WIRELESS COMMUNICATIONS RESOURCE MANAGEMENT

WIRELESS COMMUNICATIONS RESOURCE MANAGEMENT

Byeong Gi Lee

Seoul National University, Republic of Korea

Daeyoung Park

Inha University, Republic of Korea

Hanbyul Seo

LG Electronics, Republic of Korea

IEEE PRESS

John Wiley & Sons (Asia) Pte Ltd

Other Wiley Editorial Offices

John Wiley & Sons, Ltd, The Atrium, Southern Gate, Chichester, West Sussex, PO19 8SQ, UK

John Wiley & Sons Inc., 111 River Street, Hoboken, NJ 07030, USA

Jossey-Bass, 989 Market Street, San Francisco, CA 94103-1741, USA

Wiley-VCH Verlag GmbH, Boschstr. 12, D-69469 Weinheim, Germany

John Wiley & Sons Australia Ltd, 42 McDougall Street, Milton, Queensland 4064, Australia

John Wiley & Sons Canada Ltd, 6045 Freemont Blvd, Mississauga, ONT, L5R 4J3, Canada

Wiley also publishes its books in a variety of electronic formats. Some content that appears in print may not be available in electronic books.

Library of Congress Cataloging-in-Publication Data

Lee, Byeong Gi.
  Wireless communications resource management / Byeong Gi Lee, Daeyoung Park, Hanbyul Seo.
      p. cm.
  Includes bibliographical references and index.
  ISBN 978-0-470-82356-9 (cloth)
1. Wireless communication systems. 2. Radio resource management (Wireless communications) I. Park, Daeyoung. II. Seo, Hanbyul. III. Title.
  TK5103.4873.L44 2009
  621.384–dc22   2008029107

ISBN 978-0-470-82356-9 (HB)

Typeset in 10/12pt Times by Thomson Digital, Noida, India.
Printed and bound in Singapore by Markono Print Media Pte Ltd, Singapore.
This book is printed on acid-free paper responsibly manufactured from sustainable forestry in which at least two trees are planted for each one used for paper production.

To our wives

Hyeon Soon Kang
Joo Yeon Choi
Jong Soon Lee

Contents

Preface

Wireless communications, especially in its mobile form, has brought us the freedom of mobility and has changed the lifestyles of modern people. Waiting at a fixed location to receive or make a phone call, or sitting in front of a personal computer to send an e-mail or download a video program, has become an old story. Nowadays it is commonplace for people to talk over a cell phone while walking on the street, or to download and watch a movie while traveling on a train. This is the benefit made available to us by the successful evolution of wireless communications over three generations, with the fourth generation being under way.

Throughout the evolution of wireless communications, enabling technologies have continued to mature and have strived to keep up with satisfying ever increasing customer service expectations. As an example, the multiple-access technique which started from the analog *frequency-division multiple-access* (FDMA) has evolved to digital *time-division multiple-access* (TDMA) and *code-division multiple-access* (CDMA) and has advanced to the sophisticated *orthogonal frequency-division multiple-access* (OFDMA). This in turn has made possible affordable and ubiquitous multimedia services for a large majority of mobile users in both consumer and business sectors.

Notwithstanding the precipitous development of wireless technologies, customers' insatiable desire continues for faster response, larger bandwidth, and more reliable transmission. Unfortunately the desire cannot be easily satiated due to limitations in the wireless communications channels. First of all, the frequency spectrum available for wireless communications is absolutely limited. In addition, the wireless channel characteristics fluctuate due to channel loss, shadowing, and multipath fading phenomena. Furthermore, battery power is a key limitation in mobile communications.

To overcome those limitations in wireless communications and thereby enhance the spectral efficiency, significant research efforts have been exerted on the enabling technologies for the past several decades. Such technologies include *hybrid automatic repeat request* (HARQ) which combines ARQ and *forward error-correction* (FEC) techniques, *adaptive modulation and coding* (AMC) which combines modulation and coding techniques, and various diversity techniques in the time, frequency and space domains. In particular, the *multi-input multi-output* (MIMO) technique is an important multiple antenna technology that takes advantage of space diversity. OFDMA also uses the diversity effect to introduce resilience to multipath fading.

Every time a new generation of wireless communications systems has emerged, it was accompanied by the adoption of more advanced physical layer technologies to achieve enhanced system performance. In the cases of the most recent wireless systems such as Mobile WiMAX and other Advanced International Mobile Telecommunication (IMT) systems, all the above state-of-the-art communication technologies are incorporated as their air interface technologies. With that, we have consumed all the affordable physical layer technologies that communication engineers have discovered/developed to date. For further enhancement of system performance, we can only resort to intelligent and optimal utilization of the available wireless resources. Here comes the need for "wireless communications resource management." In essence, a next-generation wireless communications system can maximize overall system performance only when it adopts the most advanced physical layer technologies *and* optimizes the wireless resource management on top of those technologies.

Basically, wireless resource management determines the optimal use of wireless resources according to the wireless channel information and the *quality-of-service* (QoS) requirements of each user. Specifically, it refers to a series of processes that determine the time, order, procedure, and the amount of wireless resources to allocate to each user in the wireless communications system.

In the context of resource management, the most fundamental three wireless resources are bandwidth, transmission power, and antennas: the available frequency spectrum (and associated channel bandwidth) is limited by both nature and license fees for operators acquiring bandwidth to provision service. Battery power is limited for each mobile device and cost can be an important issue for higher power transmission. Mobile devices are limited in size for placing multiple antennas and need complicated space–time signal processing to operate.

Each available wireless resource is self-limited and the incremental improvement of a wireless resource does not necessarily increase overall system performance. The problem becomes more complicated in the multiuser environment since increasing the transmission power of one user will usually increase interference to other users or to other cells. Thus it is most desirable for wireless resource management to optimally use limited wireless resources to maximize resource efficiency and overall system performance. In effect, optimally managed wireless resources can significantly improve a channel's data rate even with low bandwidth, and also flexibly operate the wireless system, adapting to the channel characteristics and QoS requirements.

This book is intended to provide comprehensive and in-depth discussions on wireless resource management, covering a broad scope, from the provision of preliminary concepts and mathematical tools to the detailed descriptions of all the resource management techniques. It first provides preliminary discussions on the characteristics of wireless channels, basic concepts for wireless resource management problem formulation, and basic mathematical tools such as convex optimization theory useful for solving the wireless resource management problems. Then, it deals with the four different types of resource management in each chapter, namely, bandwidth management, transmission power management, antenna management, and inter-cell resource management. The topics discussed in this book include scheduling, admission control, power control, transmission power allocation, MIMO transmissions, inter-cell interference mitigation, and handoff management, and so on.

The authors believe that the book will help readers gain a thorough understanding of the overall and detailed picture of the wireless communications resource management and deepen their insight into the mechanisms of performance maximization geared by wireless resource management in next-generation wireless communication systems.

An important and useful supplement is available on the book's companion website at the following URL: www.wiley.com/go/bglee.

The authors would like to thank Seoul National University and the Institute of New Media and Communications Research for providing a comfortable environment to write the book, and the students in the Telecommunications and Signal Processing (TSP) laboratory, namely, Soomin Ko, Junho Lim, Seonwook Kim, Sunghan Ryu, Seungmin Lee, and Youngjun Ryu for assisting the writing work in various ways. In addition, the authors would like to deeply thank their wives, Hyeon Soon Kang, Joo Yeon Choi, and Jong Soon Lee, whose love and support enabled the successful completion of this year-long writing project.

About the Authors

Byeong Gi Lee

Byeong Gi Lee received the B.S. and M.E. degrees from Seoul National University, Seoul, Korea, and Kyungpook National University, Daegu, Korea, both in electronics engineering, and the Ph.D. degree in electrical engineering from the University of California, Los Angeles. He was with the Electronics Engineering Department of ROK Naval Academy as an Instructor and Naval Officer in active service from 1974 to 1979, and worked for Granger Associates, Santa Clara, CA, as a Senior Engineer responsible for applications of digital signal processing to digital transmission from 1982 to 1984, and for AT&T Bell Laboratories, North Andover, MA, as a Member of Technical Staff responsible for optical transmission system development along with the related standards works from 1984 to 1986. He joined the faculty of Seoul National University in 1986 and served as the Director of the Institute of New Media and Communications in 2000 and the Vice Chancellor for Research Affairs from 2000 to 2002.

Dr Lee was the founding chair of the Joint Conference of Communications and Information (JCCI), the chair of the Steering Committee of the Asia Pacific Conference on Communications (APCC), and the chair of the founding committee of the Accreditation Board for Engineering Education of Korea (ABEEK). He served as the TPC Chair of IEEE International Conference on Communications (ICC) 2005 and the President of Korea Society of Engineering Education (KSEE). He was the editor of the *IEEE Global Communications Newsletter*, an associate editor of *the IEEE Transactions on Circuits and Systems for Video Technology*, and the founding Associate Editor-in-Chief and the Second Editor-in-Chief of the *Journal of Communications and Networks* (JCN). He served for the IEEE Communications Society (ComSoc) as the Director of Asia Pacific Region, as the Director of Membership Programs Development, as the Director of Magazines, as a Member-at-Large to the Board of Governors, and as the Vice President for Membership Development. He served a member of the Presidential Advisory Committee of Policy Planning, the Presidential Advisory Council on Science and Technology, and the Policy Committee of the Ministry of Justice of the Korean Government. He served a Vice President of the ABEEK, the President of Korea Information and Communication Society (KICS), and the first President of the Citizens' Coalition for Scientific Society (CCSS), a non-government organization for the advancement of science and technology in Korea. He currently serves as a Commissioner of the Korea Communications Commission (KCC) and the Vice President for Member Relations of the IEEE ComSoc.

Dr Lee is a co-author of *Broadband Telecommunication Technology*, first and second editions, (Artech House: Norwood, MA, 1993 and 1996), *Scrambling Techniques for Digital Transmission* (Springer Verlag: New York, 1994), *Scrambling Techniques for CDMA Communications* (Kluwer: Norwell, MA, 2001), *Integrated Broadband Networks* (Artech House:

Norwood, MA, April 2002), and *Broadband Wireless Access and Local Networks: Mobile WiMAX and WiBro* (Artech House: Norwood, MA, 2008). He holds thirteen US patents with four more patents pending. His current fields of interest include broadband networks, wireless networks, communication systems, and signal processing.

He received the 1984 Myril B. Reed Best Paper Award from the Midwest Symposium on Circuits and Systems, Exceptional Contribution Awards from AT&T Bell Laboratories, a Distinguished Achievement Award from KICS, the 2001 National Academy of Science (of Korea) Award and the 2005 Kyung-am Academic Award. He is a Member of the National Academy of Engineering of Korea, a Member of Sigma Xi, and a Fellow of the IEEE.

Daeyoung Park

Daeyoung Park received the B.S. and M.E. degrees in electrical engineering and the Ph.D. degree in electrical engineering and computer science, all from Seoul National University, Seoul, Korea, in 1998, 2000, and 2004, respectively. He was with Samsung Electronics as a Senior Engineer from 2004 to 2007, contributing to the development of next-generation wireless systems based on the MIMO–OFDM technology. From 2007 to 2008, he was with the University of Southern California, Los Angeles, CA, as a Postdoctoral Researcher. In March 2008, he joined the faculty of the School of Information and Communication Engineering, Inha University, Korea. He received a silver award from Samsung Technical Paper Contest 2005. His research interests include communication systems, wireless networks, multiuser information theory, and resource allocation.

Hanbyul Seo

Hanbyul Seo received the B.S. degree in electrical engineering, and the M.E. and Ph.D. degrees in electrical engineering and computer science, from Seoul National University, Seoul, Korea, in 2001, 2003, and 2008, respectively. He is currently with LG Electronics and his research interests include wireless resource management, wireless MAC protocol, and wireless sensor networks.

Abbreviations

1G	First Generation
2G	Second Generation
3G	Third Generation
4G	Fourth Generation
ACK	ACKnowledgment
ALP	Active Link Protection
AMC	Adaptive Modulation and Coding
AMPS	Advanced Mobile Phone System
ARP	Autonomous Reuse Partitioning
ARQ	Automatic Repeat reQuest
AWGN	Additive White Gaussian Noise
BC	Broadcast Channel
BDCL	Borrowing with Directional Channel Locking
BER	Bite Error Rate
BLAST	Bell Labs lAyered Space–Time
BPSK	Binary Phase-Shift Keying
BR	Borrowing from the Richest
BS	Base Station
CBR	Constant Bit Rate
CDF	Cumulative Distribution Function
CDMA	Code-Division Multiple Access
CIF-Q	Channel-condition Independent packet Fair Queueing
CINR	Carrier-to-Interference-and-Noise Ratio
CP	Cyclic Prefix
CRC	Cyclic Redundancy Check
CSDPS	Channel State-Dependent Packet Scheduling
CSI	Channel State Information
CSIT	CSI available at Transmitter
CSM	Collaborative Spatial Multiplexing
DCA	Dynamic Channel Allocation
DFT	Discrete Fourier Transform
DL	DownLink
DPC	Distributed Power Control
DPC	Dirty-Paper Coding
DRC	Data Rate Control

D-STTD	Double Space–Time Transmit Diversity
DTR	Dual-Threshold Reservation
EGC	Equal-Gain Combining
EV-DO	Enhanced Version for Data Optimization
EV-DV	Enhanced Version for Data and Voice
EXP	EXP-Q and EXP-W
EXP-Q	EXPonential Queue-length rule
EXP-W	EXPonential Waiting-time rule
FA	First Available
FCA	Fixed Channel Allocation
FCC	Federal Communications Commission
FCH	Frame Control Header
FDMA	Frequency-Division Multiple Access
FEC	Forward Error Correction
FER	Frame Error Rate
FFT	Fast Fourier Transform
FGC	Fractional Guard Channel
FIFO	First-In First-Out
FRU	Flexible ReUse
GC	Guard Channel
GPS	Generalized Processor Sharing
GRP	Greedy Rate Packing
GSM	Global System for Mobile
HARQ	Hybrid ARQ
HOL	Head-Of-Line
i.i.d.	Independent and Identically Distributed
IC	Indifference Curve
IDFT	Inverse DFT
IETF	Internet Engineering Task Force
IFFT	Inverse FFT
IIR	Infinite-Impulse Response
IR	Incremental Redundancy
IS-95	Interim Standard 95
ISI	Inter-Symbol Interference
ISM	Industrial, Scientific and Medical
ITU	International Telecommunication Union
IWFQ	Idealized Wireless Fair Queueing
KKT	Karush–Kuhn–Tucker
LDPC	Low-Density Parity Check
LOS	Line-Of-Sight
LPF	Low-Pass Filter
LWDF	Largest-Weighted-Delay-First
MAC	Medium-Access Control
MAC	Multiple-Access Channel
MAI	Multiple-Access Interference
MCS	Modulation and Coding Scheme

MDP	Markov Decision Process
MIMO	Multiple-Input Multiple-Output
MISO	Multiple-Input Single-Output
ML	Maximum Likelihood
M-LWDF	Modified Largest-Weighted-Delay-First
M-LWWF	Modified Largest-Weighted-(unfinished)-Work-First
MMSE	Minimum Mean Square Error
MR	Maximal Rate
MRC	Maximal-Ratio Combining
MRS	Marginal Rate of Substitution
MS	Mobile Station
MSIR	Maximum SINR
MSQ	Mean SQuare
MU	Marginal Utility
NAK	Negative AcKnowledgment
NN	Nearest Neighbor
OFDM	Orthogonal Frequency-Division Multiplexing
OFDMA	Orthogonal Frequency-Division Multiple-Access
OSI	Open System Interconnection
OSIC	Ordered SIC
PDU	Protocol Data Unit
PF	Proportional Fairness
PFPA	Proportional-Fair Power Allocation
PG	Processing Gain
PPF	Production Possibility Frontier
PSK	Phase-Shift Keying
QAM	Quadrature Amplitude Modulation
QoS	Quality of Service
QPSK	Quadrature Phase-Shift Keying
RPA	Retransmission Power Adjustment
RT	Real Time
RTG	Receiver/transmitter Transition Gap
SDMA	Space-Division Multiple Access
SDU	Service Data Unit
SE	Spectral Efficiency
SF	Spreading Factor
SIC	Successive Interference Cancellation
SIMO	Single-Input Multiple-Output
SINR	Signal-to-Interference-and-Noise Ratio
SISO	Single-Input Single-Output
SM	Spatial Multiplexing
SNR	Signal-to-Noise Ratio
SSS	Static Service Split
STTD	Space–Time Transmit Diversity
SVD	Singular-Value Decomposition
TCP	Transport Control Protocol

TDMA	Time-Division Multiple-Access
TTG	Transmitter/receiver Transition Gap
UCC	User-Created Contents
UL	UpLink
UMTS	Universal Mobile Telecommunications System
VoD	Video-on-Demand
WCDMA	Wideband CDMA
WFQ	Weighted Fair Queueing
WiMAX	Worldwide Interoperability for Microwave Access
WMAN	Wireless Metropolitan Area Network
WRM	Wireless Resource Management
ZF	Zero-Forcing

Part One

Concepts and Background

1

Introduction

Wireless mobile communication has evolved over three generations, with the fourth generation yet to come. As a new generation stepped forward, it accompanied the implementation of new air interface technologies as well as the enhancement of system performances. In the early generations, performance enhancements stemmed mainly from physical layer technologies such as modulation, multiple access, and multiple-antenna technologies but the contribution of resource management has increased with each generation. The role of resource management is expected to become even more important in the next-generation wireless communication systems. In essence, maximized performance of a future wireless system will likely be attained by optimized resource management implemented on the basis of matured physical layer technologies.

1.1 Evolution of Wireless Communications

Recent decades have witnessed a great evolution of wireless communication systems. It was in the 1970s when the *first-generation* (1G) wireless communication systems such as the *advanced mobile phone system* (AMPS) were first deployed to support circuit-switched voice telephony based on the analog *frequency-division multiple-access* (FDMA) method.

In the early 1990s, the 1G wireless systems evolved to become *second-generation* (2G) systems by virtue of the advances in digital technologies. The 2G wireless systems also were targeted at voice telephony with circuit switching, but digital modulation and multiple access methods were incorporated as well based on digital technologies. *Time-division multiple access* (TDMA) was employed in the *global system for mobile* (GSM), and *code-division multiple access* (CDMA) was used in the IS-95 CDMA system.

The late 1990s saw an increasing demand for high data rate services such as video telephony, Web browsing, and e-mail, and this demand motivated the development of the *third-generation* (3G) systems, such as WCDMA and CDMA2000, which supported both circuit- and packet-switched services with broader bandwidths. The 3G systems were followed by enhanced systems such as CDMA2000 EV-DO (*enhanced version for data optimization*) and EV-DV (*enhanced version for data and voice*), which were developed for optimized data services, with a compatibility maintained with the earlier systems.

Wireless Communications Resource Management   Byeong Gi Lee, Daeyoung Park, and Hanbyul Seo
© 2009 John Wiley & Sons (Asia) Pte Ltd

Now, we are facing the advent of *fourth-generation* (4G) wireless communication systems. The requirements for 4G systems set by the International Telecommunication Union (ITU) include the provision of a 1 Gbps data rate in nomadic state and a 100 Mbps rate in moving state (ITU, 2003). Applications of 4G systems are expected to range very wide, from classical voice telephony to new more modern services such as the uploading of *user-created content* (UCC).

As stated above, the evolution of wireless communications is characterized by a shift from analog, small-capacity, single-target services to digital, large-capacity, integrated multiple services. For the next generation to be successfully implemented, systems should be designed to meet a number of requirements arising from that trend.

First, they should be capable of providing high transmission rates. Users want to send and receive by broadband services with very fast response times, and some applications such as *video-on-demand* (VoD) cannot be supported without broad bandwidth.

Second, they should have a flexible service architecture to integrate different types of user services on a single air-interface. If the architecture is rigidly optimized only for a single service type, the other types will suffer from poor service quality within the system.

Third, they should be equipped with the functionality of *quality-of-service* (QoS) management. The metrics of QoS may differ among different applications. For example, voice telephony has a strict delay requirement whereas e-mail is not sensitive to delays as long as data delivery is reliable. Unless the system satisfies the service-dependent QoS metric, even high throughput may fail to satisfy the users.

One might think that the above requirements could be fulfilled by enhancing the transmission technology in a physical layer. In reality, the enhancements solely due to transmission technology (or physical-layer technologies in general) are limited in various aspects.

First, transmission technologies are already matured and there is not much scope for further improvements without some unexpected breakthroughs. For example, the channel coding technologies, such as turbo code and *low-density parity-check* (LDPC) code, permit a transmission rate very close to the information-theoretic capacity in Gaussian channels (Chung *et al.*, 2001).

Second, most transmission technologies are developed to accomplish specific objectives, so each individual technology cannot yield a universally optimal performance solution. If one considers, for example, the diversity and multiplexing techniques supported by multiple transmit and receive antennas, the diversity technique is the better choice when the channel condition is hostile as it can improve the signal quality and transmission reliability. However, when the channel condition is favorable, the additional improvement of signal quality supported by the diversity technique does not contribute much, so multiplexing becomes the better choice as it can increase the overall transmission rate. (See Section 7.2 for a discussion of MIMO transmission modes.) For a system to operate successfully over a universal communication environment, it is necessary to employ a management technique that can intelligently determine the transmission technology that is optimal to the given situation.

Third, the QoS requirements of user traffic ffect the optimized solution when selecting the transmission technology. As an example, compare the water-filling and the truncated channel inversion methods, each of which can be the optimal transmission policy under its own system assumption. Interestingly, the two policies yield opposite results even though they share the common objective of maximizing channel capacity under limited battery charge. When the current channel condition is good, water-filling – the optimal policy with a loose delay requirement – can send more data with more transmission power to enjoy the hospitality of the wireless channel. In contrast, truncated channel inversion – the optimal policy with a stringent delay requirement – expends less transmission power to save the limited battery charge. (See Section 6.3.1 for a comparison of water-filling and truncated channel inversion policies.) As demonstrated in this example, a different QoS requirement leads to a different choice of transmission technology. Therefore, in future wireless systems that are expected to support various different services, it is essential to manage the transmission technologies in relation with the QoS requirement.

1.2 Wireless Resource Management

The above discussion, in effect, points to the necessity for wireless resource management. Wireless resource management refers to a series of processes that determine timing, ordering, procedures, and the amount of wireless resources to allocate to each user. This determines the optimal use of wireless resources according to the wireless channel information and the QoS requirements of each user.

The importance of resource management originates from the scarceness of wireless resources. The requirements could be met simply if we had unlimited access to the three fundamental resources – bandwidth, transmission power, and antennas, but in practice there are various limitations.

First, the frequency spectrum is limited, and licensing fees increase with bandwidth. Second, battery power is limited for each mobile device, and it is necessary to use more expensive signal amplifiers to use more transmission power. Third, the size of a mobile device limits multiple antennas, and more complicated space–time signal processing is needed to use more antennas. These problems become more complicated in the multiuser environment as an increase of transmission power of a user means an increase of interference to other users.

Thus, the quantity of each wireless resource is limited, and the use of more wireless resources does not necessarily bring a performance enhancement of the overall system. Therefore it is most desirable to manage wireless resources so that they are used as efficiently as possible, while maximizing the overall system performance.

By determining ways to use the available wireless resources optimally, resource management can lead to a significant improvement in transmission rate without using more bandwidth. In addition, it can make the system flexible enough to operate in adaptation to the channel characteristics and QoS requirements, thereby permitting a flexible service architecture for integrating various different services in a single air-interface.

1.2.1 Bandwidth Management

Bandwidth is a fundamental wireless resource which refers to the range of frequencies occupied by a transmitted signal. Since bandwidth determines the maximum symbol transmission rate, it puts a fundamental limit on the channel access rate. Broadly, bandwidth may be considered as the transmission opportunity to access the wireless medium; that is, how often we can transmit symbols over the wireless medium. If this transmission opportunity is shared by multiple users, the *scheduling* that determines the order of user access over time becomes the essential part of bandwidth management. There are various different scheduling methods having different objectives. Another important part of bandwidth is the *admission control*, which determines the admissibility of a new connection in relation with QoS. Admission control may be regarded as a long-term bandwidth management and is strongly related to scheduling.

1.2.2 Transmission Power Management

Transmission power is another important wireless resource that can determine the quality of the wireless link. A wireless link, in general, becomes more reliable as the transmission power is increased.

There are three aspects to transmission power management. The first is *power control*, which focuses on how to support a harmonized coexistence of mutually interfering transmitter–receiver pairs. The second is *power allocation*, which distributes a given amount of transmission power over multiple orthogonal subchannels with the intention of maximizing the efficiency of the transmission power. The third is *power adaptation*, which determines the transmission power adaptively according to the time-varying environments, caused by the randomness of traffic arrival or the fading process of the channel.

1.2.3 Antenna Management

Recently, antennas have become the most attractive wireless resource as they can contribute to increasing the channel capacity without requiring additional bandwidth or transmission power. However, as antennas need circuitry operating at radio frequency in the transmitting and receiving devices, increasing the number of antennas implies an increased device cost. In addition, the channel capacity heavily relies on the usage of antennas, and their optimal use varies depending on the channel that connects the transmitter–receiver pair. This implies that, for an effective operation of antennas, it is necessary to adopt an antenna management method adequate to the current channel state.

There are various *multiple-input multiple-output* (MIMO) technologies, including *diversity transmission* and *spatial multiplexing transmission*. Also, there are multiuser MIMO transmission technologies, such as *space-division multiple-access* (SDMA) and dirty paper coding.

1.2.4 Inter-cell Resource Management

For efficient spectrum use, wireless communication systems have a cellular structure in which the whole service area is divided into multiple smaller service areas, called *cells*, and deploy one *base station* (BS) at each cell. The cellular concept uses limited wireless resources very efficiently. It permits repeated use of the given frequency spectrum, reduces the distance from each BS to the users within each cell, and thus can provide the desired level of service quality using a much smaller amount of transmission power and bandwidth.

However, this cellular concept brings in some new problems to solve, including the inter-cell interference problem. The *inter-cell resource management* is intended to solve such problems arising in the multiple-cell environment. There are two aspects to inter-cell resource management. The first is *inter-cell interference management* that tries to mitigate interference from neighboring cells while maximizing reuse of the spectrum at the base stations. The second is *handoff management* that tries to maintain the QoS of an ongoing connection while the user moves across the cell boundary.

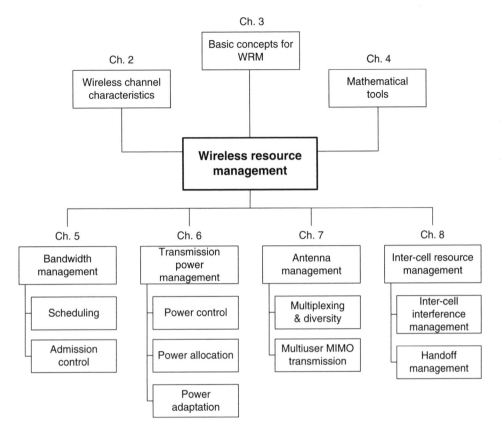

Figure 1.1   Organization of the book

1.3 Organization of the Book

This book is intended to provide comprehensive and in-depth discussions on wireless resource management, from the preliminary concepts to specific resource management techniques. Figure 1.1 depicts the overall structure of the book.

Chapters 2–4 deal with the preliminaries needed to understand and develop resource management. Chapter 2 discusses the characteristics of wireless channels and provides models that describe wireless channels mathematically. Chapter 3 covers several basic concepts needed to formulate wireless resource management problems, including the definitions of wireless resources, different multiple-access methods, and QoS-related issues. Chapter 4 introduces the basic mathematical tools useful for solving the wireless resource management problems, including convex optimization theory and dynamic programming.

Chapters 5–8 deal with the four different types of resource management discussed in the previous section. Specifically, Chapter 5 deals with bandwidth management, Chapter 6 with transmission power management, Chapter 7 with antenna management, and Chapter 8 with inter-cell resource management.

References

Chung, S.-Y., Forney, G.D., Richardson, T.J. and Urbanke, R. (2001) On the design of low-density parity-check codes within 0.0045 dB of the Shannon limit. *IEEE Communications Letters*, **6** (2), 58–60.

ITU-R (2003) Recommendation M.1643: Framework and Overall Objectives of the Future Development of IMT-2000 and Systems beyond IMT-2000.

2

Characteristics
of Wireless Channels[1]

The most salient distinction between wireless and wired communications lies in the propagation medium, or the channel for signal transmission. Whereas the media for wired communication are mostly in protected form (e.g. twisted-pair, coaxial cable, and fiber), the medium for wireless communication is unprotected – and thus subject to various kinds of noise and external interference. Such interfering factors cause attenuation of the received signal power and various distortions in the received signal waveform. Moreover, in the case of mobile communications, user movement and environmental dynamics make those impediments change over time in unpredictable ways. As a consequence, the radio propagation conditions and their statistical characteristics seriously affect the operation of wireless communication systems as well as their performance. Therefore, in preparation for the design of high-performance wireless resource management, it is crucial to understand the characteristics of wireless communication channels and, further, the technologies to reinforce the wireless channels against those impediments.

This chapter will characterize how transmitted signals are distorted while propagating through wireless channels. This will be done by establishing appropriate mathematical models that can effectively describe physical characteristics of wireless channels with a set of environmental parameters. There are several reports in the literature on developing channel models appropriate to different physical environments, but here the discussion will be limited to a fundamental level to help readers gain a solid basic understanding of wireless channel characteristics without being diverted by the details.

2.1 Channel Gain

One of the most important characteristics of wireless channels is the *channel gain* which determines what proportion of the transmission power is received at the receiver. The channel gain is defined by the ratio of the received power to the transmission power, so it dictates the level of the received signal power, which serves as the metric of link quality of the wireless

---

[1] The content of this chapter is similar to that in Lee and Choi (2008), coauthored by one of the authors, but is addressed in greater detail.

channel. The transmitter–receiver pair, in general, gets a clearer channel as the channel gain is increased.

In general, channel gain is determined by three factors – path loss, shadowing, and multipath fading.

- *Path loss* is the decay of signal power dissipated due to radiation on the wireless channels, so is determined by the channel's physical characteristics of signal propagation. In general, path loss is modeled as a function of the distance between the transmitter and receiver, on the assumption that signal loss is identical at the same distance.
- *Shadowing* is caused by obstacles such as walls and trees located in the transmitter–receiver communication path. These obstacles absorb, diffract, and reflect the transmitted signal, thereby attenuating its power. Since the locations of the obstacles and their properties are not predictable, shadowing is a random factor that can change the received power even at the same distance.
- *Multipath fading* refers to signal power variations caused by constructive and destructive additions of signal components arriving at the receiver by multiple paths. These multiple paths are generated by scattering and reflecting objects located around the transmitter–receiver communication path. As with shadowing, multipath fading is generally modeled as a statistical process.

Figure 2.1 depicts how the channel gain is determined from those three factors. The channel gain G is determined as the product of the power gains obtained from path loss, shadowing, and multipath fading. That is, for the transmission power P_T, the received power is represented by

$$P_R = P_T G = P_T G_{PL} G_S G_{MF}, \qquad (2.1)$$

where G_{PL}, G_S, and G_{MF} denote the power gains determined by path loss, shadowing, and multipath fading, respectively. Note that the power level in Figure 2.1 is plotted on a log scale, so addition in the figure is equivalent to multiplication on a linear scale.

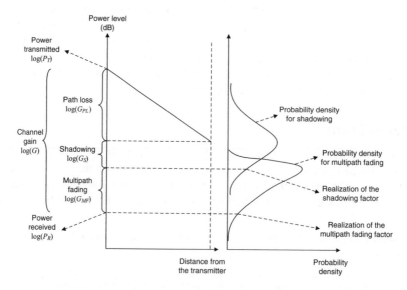

Figure 2.1   Determination of channel gain by path loss, shadowing, and multipath fading

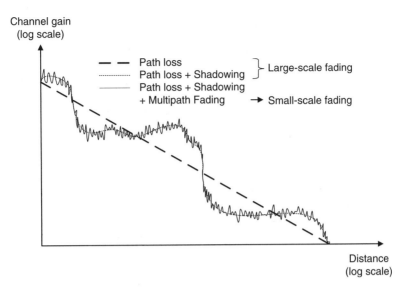

Figure 2.2 Illustration of the characteristics of channel fading

Channel gain changes depending on environmental dynamics, including the mobility of the transmitter–receiver pair. The three gain factors – path loss, shadowing, and multipath fading – are involved in the change but affect the change in different ways. The path loss factor is almost insensitive to the movement by several meters as the transmitter–receiver pair is usually assumed to be separated by hundreds of meters. The shadowing factor is affected by the lengths of obstacles which are in the range 10–100 meters. Thus, these two gain factors are affected by environmental dynamics in the scale of meters, so those variations are called *large-scale fading*.

On the other hand, the multipath fading factor is strongly related to the wavelength of the propagating signal, which in general is in the scale of micrometers. Thus, even a small movement may affect the multipath fading factor, so this variation is called *small-scale fading*.

Figure 2.2 illustrates the characteristics of the channel fading process with respect to the distance between the transmitter and receiver. Both the channel gain and the distance are plotted on a log scale.

In the following, we deal with mathematical methods to model variation of the channel gain. Some models are introduced for large-scale and small-scale fading in Sections 2.2 and 2.3, respectively. Then, several methodologies developed to mitigate the channel impediment are explained.

2.2 Large-scale Fading

Large-scale fading is caused by path loss and shadowing, which are affected only by a movement of comparatively large distance. We now consider some mathematical descriptions of large-scale fading that has been developed to date.

2.2.1 Path Loss

Path loss is determined by how the signal power decays as it propagates through the wireless channel. The critical factor is the distance between the transmitter and receiver. It is known that the signal power decreases as the distance increases, but the issue is how to determine the rate at which the signal decays. There are several path-loss models, including the free-space model, the two-ray propagation model, and a simplified path-loss model.

2.2.1.1 Free-space Model

Free space refers to the radio propagation environment where no obstruction exists between transmitter and receiver. The transmitted signal propagates along a straight line that connects the pair. According to the theory of propagation of electromagnetic waves, the signal power reduces inversely with the square of the distance in this free space, which equally applies to the power gain. To be specific, the received power P_R is related to the transmit power P_T by

$$P_R = \left(\frac{\lambda}{4\pi d}\right)^2 AP_T, \tag{2.2}$$

where d denotes the distance between the transmitter and receiver, λ the wavelength of the pass-band carrier, and A the power gain obtained at the antennas of the transmitter and receiver. This power gain is determined by radiation patterns in the signal propagating direction. From the above relation, the power gain takes the expression

$$G_{PL} = \left(\frac{\lambda}{4\pi d}\right)^2 A \tag{2.3}$$

in the free space.

2.2.1.2 Two-ray Propagation Model

Consider a path-loss model in which the signal power falls off much faster than in the free-space model. It is the *two-ray propagation model* in which the received signal consists of two propagation components. One is the *line-of-sight* (LOS) ray which directly arrives at the receiver, and the other is the *reflected ray* which is reflected on the ground. This is the simplest model that takes into consideration the effects of reflection, diffraction, and scattering, so it is widely used in evaluating the performance of wireless systems.

Figure 2.3 is a schematic of signal propagation in the two-ray model. The ground distance between the transmitter and receiver is d, and the altitudes of the transmitter and receiver antennas are h_t and h_r, respectively. The LOS ray arrives at the receiver after propagating through a distance a. The reflected ray travels a distance b and a distance c before and after the reflection, respectively, whose projections on the ground are d' and d'', respectively. The angle of incidence as well as the angle of reflection is θ.

Assume that the signal power of each ray is determined according to the free-space model. Assume also that the ground reflection causes no signal power loss but just adds a

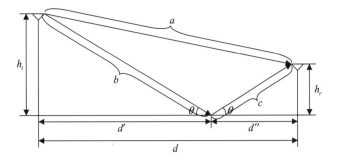

Figure 2.3   Signal propagation in the two-ray model

phase shift of π. Then the received power of the LOS and the reflected rays are respectively given by

$$P_1 = \left(\frac{\lambda}{4\pi a}\right)^2 A P_{\mathrm{T}}, \qquad P_2 = \left(\frac{\lambda}{4\pi(b+c)}\right)^2 A' P_{\mathrm{T}}, \tag{2.4}$$

where A and A' denote the antenna gains in the direction of the LOS and the reflected ways, respectively. The two rays arrive at the receiver with the phase difference $\phi' = \phi + \pi = 2\pi (b + c - a)/\lambda + \pi$. The first term of ϕ' is the phase difference caused by the difference of the two propagating paths, and the second term (π) is the additional phase shift by the ground reflection. On superposing the two rays, the received power is

$$P_{\mathrm{R}} = \left| \sqrt{P_1} + \sqrt{P_2}\, e^{-j\phi'} \right|^2 = P_{\mathrm{T}} \left(\frac{\lambda}{4\pi}\right)^2 \left| \frac{\sqrt{A}}{a} - \frac{\sqrt{A'}\, e^{-j\phi}}{b+c} \right|^2 . \tag{2.5}$$

Starting from this equation, we can now derive the received power in terms of d, the ground distance between the transmitter and receiver. First, by relating the basic relations

$$b^2 = d'^2 + h_t^2, c^2 = d''^2 + h_r^2, d' + d'' = d, \tag{2.6}$$

$$\sin\theta = \frac{h_t}{b} = \frac{h_r}{c}, \tan\theta = \frac{h_t}{d'} = \frac{h_r}{d''}, \tag{2.7}$$

we can easily derive

$$b + c = b\left(1 + \frac{h_r}{h_t}\right) = \sqrt{(d'^2 + h_t^2)}\left(1 + \frac{h_r}{h_t}\right)$$

$$= \sqrt{\left(d' + d'\frac{h_r}{h_t}\right)^2 + (h_t + h_r)^2} = \sqrt{d^2 + (h_t + h_r)^2}. \tag{2.8}$$

From the figure we get the relation

$$a = \sqrt{d^2 + (h_t - h_r)^2}. \tag{2.9}$$

So, the difference between the two propagating paths is given by

$$b + c - a = d\left(\sqrt{1 + \left(\frac{h_t + h_r}{d}\right)^2} - \sqrt{1 + \left(\frac{h_t - h_r}{d}\right)^2}\right). \tag{2.10}$$

Here, it is assumed that the distance d is much larger than the antenna heights h_t and h_r. Then, by applying the Taylor series approximation to (2.10):

$$\phi = \frac{2\pi(b + c - a)}{\lambda} \approx \frac{4\pi h_t h_r}{\lambda d}. \tag{2.11}$$

Now additionally assume that the distance d is long enough to approximate $A \approx A'$, $a \approx b + c \approx d$, and $\phi \approx 0$. Then by inserting these approximations into (2.5) we get

$$P_R \approx P_T\left(\frac{\lambda\sqrt{A}}{4\pi d}\right)^2 |1 - e^{-j\phi}|^2 = P_T\left(\frac{\lambda\sqrt{A}}{4\pi d}\right)^2 |1 - \cos\phi + j\sin\phi|^2. \tag{2.12}$$

Since $\cos\phi \approx 1$ and $\sin\phi \approx \phi$ for $\phi \approx 0$, this equation is simplified to

$$P_R \approx P_T\left(\frac{\lambda\sqrt{A}}{4\pi d}\right)^2 (\phi)^2 \approx P_T A \frac{(h_t h_r)^2}{d^4}. \tag{2.13}$$

From this, the power gain is derived as

$$G_{PL} = A\frac{(h_t h_r)^2}{d^4} \tag{2.14}$$

for the two-ray propagation model.

Observe in (2.14) that the received power decreases at the rate of $1/d^4$. This happens because the reflected ray is added in a destructive manner when the distance d is very large. To be more specific, since $A \approx A'$, $a \approx b + c \approx d$ and $\phi \approx 0$ in this case, the direct and reflected rays are almost the same except for the phase difference of π, which results in a destructive interference. The rate of decrease with $1/d^4$ implies that, in real communication environments where reflecting objects exist, the path loss is much higher than in free space. We also observe that the path loss is independent of wavelength, and the channel gain increases if the transmitter and receiver antennas are installed at higher altitudes.

2.2.1.3 Simplified Path-loss Model

The two path-loss models discussed above – free-space and two-ray – may not be suitable for practical use, due to their unrealistic assumptions. The free-space model assumes that no objects scatter or reflect the transmitted signal, and the two-ray model assumes a very large distance between the transmitter and receiver. As a result, the channel gain predicted by these models deviates from empirical results.

However, the two models commonly establish the fact that the power gain from path loss is proportional to $1/d^\alpha$ for a *path-loss exponent* α. This means that the signal power decays linearly on a log scale and the slope is equal to α. The path-loss exponent α is 2 in the free space model and is 4 in the two-ray model. Thus, a simple way to formulate path loss is to derive the appropriate path-loss exponent α from empirical results.

A widely used practical model is the *two-slope path-loss model* which employs two different path-loss exponents according to the distance between the transmitter and receiver. To be specific, the channel gain in this model is given by

$$
G_{\mathrm{PL}} =
\begin{cases}
A \left(\dfrac{d_0}{d} \right)^{\alpha_1}, & d_0 \leq d \leq d_c, \\[4mm]
A \left(\dfrac{d_0}{d_c} \right)^{\alpha_1} \left(\dfrac{d_c}{d} \right)^{\alpha_2}, & d > d_c,
\end{cases}
\tag{2.15}
$$

for the reference distance d_0 at which the channel gain becomes A. The distance d_c is called the *critical distance* at which the path-loss exponent changes. In general, the received power decreases more slowly for a smaller distance since the reflected rays are more likely to be added constructively. Thus, the first path-loss factor is usually smaller than the second one; that is, $\alpha_1 \leq \alpha_2$. With these path-loss exponents, the signal power begins to decrease drastically after the critical point.

Figure 2.4 illustrates the two-slope path-loss model in the log scale of power and distance. The parameters in (2.15) are heavily dependent on the communication environment. For example, a set of parameters for microcellular systems with a carrier frequency of 1.9 GHz is given by (Feuerstein *et al.*, 1994)

$$
\alpha_1 = 2.07, \alpha_2 = 4.16, d_c = 573 \, \mathrm{m}, \tag{2.16}
$$

for an antenna height 13.3 m. In this example, the path-loss exponent is close to 2 before the critical distance. This means that the free-space model well approximates the path loss with a short distance. But when the distance increases beyond the critical distance, the path-loss exponent approaches the value of the two-ray model, which is 4.

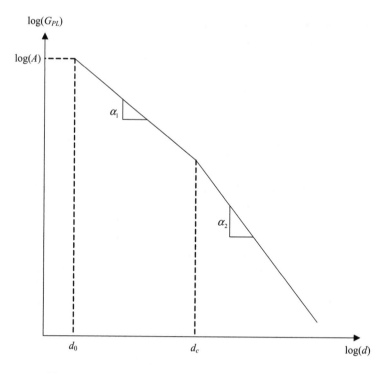

Figure 2.4   Characteristics of the two-slope path-loss model

A more simplified path-loss model uses only a single path-loss exponent; that is, setting $\alpha_1 = \alpha_2$. This model is useful in evaluating macrocellular systems since most users are located further away than the critical distance from the base station. An example of this model in 3GPP (2005), which models the path loss of macro-cell systems with a carrier frequency of 2 GHz, is

$$10 \log_{10} G_{PL} = -15.3 - 37.6 \log_{10} d, \tag{2.17}$$

where d denotes the transmitter–receiver distance in meters. This model is equivalent to setting the path-loss exponent to 3.76. In this model, the minimum distance between the transmitter and receiver is constrained to be larger than 35 meters, which may be the reference distance d_0 in (2.15).

2.2.2 Shadowing

Shadowing refers to the additional attenuation of signal power caused by blocking objects between the transmitter and receiver. Since the locations and effects of these obstacles are unpredictable, shadowing brings in variations of the channel gain even at a fixed transmitter–receiver distance. For this reason, the shadowing effect is formulated by a statistical model.

The most common model is *log–normal shadowing*, in which the channel gain is assumed to have a log–normal distribution. That is, the logarithm of the channel gain is a zero-mean Gaussian random variable given by

$$10 \log_{10} G_s \sim N(0, \sigma_s^2) \tag{2.18}$$

for the standard deviation σ_S of the log–normal shadowing whose value ranges from 6 to 10 dB, with a larger σ_S rendering more fluctuation of the channel gain.

The shadowing effect is dictated by obstacles such as walls and buildings. Thus, the amounts of signal power attenuation are not independent but correlated for two different locations not much separated. The shadowing effect is nearly the same at two closely located points, but signal power attenuation becomes more uncorrelated as the locations are separated further. In formulating this correlated property of the shadowing effect, it is sufficient to consider the covariance between two different points. This is because the joint distribution of any multivariate Gaussian random variable can be determined by its covariance.

A simple method to formulate the correlation of the shadowing effect is to assume that the correlation decreases exponentially with the distance between two different locations. The *decorrelation distance* d_{dc} is defined by the distance at which the correlation becomes $1/e$ of the variance of the log–normal shadowing. With this method, the power gains $G_{S,1}$ and $G_{S,2}$, determined by the shadowing effect at two points separated by distance δ, is formulated by

$$\mathbb{E}[(10 \log_{10} G_{S,1})(10 \log_{10} G_{S,2})] = \sigma_S^2 e^{-\delta/d_{\mathrm{dc}}}. \tag{2.19}$$

The decorrelation distance is also a system parameter dependent on communication environments. For example, the standard deviation σ_S and the decorrelation distance d_{dc} are set to 8 dB and 50 m in 3GPP (2005).

2.3 Small-scale Fading

In general, there exist a large number of objects between the transmitter and receiver, which create multiple propagation paths by reflecting and scattering the transmitted signal. The transmitted signal arrives at the receiver along multiple paths as illustrated in Figure 2.5.

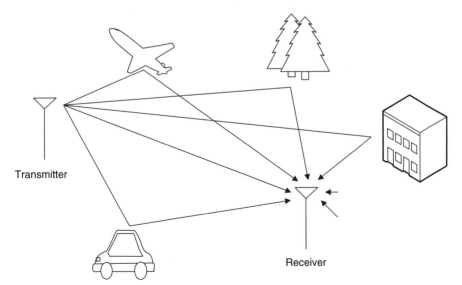

Figure 2.5 Illustration of a multipath environment (Reproduced with permission from B.G. Lee and S. Choi, *Broadband Wireless Access and Local Networks: Mobile WiMAX and WiFi*, Artech House, Norwood, MA. © 2008 Artech House, Inc.)

These paths have different attenuation properties and add different phase shifts to the signal. The multipath components may be added in a constructive or destructive manner. In other words, the aggregate channel response consists of multiple replicas of the transmitted signal, and the "effective" channel seen by the receiver at a time instance is determined by how the multipath components are added – constructively or destructively. Since the constructive/ destructive addition is sensitive to small phase shifts which are caused by wavelength-scale movements, a slight change in the dynamics of the transmitter–receiver pair or of reflecting objects may cause a large variation in the "effective" wireless channel. Consequently, the multipath environment induces rapid channel fluctuations and, therefore, small-scale fading.

In addition to this time-variance of wireless channels, the multipath environment also induces frequency-domain channel fading. Each multipath component of the signal arrives at a different time. As a result, even though the transmitter sends a single pulse with very short duration, the receiver sees a dispersed and distorted version in which the transmitted pulse is spread in the time domain. In view of circuit theory, the wireless channel in this multipath environment is like a *time-varying, memory channel*. That is, the impulse response of the channel is not an impulse but a dispersed waveform with non-zero duration. This implies that the output of the channel is expressed by a composition of input signals which had been transmitted at different time instances in the past. This phenomenon brings the channel fading into the frequency domain, as the frequency response of the wireless channel may not be a constant: The channel may suffer from deep fading at a specific frequency band.

An example of multipath fading is given in Figure 2.6 which illustrates two impulses transmitted at times τ_1 and τ_2 as well as their received versions. The notations d_1 and d_2 correspond to the smallest propagation delay among the multipath components for each transmission. The channel impulse response in this figure is, in fact, a superposition of the multipath components that have different attenuations and phase shifts. This means that if we

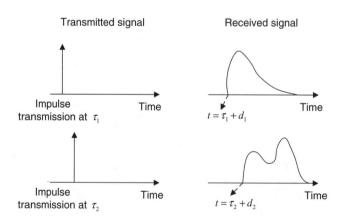

Figure 2.6 Example of multipath fading channels (Reproduced with permission from B.G. Lee and S. Choi, *Broadband Wireless Access and Local Networks: Mobile WiMAX and WiFi*, Artech House, Norwood, MA. © 2008 Artech House, Inc.)

take a sample at a point of the channel response, it consists of multiple replicas of the transmitted signal that have traveled different paths. Thus, different channel responses are observed at different observation times, τ_1 and τ_2. This variation of channel is referred to as *fading in the time domain*. On the other hand, we observe in this figure that a dispersed version of the transmitted signal arrives at the receiver even though a narrow pulse is transmitted. This phenomenon causes channel fading in the frequency domain as the time-domain response of the channel is not an impulse but a function of time t that has duration larger than zero. This is referred to as *fading in the frequency domain*.

The channel variations in time and frequency domains appear unpredictable to the transmitter–receiver pair as it is hardly possible to estimate the multipath components and their interoperations. Therefore, it is reasonable to characterize multipath fading channels in a statistical way. In order to describe channel fading in both the time and frequency domains, we denote by $h(t; \tau)$ the impulse response of the multipath fading channel observed at time τ. Here, the variable t indicates the time delay elapsed after the impulse is transmitted. We characterize the channel response $h(t; \tau)$ as a complex-valued random process, and we assume that $h(t; \tau)$ is wide-sense stationary. The channel response $h(t; \tau)$ for a fixed observation time τ describes how the transmitted narrow pulse is dispersed while traveling through the multipath fading channel. On the other hand, the response $h(t; \tau)$ for a fixed time delay t describes how the multipath components corresponding to the time delay t vary as the observation time passes. Figure 2.7 depicts the channel response $h(t; \tau)$ and its frequency response $H(f; \tau)$ for the examples in Figure 2.6. The time-domain fading refers to the channel variance with respect to the observation time τ, while the frequency-domain fading is dictated by the time delay t. How

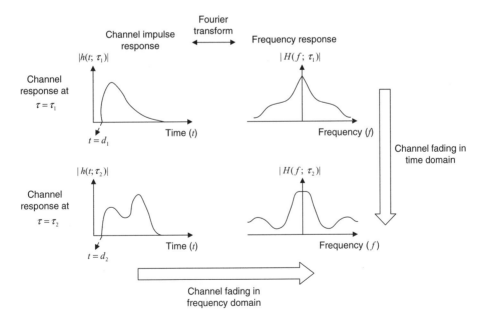

Figure 2.7 Illustration of channel response in time and frequency domains (Reproduced with permission from B.G. Lee and S. Choi, *Broadband Wireless Access and Local Networks: Mobile WiMAX and WiFi*, Artech House, Norwood, MA. © 2008 Artech House, Inc.)

to characterize and formulate the multipath fading in time and frequency domains is discussed in the following subsections.

2.3.1 Fading in the Time Domain

Formulating channel fading in the time domain involves finding a mathematical method that describes how the channel $h(t; \tau)$ changes as the observation time τ elapses. In other words, we characterize $h(t; \tau)$ by fixing the time delay t to a certain value and changing the observation time τ.

2.3.1.1 Marginal Distribution of Wireless Channels

A channel realization with fixed time delay and observation time is a superposition of several multipath components. Since the channel $h(t; \tau)$ is a complex number, the multipath components are added in both in-phase and quadrature-phase. If it is assumed that there are a large number of scatterers in the channel, the number of multipath components also increases to a large value. This assumption is valid in urban or suburban areas where there exist many obstacles such as walls and buildings around the transmitter–receiver pair. In addition, it can be assumed that each multipath component is independent of each other as each component arrives at the receiver after undergoing its own propagation path. These assumptions permit application of the central limit theorem which leads to the result that both the in-phase and quadrature-phase components are independent Gaussian random variables. This means that the channel can be written as

$$h(t; \tau) = x(t; \tau) + jy(t; \tau) \tag{2.20}$$

for independent Gaussian random variables $x(t; \tau)$ and $y(t; \tau)$. Note that these two random variables are real numbers.

First consider the case where the expectations of the in-phase and quadrature-phase components are zero, so that $\mathbb{E}[x(t; \tau)] = \mathbb{E}[y(t; \tau)] = 0$, and their variances are the same, so that $\mathbb{E}[x^2(t; \tau)] = \mathbb{E}[y^2(t; \tau)] = \sigma_R^2$. This means that the signal arrives uniformly from all directions and there exists no component that directly comes from the transmitter. In this case, the channel gain $g(t; \tau) = |h(t; \tau)|^2 = x^2(t; \tau) + y^2(t; \tau)$ is exponentially distributed with a probability density function (Viniotis, 1998)

$$f_G(g) = \frac{1}{2\sigma_R^2} e^{-g/2\sigma_R^2}. \tag{2.21}$$

This means that the channel gain is an exponential random variable with mean $\bar{g} = 2\sigma_R^2$. Then, the envelope of the channel $r(t; \tau) = |h(t; \tau)| = \sqrt{x^2(t; \tau) + y^2(t; \tau)}$ is a Rayleigh random variable with distribution

$$f_R(r) = \frac{2r}{\bar{g}} \exp\left[-\frac{r^2}{\bar{g}}\right]. \tag{2.22}$$

The phase of the channel $\theta(t; \tau) = \arctan(y(t; \tau)/x(t; \tau))$ is uniformly distributed in the interval $[0, 2\pi]$ and is independent of the channel envelope. Since the channel envelope is Rayleigh distributed, the channel which is modeled as above is called a *Rayleigh fading channel*.

Now consider the case where one of the multipath components comes directly from the transmitter, not experiencing reflection nor scattering. This *line-of-sight* (LOS) component is not affected by the small-scale dynamics, so it is a deterministic component, not a random one. We call the multipath components other than the LOS one *diffused components*. In cases where LOS and diffused components coexist, the in-phase and quadrature-phase components are still Gaussian random variables by the central limit theorem applied to the diffused component, but their means are not zero any longer owing to the deterministic nature of the LOS component. Since any additional phase shift to the channel does not affect the distribution of the channel gain and envelope, we subtract the phase shift of the LOS component without touching the channel gain. Thus, without loss of generality, it can be assumed that the in-phase component takes a non-zero mean which is denoted by A, while the quadrature-phase component has zero mean. The variances of the two phase components are assumed to be σ_R^2 again since they are determined by the diffused components. In a similar way, we can obtain that the envelope of the channel is a Rice random variable with probability density function (Viniotis, 1998)

$$f_R(r) = \frac{r}{\sigma_R^2} \exp\left[-\frac{(r^2 + A^2)}{2\sigma_R^2}\right] I_0\left(\frac{Ar}{\sigma_R^2}\right), \tag{2.23}$$

where $I_0(z)$ is a modified Bessel function of the first kind and zero order, defined by

$$I_0(z) = \frac{1}{2\pi} \int_0^{2\pi} e^{z \cos\theta} d\theta. \tag{2.24}$$

We call this type of channel model a *Rician fading channel*. The average channel gain of a Rician fading model is given by $\bar{g} = A^2 + 2\sigma_R^2$, which is equal to the sum of channel gains of the LOS and the diffused components. The Rician fading model is usually parameterized by a parameter K, defined by

$$K = \frac{A^2}{2\sigma_R^2}, \tag{2.25}$$

which is equal to the ratio of the power in the LOS component to that in the diffused component. In terms of the parameter K, the Rician distribution in (2.23) may be rewritten as

$$f_R(r) = \frac{2r(K+1)}{\bar{g}} \exp\left[-K - \frac{(K+1)r^2}{\bar{g}}\right] I_0\left(2r\sqrt{\frac{K(K+1)}{\bar{g}}}\right). \tag{2.26}$$

A small K implies severe fluctuation in the channel fading as the diffused component (i.e., the random one) dominates the LOS component (i.e., the deterministic one). For $K = 0$, we get Rayleigh fading where there is no LOS component; and for $K = \infty$, we get a time-invariant

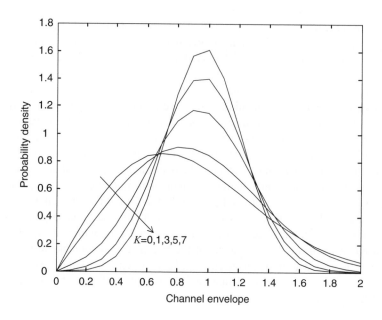

Figure 2.8   Distribution of Rician fading channels for different values of K

channel where there is no multipath component except the LOS one. Figure 2.8 compares the distributions of Rician fading channel for different values of K but with the same average channel gain 0 dB (unity on linear scale). Note that the distribution converges to the average value as K increases.

Sometimes, neither the Rayleigh nor the Rician model may match with the experimental results due to the application of the central limit theorem, especially when the number of multipaths is not large enough. Thus, an alternative model is necessary to match the fading channel statistics. The Nakagami-m distribution is often used to characterize the statistics of a channel envelope, whose probability density is given by

$$f_R(r) = \frac{2m^m r^{2m-1}}{\Gamma(m)\bar{g}^m} \exp\left[-\frac{mr^2}{\bar{g}}\right] \tag{2.27}$$

for a parameter $m \geq 0.5$ and the average channel gain $\bar{g}$. Here, $\Gamma(m)$ is the gamma function which is defined by

$$\Gamma(m) = \int_0^\infty x^{m-1} e^{-x} dx, \quad \text{for} \quad m > 0. \tag{2.28}$$

The Nakagami distribution reduces to the Rayleigh distribution when $m = 1$. In addition, we obtain the PDFs whose tails are heavier than that of a Rayleigh-distributed one for m in the range $0.5 \leq m \leq 1$. This implies that the Nakagami model renders a severer channel fluctuation than both the Rayleigh and Rician models. When the number of multipaths is not large, the channel envelope, in general, fluctuates more severely as it is more probable that all the multipath components are added to one direction, constructively or destructively. The Nakagami distribution is a useful model in characterizing this type of channel fading.

2.3.1.2 Doppler Spread and Coherence Time

Consider now the channel variation during the observation time difference $\Delta\tau$. To characterize this variation, consider the autocorrelation function of $h(t; \tau)$ for a fixed time delay t by

$$A_t(\Delta\tau) = \mathbb{E}[h^H(t; \tau)h(t; \tau + \Delta\tau)], \tag{2.29}$$

where h^H is the hermitian of h. The autocorrelation function in (2.29) defines how the channel response is correlated over time. It is strongly related with the Doppler effects caused by the dynamics of the transmitter–receiver pair or the scattering objects. Denote by f_{D_n} the Doppler frequency of the nth multipath component. A different multipath component, in general, has a different Doppler frequency as the direction of the scatterer mobility and the angle of arrival differ for different components. We can get the power spectrum of the Doppler effect in the channel by applying the Fourier transform to (2.29):

$$S_t(\lambda) = \int_{-\infty}^{\infty} A_t(\Delta\tau)e^{-j2\pi\lambda\Delta\tau} d\Delta\tau. \tag{2.30}$$

The function $S_t(\lambda)$ calculates the average power output of the channel experienced by the components having the Doppler frequency λ. Hence, we may call this function the *Doppler power spectrum* of the channel. Figure 2.9 illustrates the relation existing between the autocorrelation function $A_t(\Delta\tau)$ and the Doppler power spectrum $S_t(\lambda)$.

The Doppler power spectrum implies how the bandwidth of the output signal is spread. A multipath component with Doppler frequency f_{D_n} makes that amount of frequency shift specified by the spectrum. As a result, the receiver with the maximum Doppler frequency $f_D = \max_n |f_{D_n}|$ experiences a spread of signal bandwidth in the interval $[-f_D, f_D]$. This

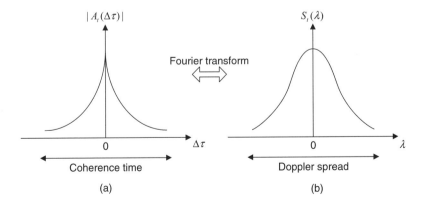

Figure 2.9 Illustration of (a) autocorrelation function $A_t(\Delta\tau)$ and (b) Doppler power spectrum $S_t(\lambda)$ (Reproduced with permission from J. Proakis, *Digital Communications*, 3rd ed., McGraw-Hill, New York, NY. © 1995 McGraw-Hill)

implies that the bandwidth of the received signal becomes slightly larger than that of the transmitted signal. The range of the values of λ over which the Doppler power spectrum is non-zero is called the *Doppler spread*, B_D, of the channel. The Doppler power spectrum is spread over the interval $[-f_D, f_D]$, so the Doppler spread can be approximated by $B_D \approx f_D$ where the maximum Doppler frequency is given by $f_D = f_c v/c$ for the carrier frequency f_c, the velocity of the transmitter-receiver pair v, and the speed of light c. Note that the Doppler spread is proportional to the carrier frequency and to the mobility of the transmitter–receiver pair.

The Doppler spread is closely related to the speed of channel variation. From the reciprocal relation of the Fourier transform, the range over which the autocorrelation function $A_t(\Delta\tau)$ is non-zero is inversely proportional to the Doppler spread B_D. Specifically, if we define by the *coherence time* T_C the maximum length of the observation time within which the channel is correlated, it can be approximated by

$$T_C \approx \frac{1}{B_D}. \tag{2.31}$$

That is, the coherence time is inversely proportional to the Doppler spread. This implies that the channel seen at the second observation can be regarded as an uncorrelated one if the coherence time has elapsed since the first observation. The relation between the coherence time and Doppler spread is illustrated in Figure 2.9.

It is obvious that more dynamics in the communication environment renders a higher Doppler frequency, which implies a larger Doppler spread. Also, this gives a smaller coherence time, which implies that the channel condition changes more rapidly. Note that a time-invariant channel renders a constant autocorrelation function (that is, $A_t(\Delta\tau) = c$)), and in this case the Doppler power spectrum becomes a delta function: $S_t(\lambda) = c\delta(\lambda)$. This implies that there is no Doppler spread in a time-invariant channel. Therefore, when the carrier frequency or the mobility of the transmitter–receiver pair increases, the channel response estimated by the transmitter–receiver pair becomes invalid sooner and thus a more prompt adaptation is required to the channel variation.

The coherence time plays an important role in relation to the delay requirement of the applied traffic. If the coherence time is much shorter than the delay requirement, a transmitting data unit is much more likely to experience several independent channel states. In such a *fast-fading* case, it is permissible to transmit the data unit adaptively to different channel fades it undergoes. On the other hand, if the coherence time is much longer than the delay requirement, the channel state hardly changes while transmitting a data unit. In such a *slow-fading* case, the data unit can be transmitted under the given fixed channel state. It is noteworthy that the speed of channel fading is a relative factor and the type of application is also an important factor in determining whether the channel fading is fast or slow.

2.3.2 Fading in the Frequency Domain

Formulating channel fading in the frequency domain involves determining a mathematical method that describes how the channel response $h(t; \tau)$ appears as the time delay t passes for a fixed observation time τ. For this, we define the autocorrelation function of $h(t; \tau)$ for a fixed observation time τ by

$$A_\tau(t_1, t_2) = \mathbb{E}[h^H(t_1; \tau)h(t_2; \tau)]. \tag{2.32}$$

Since it is reasonable to assume that multipath components with different path delays are uncorrelated, (2.32) can be written as

$$A_\tau(t_1, t_2) = A_\tau(t_1)\delta(t_1 - t_2), \tag{2.33}$$

where $A_\tau(t) = \mathbb{E}[h^H(t; \tau)h(t; \tau)]$. This relation implies that the autocorrelation can have a non-zero value only when $t_1 = t_2$. Thus, in characterizing channel fading in the frequency domain, it suffices to consider the function $A_\tau(t)$ which gives the average channel gain seen by the receiver with the time delay t. In other words, $A_\tau(t)$ is equal to the average channel gain of the multipath component that has the path delay t. Hence, $A_\tau(t)$ is called the *multipath intensity profile* of the channel. The range of the values of t over which the multipath intensity profile is non-zero is called the *delay spread*, T_M, of the channel.

The multipath intensity profile is determined by various factors of the communication environment. The major factor is whether there exist some objects that can make multipaths with relatively large time delays. Usually, a rich scatterer environment renders a longer delay spread. Table 2.1 lists examples of multipath intensity profiles in various communication environments.

The multipath intensity profile with a non-zero delay spread necessarily renders channel fading in the frequency domain. This means that signal components carried by different tones may experience different channel gains. If we consider the channel correlation between two tones with different frequencies f_1 and f_2 in the form

$$S_\tau(f_1, f_2) = \mathbb{E}[H^H(f_1; \tau)H(f_2; \tau)], \tag{2.34}$$

we get

$$S_\tau(f_1, f_2) = \int_{-\infty}^{\infty} \mathbb{E}[h^H(t; \tau)h(t; \tau)] e^{-j2\pi(f_2 - f_1)t} dt \tag{2.35}$$

Table 2.1 Example of multipath intensity profiles for different communication environments (© 2006. 3GPPTM TSs and TRs are the property of ARIB, ATIS, CCSA, ETSI, TTA and TTC who jointly own the copyright in them. They are subject to further modifications and are therefore provided to you "as is" for information purposes only. Further use is strictly prohibited.)

Multipath number	Rural area		Hilly terrain		Urban area	
	Relative time (μs)	Average power (dB)	Relative time (μs)	Average power (dB)	Relative time (μs)	Average power (dB)
1	0.0	0.0	0.0	0.0	0.0	−3.0
2	0.1	−4.0	0.1	−1.5	0.2	0.0
3	0.2	−8.0	0.3	−4.5	0.5	−2.0
4	0.3	−12.0	0.5	−7.5	1.6	−6.0
5	0.4	−16.0	15.0	−8.0	2.3	−8.0
6	0.5	−20.0	17.2	−17.7	5.0	−10.0

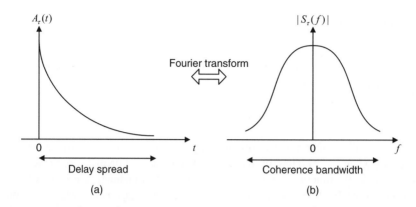

Figure 2.10 Relation between (a) multipath intensity profile $A_\tau(t)$ and (b) frequency spectrum $S_\tau(f)$ (Reproduced with permission from J. Proakis, *Digital Communications*, 3rd ed., McGraw-Hill, New York, NY. © 1995 McGraw-Hill)

by applying the relation $\mathbb{E}[h^{\mathrm{H}}(t_1; \tau)h(t_2; \tau)] = 0$, for $t_1 \neq t_2$. On putting $f = f_2 - f_1$ and applying (2.33), equation (2.35) can be rewritten as

$$S_\tau(f) = \int_{-\infty}^{\infty} A_\tau(t)e^{-j2\pi ft}\mathrm{d}t, \qquad (2.36)$$

which is the Fourier transform of the multipath intensity profile $A_\tau(t)$. The function $S_\tau(f)$ represents the correlation between two sinusoidal signals with frequency difference f. Hence, we may call it the *frequency spectrum* of the channel. Figure 2.10 illustrates the relation existing between the multipath intensity profile $A_\tau(t)$ and the frequency spectrum $S_\tau(f)$.

The range of f over which the frequency spectrum is non-zero is called the *coherence bandwidth*, B_C, of the channel. Two sinusoidal frequencies separated further than B_C experience uncorrelated channels. Similarly to the relation between the Doppler spread and the coherence time, the delay spread and the coherence bandwidth are also related reciprocally; that is

$$T_M \approx \frac{1}{B_C}. \qquad (2.37)$$

The relation between the delay spread and the coherence bandwidth is illustrated in Figure 2.10.

The coherence bandwidth gives an important interpretation about the distortion of the transmitted signals. If the coherence bandwidth is small compared with the bandwidth of the transmitted signal, some sinusoidal components of the signal may undergo channel fades differently from the others. As a result, the signal is severely distorted by the channel and it becomes difficult for the receiver to recover the original signal. We call this case the *frequency-selective fading channel*, in the sense that the channel fluctuates over the bandwidth of the transmitted signal. On the other hand, if the coherence bandwidth is large, all the frequency components of the transmitted signal experience the same channel state. In this case, the channel seen by the receiver is flat in the frequency domain, so we call this case the *frequency-flat fading channel*. Note that when the multipath intensity profile is a delta function, the

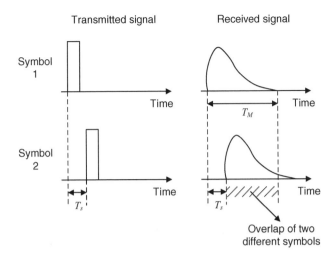

Figure 2.11   Illustration of inter-symbol interference

frequency spectrum becomes constant regardless of the frequency by the relation in (2.35); that is, the channel has zero delay spread.

As noted above, the frequency selectivity of the channel distorts the transmitted signals as illustrated below. From the reciprocal relation between the coherence bandwidth and delay spread, a signal bandwidth larger than the coherence bandwidth (i.e., $B > B_C$) is equivalent to a symbol time shorter than the delay spread (i.e., $T_s = 1/B < T_M$). Thus, in a frequency-selective fading channel, a new symbol is transmitted into the channel before the previously transmitted symbol finishes its arrival at the receiver. As a result, a symbol is interfered with by the adjacent symbols as illustrated in Figure 2.11. This is called *inter-symbol interference* (ISI). The effect of ISI becomes more severe for high-speed communication systems which employ wider bandwidth; that is, shorter symbol time.

Inter-symbol interference can be eliminated, or at least reduced, by employing an equalizer at the receiver. The basic principle of equalization is to pass the received signal through a filter with a properly designed response. Figure 2.12 shows an example of a commonly used equalizer structure that consists of a linear non-recursive N-tap filter. The coefficient connected with each tap is determined through an adaptive processing such that the desired performance objective is achieved based on the channel estimation helped by a training sequence. Ideally, ISI would be eliminated if the frequency response of the equalizer is made equal to the inverse of that of the channel. However, this is hardly possible as the number of filter taps is limited and exact channel estimation is required for each multipath component. Thus, an equalizer is usually designed based on more practical criteria such as the minimization of the mean-squared error. From the perspective of the frequency domain, the equalizer tries to make the frequency response of the channel as flat as possible by determining its response. From the perspective of the time domain, the equalizer tries to cancel the ISI by subtracting the previously transmitted symbols from the currently received signal.

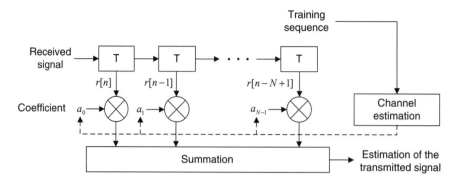

Figure 2.12   Example of equalizer structure

2.4 Technologies against Channel Fading

As described in the previous sections, wireless channels generally fluctuate in both time and frequency domains. Channel fading makes more errors in detecting a given transmitted signal. Suppose that the error probability of a data unit is given by $f_e(\gamma)$ for a *signal-to-noise ratio* (SNR) γ.[2] In many cases, the error probability is assumed to be a convex function of the SNR in the region of interest, such as $f_e(\gamma) < 10^{-1}$. This means that an increase in SNR does not contribute much in reducing the error probability but the error probability rapidly increases as the SNR decreases.

Now consider the error probability of a fading channel which incites fluctuation in the SNR around the average value $\bar{\gamma}$. In this case, the average error probability is given by $\mathbb{E}[f_e(\gamma)]$ but, by Jensen's inequality and the convexity assumption of the error probability, we get

$$\mathbb{E}[f_e(\gamma)] \geq f_e(\mathbb{E}[\gamma]) = f_e(\bar{\gamma}). \tag{2.38}$$

This indicates that more errors occur in a fading channel than in a non-fading channel with the same average SNR. The error probability curve in Figure 2.13 explains this phenomenon. The figure demonstrates that the average error probability becomes higher as the SNR fluctuates more around the fixed average value. It happens due to the convexity of the curve: That is, the error probability increases rapidly in the low SNR region but decreases slowly in the high SNR region. As a consequence, channel fading becomes a major source that makes wireless communications unreliable.

Various techniques have been introduced to overcome such unreliability and reinforce the performance of wireless systems in fading channels. There are two major approaches that mitigate the negative effect of channel fading. One is to reduce the channel fluctuation and make the channel more reliable. A representative method is the *diversity technique* which diversifies the channel into multiple independent channels such that at least one of them falls in a good channel state. Besides this, the *equalization technique* discussed in Section 2.3.2 and the *orthogonal frequency-division multiplexing* (OFDM) technique which divides the whole bandwidth into multiple narrow subchannels may be included in this category.[3] The

---

[2] The data unit may be a single bit or a frame which consists of multiple data bits.
[3] See Section 3.3.4 for detailed discussion of OFDM.

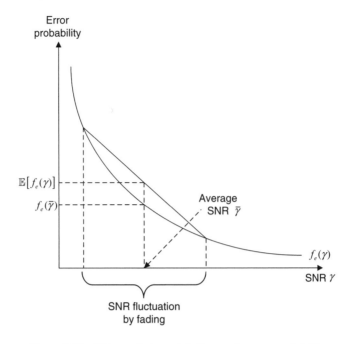

Figure 2.13   Effects of channel fading on the error probability

other approach is to operate the communication system adaptively to the channel fluctuation. This approach includes the hybrid ARQ technique which is a hybridization of *forward error correction* (FEC) and *automatic repeat request* (ARQ) and the *adaptive modulation and coding* (AMC) technique which adjusts the transmission rate according to the state of the fading channel. In the following, the diversity, hybrid ARQ, and AMC techniques are discussed.

2.4.1 Diversity

Diversity techniques mitigate the effect of channel fading by using multiple independent channels. If multiple replicas of the same information signal are transmitted through independent fading channels, the probability that all the channels undergo deep fading at the same time substantially decreases. In this situation it is much more likely that at least one channel stays in a favorable condition and the information can be delivered more reliably through that favorable channel. The effect of diversity may be described by using probability as follows: If the probability that one channel suffers from deep fading is p, then the probability that the N independent channels provided by the diversity technique go into deep fading all together drops to p^N. The number of independent channels, N, is called the *diversity order* and is represented as the slope of the error probability curve plotted on a log scale.

There are several ways to implement the diversity technique. With *time diversity* the same signal is transmitted in N different time slots. In order to secure the independency of each time slot, we should separate out two successive time slots by more than the coherence time of the channel. With *frequency diversity* the same signal is transmitted over N (sub)carriers, with

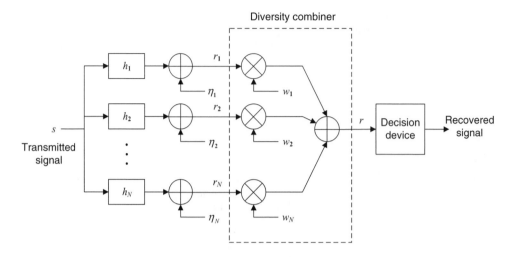

Figure 2.14   Functional block diagram of an Nth-order diversity combiner

the separation of the (sub)carrier frequencies made larger than the coherence bandwidth of the channel. *Space diversity* achieves the diversity effect by using multiple transmitter and/or receiver antennas. In this case the multiple antennas are placed sufficiently apart from each other (usually more than 10 wavelengths) such that the multipath fading in each antenna becomes nearly independent. There may be other ways to achieve diversity, such as *angle diversity* and *polarization diversity*, but they are not used as widely as the above three methods.

Now we need to consider how to combine the N independent diversity components for an effective detection of the transmitted signal. Whichever diversity scheme may be used, all can be modeled by a combination of N replicas of the same signal transmitted over independent channels. Write the nth diversity sample of the common transmitted signal s by

$$r_n = h_n s + \eta_n \tag{2.39}$$

for the diversity channel function h_n and the additive noise η_n of the nth channel, both of which are complex variables. The noise term η_n is assumed to be a zero-mean white Gaussian process with an identical variance σ^2. A single reliable signal r is generated by combining these N samples linearly; that is

$$r = \sum_{n=1}^{N} w_n r_n = hs + \eta \tag{2.40}$$

for a proper weighting factor w_n. Then the decision device restores the transmitted signal from r. Figure 2.14 depicts this combining process schematically based on (2.39) and (2.40).[4]

---

[4] Actually, this formulation corresponds to the receive diversity where multiple diversity samples arrive at the receiver and are combined into a single decision variable. We can also consider the transmit diversity where the same signal is transmitted through multiple transmitting antennas and their superposition arrives at the single receiving antenna. Here, we omit the formulation and solution of the transmit diversity as they are analogous to the receive diversity case. Detailed discussion of the transmit diversity is provided in Chapter 7 in the context of multiple antenna management.

There are three different approaches to diversity combining. First, *selection combining* is the simplest method that arranges the weighting factors w_n such that the sample with the largest channel gain is selected; that is

$$w_n = \begin{cases} 1, & \text{if} \quad n = \arg\max_m |h_m|^2 \\ 0, & \text{otherwise.} \end{cases} \tag{2.41}$$

Second, *equal-gain combining* (EGC) arranges w_n such that the phase differences of the samples are eliminated and thus all the diversity components can be added constructively. This means that the weighting factors are given by

$$w_n = \exp[-j\angle h_n]. \tag{2.42}$$

Third, *maximal-ratio combining* (MRC) arranges w_n such that the SNR of the combined signal r in (2.40) can be maximized. As the MRC maximizes the SNR which governs the reliability of the transmission, it generally performs better than the other combining techniques.

Now consider the weight factors of the MRC in more detail. The SNR of the combined signal is defined by

$$\gamma = \frac{|hs|^2}{\mathbb{E}|\eta|^2} = \frac{\left|\sum_{n=1}^{N} h_n w_n\right|^2 |s|^2}{\mathbb{E}\left|\sum_{n=1}^{N} w_n \eta_n\right|^2}. \tag{2.43}$$

Since the noise process η_n is assumed to have zero mean and be independent of the at other channels, we get

$$\gamma = \frac{\left|\sum_{n=1}^{N} h_n w_n\right|^2 |s|^2}{\mathbb{E}\left[\sum_{n=1}^{N} |w_n|^2 |\eta_n|^2\right]} = \frac{\left|\sum_{n=1}^{N} h_n w_n\right|^2 |s|^2}{\sigma^2 \sum_{n=1}^{N} |w_n|^2}. \tag{2.44}$$

By applying the Schwarz's inequality, we get

$$\left|\sum_{n=1}^{N} h_n w_n\right|^2 \le \left(\sum_{n=1}^{N} |h_n|^2\right)\left(\sum_{n=1}^{N} |w_n|^2\right), \tag{2.45}$$

where the equality holds when $w_n = ch_n^H$ for a constant c. By applying (2.45) to (2.44), we get

$$\gamma \le \sum_{n=1}^{N} \frac{|h_n|^2 |s|^2}{\sigma^2}, \tag{2.46}$$

and the maximal SNR yields when

$$w_n = ch_n^H. \tag{2.47}$$

Note in (2.47) that the MRC first removes the phase differences among the diversity samples by multiplying the complex conjugate of each diversity channel. Note also that the MRC gives

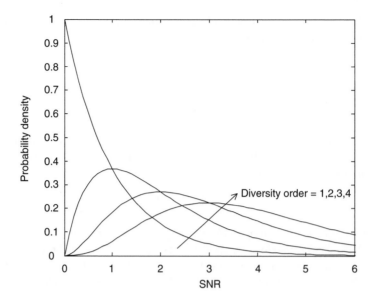

Figure 2.15 Distribution of signal-to-noise ratio (SNR) after applying the MRC

more weight on the sample with better channel condition by determining the weighting factor in proportional to the amplitude of each channel. One interesting result is that the maximal SNR achieved by (2.47) is the summation of the SNRs of each channel (i.e., diversity component). As a consequence, a deep fading in a channel can be compensated by another channel with a high SNR. In this sense, the MRC may be said to fully utilize all the signal power contained in the diversity samples. This performance improvement can be achieved on the condition that both the phase and amplitude of each diversity channel are exactly known.

As the signal-to-noise ratio achieved by the MRC becomes the sum of the SNR at each channel, the SNR fluctuation reduces significantly as the diversity order N increases. Figure 2.15 shows how the probability density of the SNR changes after applying the MRC to N i.i.d. Rayleigh fading channels with the average SNR $\bar{\gamma}_n = 1$. Under this assumption, the distribution of the combined SNR is χ^2 with $2N$ degrees of freedom, with the probability density (Viniotis, 1998)

$$p(\gamma) = \frac{\gamma^{N-1} e^{-\gamma/\bar{\gamma}}}{\bar{\gamma}^N (N-1)!}, \tag{2.48}$$

whose expected value is $N\bar{\gamma}_n$. It is noteworthy that the average SNR of the combined signal increases in proportion to the diversity order N. Figure 2.15 reveals that the probability of suffering from deep fading decreases drastically as the diversity order increases.

2.4.2 Hybrid ARQ

Hybrid ARQ is a technique that combines two representative transmission error recovery techniques, ARQ and FEC.

ARQ is a technique that recovers transmission error by requesting a retransmission of the same data to the transmitter. An ARQ scheme requires an error-detecting mechanism such as the *cyclic redundancy check* (CRC) and a feedback channel that reports the successful/failed reception of a data block. If the reception is successful, the receiver sends an *acknowledgement* (ACK) message and, otherwise, a *negative acknowledgement* (NAK) to signal a retransmission request. Since ARQ confirms whether or not a data block is reliably delivered, it is an essential technique in wireless data networks.

FEC is a technique that recovers bit errors in the receiver by utilizing the redundancy carried over the channel-coded data. In essence, the channel coding technique helps to overcome the unreliability of wireless channels at the cost of added redundancy. The redundancy part, called *parity bits*, does not contain any new information but increases the dimension of the signal space and, as a result, increases the distance among different encoded sequences of the information bits. According to information theory, the transmission errors can be corrected if the number of erred bits does not exceed half the minimum distance of the employed channel coding scheme. Such a coding gain is, therefore, achieved at the cost of a lower information rate. Coding gain increases as the information rate decreases; that is, as the number of added redundancy bits increases.

Channel coding schemes can be categorized based on how to generate a codeword from the information bits. One category of the channel coding schemes is the *linear block code*. This uses several parity bits in a block of data bits to be able to detect and correct transmission errors. The parity bits are determined by a linear combination of the data bits in a finite field. By multiplying a proper parity-check matrix to the received bits, linear block codes can identify the error pattern as long as it is detectable. Another category is the *convolutional code* which generates the coded bit sequence by passing the data bits through a linear finite-state shift register. Since each bit of the coded sequence is a linear combination of the data bits in the shift register, the original data bits can be recovered by tracing which state of the shift register generates the received sequence. This operation can be easily done by the trellis diagram which illustrates the relation between the shift register state and the generated sequence. Yet another category is the *concatenated code*. This uses two levels of channel coding (i.e., a linear block code and a convolutional code) with an interleaver intervening between them for the randomization of the coded sequence. One advantage of the concatenated code is that it can employ an iterative decoding where a decoder can utilize the output of the other decoder as a pre-knowledge. Owing to this merit, concatenated codes such as turbo codes generally achieve very low error probability with reasonable complexity.

The ARQ technique can operate more efficiently when used in conjunction with the FEC technique. This hybridization of FEC and ARQ is called *hybrid ARQ* (HARQ). The simplest version of HARQ, known as type I, simply combines FEC and ARQ by encoding the data block with a channel coding scheme prior to transmission. When the coded data block is received, the receiver first decodes the error-correction code. If all transmission errors are correctable by the employed channel coding, and the receiver can obtain the correct data block, it feeds back an ACK message. If all transmission errors are not correctable, the received coded data block is discarded and a retransmission is requested by the receiver via a NAK message, similar to ARQ. Since a small amount of errors can be corrected by FEC without a retransmission, this type of HARQ is more effective to the fading channels than ordinary ARQ mechanisms. Figure 2.16 plots the operational block diagram of the HARQ mechanism in general.

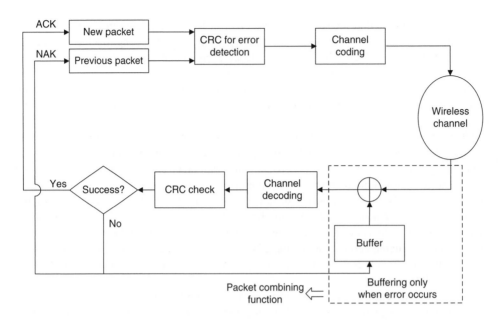

Figure 2.16   Operational block diagram of HARQ in general

A simple hybridization of FEC and ARQ can render considerable performance improvement, but the real advantage of HARQ comes from the *packet combining function*, which
utilizes the information contained in the erroneously received block. In the packet combining
function, the erred data blocks are stored at the receiver rather than discarded and are combined
with the retransmitted block before being fed to the decoder of the error-correction code. When
the transmitter repeats sending the same coded data block in retransmissions as in type I HARQ,
a newly received block can be combined with the previous ones by applying the MRC to their
soft value (i.e., the amplitude and phase of each modulated symbol before entering the hard
decision process). This kind of data block combining is called *Chase combining* and can be
regarded as type I HARQ with soft combining (Chase, 1972). As it is probable that a relatively
small number of bits are erred in an erroneously received block when compared with the total
number of bits transmitted, an erred data block still has much useful information even though
the number of bit errors in it exceeds the error-correction capability. Thus, combining the
received data blocks helps in reinforcing the reliability of each bit of the transmitted codeword.
Obviously, this improvement is achieved at the cost of additional memory deployment at the
receiver, as shown in Figure 2.16.

The performance of the packet combining function can be improved further, by employing
type II/III HARQ. In this scheme, different (re)transmissions are encoded differently rather
than simply repeating the same coded bits as in Chase combining. The received data blocks are
combined into a single codeword and this combined codeword can be decoded more reliably
since coding is effectively done across retransmissions. A typical operation of type II HARQ is
illustrated in Figure 2.17. The information bits are first encoded by a low-rate channel coder
(usually called the "mother code"). At the initial transmission, the information bits (sometimes
called the "systematic bits") and a selected number of parity bits (Parity 0 in the figure) are

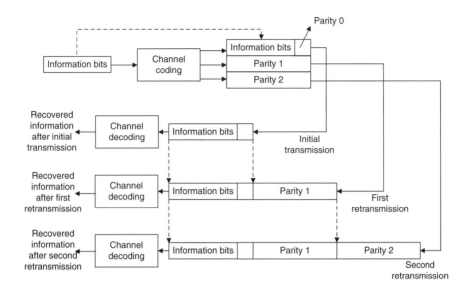

Figure 2.17 Operational block diagram of type II HARQ

transmitted. If the transmission is not successful, the transmitter sends additional selected parity bits (Parity 1, 2) in the following retransmissions. The receiver puts together the newly received parity bits with those previously received. This operation produces a new codeword with a lower code rate (i.e., with more parity bits). Thus, the receiver can decode a stronger codeword as more retransmissions are received. As new parity bits are added after each retransmission, type II HARQ is sometimes called *incremental redundancy* (IR), in contrast to the Chase combining for which the number of parity bits does not change. Usually, the frame used in each (re)transmission is obtained by puncturing the output of the mother code. The punctuation pattern used during each (re)transmission is different, so different coded bits are sent at each time.

The difference of type III HARQ from type II is that the retransmission packets are made decodable by themselves. This means that each retransmission in type III contains the same information bits with different parity bits, in contrast to the case of type II in which no information bits were retransmitted. In this sense, each retransmission in type III can be said to be *self-decodable*.

Figure 2.17 shows how type III HARQ operates. One can easily recognize the difference from type II by comparing this figure with Figure 2.18. Type III HARQ may obtain a weaker channel coding gain than in type II but, due to the self-decodable property, it is advantageous in situations where the initial transmission was so severely damaged that the erred data block cannot be recovered by combining it only with the redundancy information delivered in the retransmissions of type II HARQ. It is noteworthy that the Chase combining can be regarded as a special case of type III HARQ where the parity part remains the same in each retransmission.

The HARQ technique can be interpreted as an *implicit* channel adaptation method. The unreliability of poorly conditioned channels can be overcome by using a stronger channel coding with a lower coding rate. When the channel suffers from deep fading, a HARQ protocol

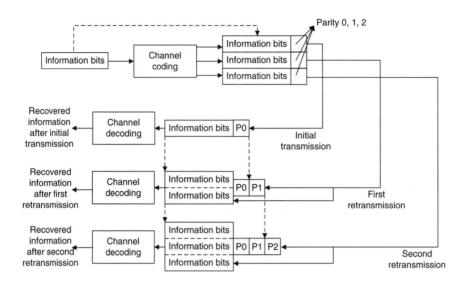

Figure 2.18   Operational block diagram of type III HARQ

automatically reduces the coding rate by requesting additional information to be delivered via retransmissions. Through several retransmissions, the receiver obtains a stronger codeword which can be successfully decoded against high level of noise and interference. On the other hand, a single initial transmission is sufficient to deliver the data block when the channel condition is favorable. Consequently, the transmitter–receiver pair equipped with an HARQ adjusts its transmission rate adaptively to the current channel condition. This adaptation is performed without any explicit message that indicates the channel condition or the desirable transmission rate, but this information is implicitly contained in the ACK/NAK message fed back from the receiver to the transmitter.

2.4.3 Adaptive Modulation and Coding

AMC refers to the transmission technique which adjusts the transmission rate based on the current channel condition. If the channel is in good condition, AMC increases the transmission rate by using a higher order modulation and/or a higher rate channel coding. A poorly conditioned channel is overcome by reducing the modulation order and adding more redundancy bits in channel coding. This enables a robust and spectrally efficient transmission over time-varying fading channels: It brings in higher throughput when the channel is in a good state and brings in reliable communication when the channel suffers from deep fading.

The AMC technique inherently requires a channel estimation process at the receiver and a feedback mechanism to report the estimated channel condition to the transmitter. So in implementing AMC it is important to arrange that the channel information is kept as accurate as possible. This requires making the delay caused by the channel-estimating and reporting process less than the coherent time of the channel. Any AMC technique will perform poorly if the channel is changing faster than this delay. Once the transmitter receives the channel status

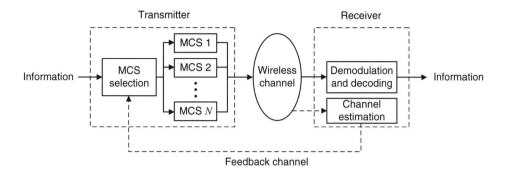

Figure 2.19   Operational block diagram of AMC techniques

information, it selects the *modulation and coding scheme* (MCS) best fitted to the received channel condition. Since the transmission scheme is adapted based on the reported current channel condition, AMC can be interpreted as an *explicit* channel adaptation method in contrast to the HARQ case where simple ACK/NAK messages are used. Figure 2.19 illustrates the overall operation of the AMC technique.

Now we consider how to determine the appropriate MCS (i.e., the transmission rate) from the reported channel condition. In general, the MCS scheme is determined such that the transmission rate can be maximized while providing the required level of reliability. This means that the *spectral efficiency* (SE), which is defined by the amount of information bits delivered over one single transmission symbol, should be maximized while the error probability is kept below a certain level. Ideally, the maximum spectral efficiency achievable under a given *additive white Gaussian noise* (AWGN) channel is given by the well-known Shannon's capacity equation

$$\text{Spectral efficiency, } SE = \log_2(1 + \gamma)(\text{bits}/\text{sec}/\text{Hz}) \tag{2.49}$$

for the SNR γ of the channel. This equation gives the maximum of information-theoretically achievable spectral efficiency with arbitrarily small error probability. However, it is nearly impossible to attain this capacity in practical communication as it requires an extremely long codeword.

A more practical model is developed by using an upper bound for the bit error rate in *M*-ary modulation schemes (Goldsmith and Chua, 1997). For an AWGN channel with *M*-ary QAM, the bit error rate under ideal coherent phase detection is bounded by

$$P_b \leq 2\exp\left[\frac{-1.5\gamma}{M-1}\right], \tag{2.50}$$

where the *M*-ary QAM implies the quadrature amplitude modulation which uses M signal points in the signal constellation. If we restrict the system environment by $M \geq 4$ and $0 \leq \gamma \leq 30\,\text{dB}$, we can obtain a tighter upper bound; that is

$$P_b \leq 0.2\exp\left[\frac{-1.5\gamma}{M-1}\right]. \tag{2.51}$$

Table 2.2 Data rate set of the forward link of CDMA2000 1xEV-DO system (Reproduced with permission from TTAT.3G-C.S0024-A, IMT-2000 3GPP2—CDMA2000 high rate packet data air interface specification, ver.3.0, December, 2007. © 2007 TTA)

Data rate (kbps)	Packet length (bits)	Number of slots	FEC rate	Modulation	Average repetition
38.4	1024	16	1/5	QPSK	9.6
76.8	1024	8	1/5	QPSK	4.8
153.6	1024	4	1/5	QPSK	2.4
307.2	2048	4	1/3	QPSK	1.2
614.4	2048	2	1/3	QPSK	1.02
921.6	3072	2	1/3	8PSK	1.02
1228.8	2048	1	1/3	QPSK	0.5
1843.2	3072	1	1/3	8PSK	0.5
2457.6	4096	1	1/3	16QAM	0.5

The achievable spectral efficiency can be related with the channel condition by inverting the above upper bounds. Consider first the case of (2.50). Suppose that the reliability requirement is given by providing a *bit error rate* (BER) lower than the target level P_b. Then, this requirement can be satisfied by using the modulation order M which meets the upper bound in (2.50) with equality. By inverting (2.50), we get

$$M = 1 + \frac{-1.5\gamma}{\ln(0.5P_b)}. \tag{2.52}$$

We can deliver $\log_2 M$ information bits with M-ary modulation, so the spectral efficiency becomes

$$SE = \log_2\left(1 + \frac{-1.5\gamma}{\ln(0.5P_b)}\right) = \log_2\left(1 + \frac{\gamma}{\Gamma}\right) \tag{2.53}$$

for $\Gamma = -\ln(0.5P_b)/1.5$. By comparing (2.53) with (2.49), note that the spectral efficiency that is achievable practically is equal to the information-theoretical capacity with the SNR degraded by Γ. Thus, we may interpret Γ as the *SNR gap* which is caused by the gap between the practical implementation and the information-theoretic result. This SNR gap is a decreasing function of the target BER P_b. This means that, as the reliability requirement gets stringent and the target BER decreases, the transmitter has to reduce the transmission rate by securing more SNR gap. If the spectral efficiency is obtained from (2.51), only the SNR gap changes to $\Gamma = -\ln(5P_b)/1.5$. Note that the service rate becomes the information-theoretical capacity when no SNR gap is assumed by letting $\Gamma = 1$.

As an example of practical implementations of the AMC technique, Table 2.2 lists the data rate set of the forward link of the CDMA2000 1xEV-DO system (3GPP2, 2001).[5] This system provides nine different data rates by using different FEC rates, modulation orders and repetition factors as listed in the table. The MCS with the lowest data rate, 38.4 kbps, delivers 1024

---

[5] The system in 3GPP2 (2001) has two different MCS modes for each of the 307.2, 614.4, and 1228.8 kbps, respectively. The table shows only one MCS for each transmission rate.

information bits over 16 slots where each slot time is 1.6 ms. The frame format for this MCS provides 24 576 modulation symbols and 1024 information bits are mapped into the provided symbol space as follows. A channel coding with the rate of 1/5 renders $1024 \times 5 = 5120$ bits and the QPSK modulation makes $5120/2 = 2560$ symbols. The whole 2560 symbols are repeated nine times and 1536 symbols among them are repeated once more. Then, the average repetition of a modulation symbol is equal to $9 + (1536/2560) = 9.6$. So we finally get $(2560 \times 9) + 1536 = 24\,576$ modulation symbols which match with the provided frame format. The MCS with the highest data rate, 2457.6 kbps, employs the 1/3 rate coding and 16QAM, and only half of the provided modulation symbols are transmitted (i.e., the average repetition is 0.5).

References

3GPP (2005) TR 25.814 V 1.0.1: *Physical Layer Aspects for Evolved UTRA*.

3GPP (2007) TS 45.005 V 7.9.0: *Radio Transmission and Reception*.

3GPP-2 (2000) C.S0024: CDMA2000: High Rate Packet Data Air Interface Specification, version 3.0.

Chase, D. (1972) A class of algorithms for decoding block codes with channel measurement information. *IEEE Transactions on Information Theory*, **IT-18**, 170–182.

Feuerstein, M.J., Blackard, K.L., Rappaport, T.S., Seidel, S.Y. and Xia, H.H. (1994) Path loss, delay spread, and outage models as functions of antenna height for microcellular system design. *IEEE Transactions on Vehicular Technology*, **43**, 487–498.

Goldsmith, A.J. and Chua, S.-G. (1997) Variable-rate variable-power MQAM for fading channels. *IEEE Transactions on Communications*, **45**, 1218–1230.

Lee, B.G. and Choi, S. (2008) *Broadband Wireless Access and Local Networks: Mobile WiMAX and WiFi*, Artech House.

Proakis, J.G. (1995) *Digital Communications*, McGraw-Hill.

Viniotis, Y. (1998) *Probability and Random Processes for Electrical Engineers*, McGraw-Hill.

3

Basic Concepts for Resource Management

Spurred by recent advances in portable multimedia devices, the demand for high-quality, high-speed wireless communications has increased drastically. The traffic carried over wireless communication systems has diversified in type, data rate, service duration, and service quality. In order to accommodate such a diverse set of requirements, systems should be capable of supporting very high data rates at very low latency and in very reliable ways. This may possibly be attained to some extent by developing new technologies in the physical and network layers. However, it is basically impossible to fully attain the goals owing to limited resources, so resource management has to get involved to help with efficient utilization.

Resource management in wireless communications refers to a series of processes that determine the timing, ordering, procedures, and the amount of wireless resources to allocate to each user. The wireless link tends to become a bottleneck, but wireless resource management aims at delivering the required service quality of each user as far as possible. Resource management is necessary for every wireless network, no matter what capability it may have. This is why wireless resource management has attracted so much research interest.

For a resource management technique to be effective, it is necessary to define certain factors. First, there is a set of available wireless resources to share among the constituent users. Second, there is the information available at the resource manager and the methods of information exchange among the users or protocol layers. Third, there are the service requirements of each user, which may be differently determined depending on the traffic characteristics and the performance metric. Fourth, there are the objectives to optimize in relation to the performance metric of the service provider.

This chapter discusses various concepts essential to formulating and solving the resource management problems. It first examines types of wireless resources and their characteristics and considers a generic formulation of resource management problems. Then it investigates various multiple-access methods that concretize the wireless resources into different forms of allocation. Next it discusses how to reflect each user's quality of service in formulating resource management problems. Finally, it considers the operation of wireless resource management in terms of the layered architecture of communication protocols.

3.1 Definition of Resource Management

Wireless resource management is a series of processes needed to determine the timing and the amount of relevant resources to allocate to each user. It is necessary to define first what types of wireless resource are to be allocated, and then define the objectives that the resource management tries to optimize and the constraints that restrict the degree of freedom in allocating resources.

3.1.1 Wireless Resources

In a transmitter–receiver pair equipped with multiple antennas, the information-theoretical capacity, which is the upper limit on the transmission rate supporting reliable information delivery, is given by (Telatar, 1999)

$$C = W \max_{\mathbf{Q}:\mathrm{Tr}(\mathbf{Q})=1} \log_2 \left[\det \left(\mathbf{I}_N + \frac{E_s}{N_0} \mathbf{HQH}^{\mathrm{H}} \right) \right] (\text{bits/sec}), \tag{3.1}$$

where W is the bandwidth, $\mathbf{Q}$ the covariance matrix of the transmitted signal, E_s the symbol energy, N_0 the noise spectral density, and $\mathbf{H}$ the $N_R \times N_T$ channel matrix for the channel with N_T transmit and N_R receive antennas. According to this equation, there are three different kinds of wireless resource – bandwidth, transmission power, and antennas.

3.1.1.1 Bandwidth

Bandwidth, denoted by W (in Hz) in (3.1), is the width of the spectrum band that the transmitter–receiver pair occupies. The bandwidth allocated to a transmitter–receiver pair determines the access rate to the channel, or the data rate for signal transmission.

Consider the maximum signal transmission rate constrained by limited bandwidth and the methods of achieving the maximum transmission rate. One could try to transmit signals at an arbitrarily fast rate in a limited bandwidth, but there would appear interference between the transmitted symbols; that is, *inter-symbol interference* (ISI). For example, consider a system that transmits 1 to send information bit 1 and −1 to send information bit 0 for a very long time duration. The bandwidth occupied by this signaling method is very narrow due to the reciprocal relation existing between the time duration and frequency bandwidth. Thus, by this method, a high symbol transmission rate can be achieved with very narrow bandwidth. Practically, however, the transmission rate of this method is limited by ISI. The symbols that were transmitted in the past affect the currently transmitting symbol, so the receiver cannot decode the transmitted symbols correctly.

In general, there exists an upper bound to the symbol transmission rate that does not suffer from ISI for a given bandwidth. This upper bound is determined by the *Nyquist criterion* as follows.

Suppose that a transmitter–receiver pair is given with a pass-band spectrum with a bandwidth W (Hz) and a center frequency of f_c (Hz). If a base-band signal $x(t)$ with frequency spectrum $X(f)$ is transmitted over this bandwidth, then its modulated signal $x_c(t)$ takes the expression

$$x_c(t) = x(t)\cos 2\pi f_c t, \tag{3.2}$$

with the frequency spectrum

$$X_c(f) = [X(f - f_c) + X(f + f_c)]/2. \tag{3.3}$$

Figure 3.1 depicts the spectra of the base-band and modulated signals. The figure confirms that if the given pass-band spectrum has a bandwidth W then the bandwidth of the base-band signal that can be transmitted over it is $W/2$. In this case, the maximum symbol rate that can be transmitted on this spectrum without suffering from ISI is given by

$$\max(R_s) = 2 \times W/2 = W(\mathrm{Hz}) \tag{3.4}$$

by the sampling theorem (Haykin, 2001).

Now consider a system that transmits the information symbol stream $x[n]$ with symbol rate $R_s = 1/T_s = W$. This implies that the system transmits one symbol in every T_s seconds. We modulate each symbol by the synch pulse $\mathrm{sinc}(Wt) = \sin(\pi Wt)/\pi Wt$, whose frequency response is $\mathrm{rect}(f/W)/W$ for

$$\mathrm{rect}(f) = \begin{cases} 1, & \text{if} \quad |f| \le 1/2, \\ 0, & \text{otherwise.} \end{cases} \tag{3.5}$$

Note that the above modulation function is the time response of an ideal low-pass filter as depicted in Figure 3.2. Then, the nth transmission is represented by $x[n]\mathrm{sinc}(W(t - nT_s))$ and

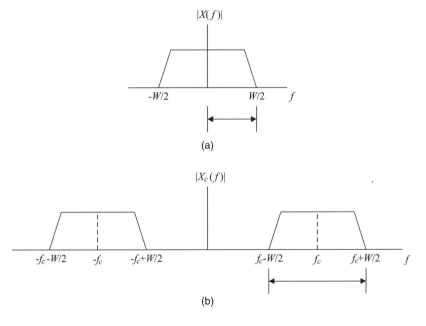

(a)

(b)

Figure 3.1   Spectra of (a) base-band signal, and (b) modulated (pass-band) signal

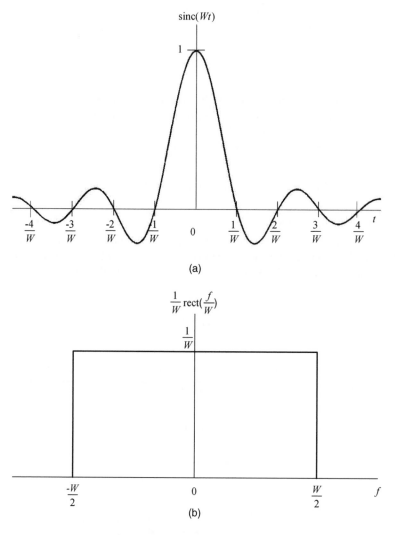

Figure 3.2 Response of the ideal low-pass filter in (a) time domain, and (b) frequency domain

hence the transmitted signal $x(t)$ takes the form

$$x(t) = \sum_{n=-\infty}^{\infty} x[n]\mathrm{sinc}(W(t-nT_s)) \tag{3.6}$$

as illustrated in Figure 3.3. If the channel is ideal (i.e., the channel impulse response is $\delta(t)$) and the receiver is perfectly synchronized to the transmitter, then the mth sample of the received signal with the sampling rate $R_s = 1/T_s = W$ is given by

$$x(mT_s) = \sum_{n=-\infty}^{\infty} x[n]\mathrm{sinc}(W(mT_s-nT_s)) = x[m] + \sum_{\substack{n=-\infty, \\ n\neq m}}^{\infty} x[n]\mathrm{sinc}(W(mT_s-nT_s)) = x[m].$$

$$\tag{3.7}$$

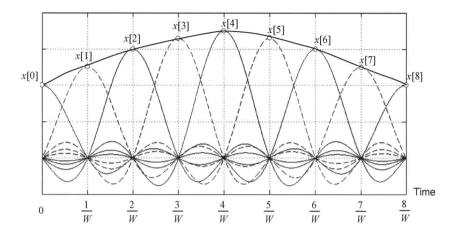

Figure 3.3 Example of transmitted signal using the time response of the ideal low-pass filter as the modulating pulse

The last equality is valid since sinc(n) = 0 for $n = \pm1, \pm2, \ldots$. This relation implies that the transmitted symbol stream can be detected by using an ideal low-pass filter, without being affected by ISI.

Now, we derive the bandwidth of the signal in (3.6). The Fourier transform of the transmitted signal is given by

$$X(j2\pi f) = \sum_{n=-\infty}^{\infty} x[n](e^{-j2\pi nT_s f}) \frac{1}{W} \text{rect}\left(\frac{f}{W}\right), \tag{3.8}$$

which becomes 0 for $|f| > W/2$ by (3.5). This implies that the bandwidth needed to send the signal in (3.6) in the pass-band is W. This shows that if an adequate form of symbol modulation is used, it is possible to achieve the symbol transmission rate W without suffering from the ISI in the system having a limited bandwidth W.

Consequently, the bandwidth W in (3.1) becomes the channel access rate – that is, the speed of transmitting the symbols. It implies that the transmitter can send a data symbol in every $1/W$ seconds. A larger bandwidth means a greater chance to access the wireless channel during a given time, and thus the transmission rate of a user increases in proportion to the bandwidth allocated to it. This proportional property makes it relatively easy to estimate the contribution of the bandwidth allocation to user performance. For example, the transmission of a user would be doubled if the allocated bandwidth is doubled.

In general, the whole radio spectrum is divided into a large number of frequency bands, each of which is dedicated to a specific wireless communication service prescribed by the licensing authority. An increment in bandwidth requires an increase in spectrum, so an increased user transmission rate requires an increased license fee. There is another choice of using the license-exempt band such as the *industrial, scientific and medical* (ISM) band at 2.4 GHz. However, no exclusive utilization is guaranteed in this band, so the transmitted signal in this band is subject to collision with other signals. Therefore, the cost of utilizing the frequency spectrum is the major limiting factor on the bandwidth. Chapter 5 provides detailed discussions on bandwidth.

3.1.1.2 Transmission Power

The symbol energy E_s in (3.1) is the energy of each symbol delivered to the receiver. In general, a higher received energy better mitigates the effect of noise and improves the quality of the wireless link. With this improved quality, the transmitter can put more information onto each single transmission symbol, thereby increasing the transmission rate. Since the symbol energy can be increased by increasing the transmission power, the transmission power is another important wireless resource. However, the power contributes to the transmission rate in a nonlinear manner: The same increment in transmission power may result in a different increment in transmission rate depending on the communication environment, such as the noise density. This property indicates that it is necessary to develop intelligent transmission power management schemes that can dictate how to allocate transmission power according to the environment. Chapter 6 provides detail on the management of transmission power.

There are several limiting factors associated with transmission power. First, using more power accelerates depletion of the battery and thus reduces the lifetime of battery-powered portable devices. Second, using excessive transmission power is limited even for power-supplied devices, due to the extra cost of implementing high-output power amplifiers. Third, using too much transmission power may be detrimental to overall performance as it causes interference with the transmitted signals of other wireless devices. For this reason, transmission power is usually regulated by the spectrum licensing organizations, such as the Federal Communications Commission (FCC) in the United States (FCC, 2007).

3.1.1.3 Antennas

According to (3.1), the capacity also increases as the number of antennas increases at transmitter or receiver. In the equation, the $N_T \times N_T$ matrix $\mathbf{Q}$ is the covariance matrix of the transmitted signal at the transmitter antennas and the constraint $\mathrm{Tr}(\mathbf{Q}) = 1$ limits the total transmission power consumed at each antenna to be the unit power. The covariance matrix $\mathbf{Q}$ describes how the information to be transmitted is mapped to each antenna and is determined such that the capacity is maximized under the power constraint. Thus, the capacity is an increasing function of the number of antennas, as more antennas make more covariance matrices feasible.

In fact, the antenna has recently become the most attractive wireless resource as it can contribute to increasing the channel capacity without requiring additional bandwidth or transmission power. However, as antennas necessitate the circuitry operating at radio frequency in the transmitting and receiving devices, increasing the number of antennas means an increased device cost. In addition, as indicated in (3.1), the channel capacity heavily relies on the usage of antennas and their optimal use varies depending on the channel which connects the transmitter–receiver pair. This implies that, for effective operation of antennas, it is necessary to adopt an intelligent antenna management method adequate to the channel state. However, the complexity of searching for optimal management grows as the number of antennas increases, so this is another limiting factor in using multiple antennas. Nevertheless, multiple antenna management is expected to be the key technology to develop for broadband wireless communications, since antennas are considered a less expensive resource than bandwidth or transmission power. Chapter 7 deals with various antenna management schemes.

3.1.2 Problem Formulation

Resource management problems are usually formulated in mathematical expressions. The problems then take the form of constrained optimizations: a predetermined objective is optimized under constraints dictating the feasibility of the solution. Formulation of resource management should reflect the policies of the service provider. The formulation may take various different forms depending on the resource management policies and each problem may be solved by a unique method.

3.1.2.1 Objectives and Constraints

The objective to maximize is a capacity-related performance metric, such as the total throughput and the number of admitted users; and the cost to be minimized is the amount of the resources to be consumed in supporting the service quality.

As an objective in the resource management problem, the system capacity itself is an important performance metric from the network operator's viewpoint but it is not directly related to the *quality of service* (QoS) that each individual user would like to get. In order to fill in this gap, many researches have employed the concept of *utility* which quantifies the satisfaction of each user out of the amount of the allocated resources, thereby transforming the objective to the maximization of the sum of all users' utility. The utility function is determined differently depending on the characteristics of the application, but the qualitative properties of typical utility functions have been studied intensively.

Figure 3.4 shows three types of application for which different utility functions are defined with respect to the allocated transmission rate (Shenker, 1995). First, data applications – such as file transfer and e-mail – are known to have an elastic nature, which is characterized by a strong adaptivity to delay and bandwidth. An elastic application has decreasing marginal improvement in utility with respect to increased bandwidth, as shown in Figure 3.4(a). Second, hard real-time applications – such as traditional telephony which provides delay-sensitive data services – perform very poorly if the data packets arrive later than the delay bound. Consequently, the utility function of hard real-time applications takes the shape of a step function which sharply increases to a saturated satisfaction beyond a threshold level of the allocated transmission rate, as

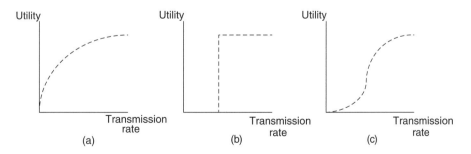

Figure 3.4   Utility of various application types with respect to transmission rate: (a) elastic application, (b) hard real-time application, and (c) adaptive real-time application

shown in Figure 3.4(b). Third, some real-time applications – such as video streaming services equipped with scalable coding[1] – have a certain level of adaptivity in delay and bandwidth, as appears in the utility function in Figure 3.4(c). They are more delay-tolerant than the hard real-time applications and are able to adjust the data-generating rate in response to network congestion by employing layered encoding or other means. The performance is very poor when the allocated transmission rate is smaller than the intrinsic data generation rate but increases sharply as the transmission rate increases above the intrinsic generation rate. Beyond that, the increase of transmission rate yields diminishing returns in utility because the additional performance improvement can be only marginal once the indispensable information is readily received.

The constraints in problem formulation mainly come from limitations in available resources, as the amount of bandwidth, transmission power, and the number of antennas are limited. In addition, the user performance is also constrained by the effort to support a certain level of service quality to each user. This performance-related constraint may be an absolute one (e.g., requirement on the minimum transmission rate) or a relative one (e.g., relative throughput ratio among the constituent users), depending on the service provider's policy. The performance-related constraint may not be necessary in the formulation with a utility-based objective, as each user's QoS is already considered in defining the utility function.

The objective and constraints should be formulated properly by the employed multiple-access method and the QoS provisioning policy. The multiple-access method determines the form of available resources, the performance of each user obtained from the allocated resources, and the interactions among the neighboring users. Thus, it plays the key role in formulating the performance metric and the resource-related constraint. This point will be discussed in more detail in Section 3.2 where different multiple-access methods are introduced. The QoS provisioning policy renders a target state of each user's performance, thereby determining the share of each user in occupying wireless resources. In this sense, it *balances* the amount of resources allocated to each user and the resulting performance. Thus, it is of great importance in formulating the performance-related constraint. There is further discussion of QoS provision in Section 3.3.

In order to illustrate mathematically formulated resource management, consider the optimization problem below. This problem is formulated for the allocation of bandwidth and transmission power in a multichannel system having single transmit and single receive antennas.[2]

$$\underset{p_{k,n}, \Omega_k}{\text{maximize}} \sum_{k=1}^{K} R_k = \sum_{k=1}^{K} \sum_{n \in \Omega_k} \frac{W}{N} \log_2 \left(1 + \frac{p_{k,n} g_{k,n}}{\sigma_{k,n}^2} \right) \qquad (3.9a)$$

$$\text{subject to} \bigcup_{k=1}^{K} \Omega_k \subseteq \{1, 2, \ldots, N\}, \qquad (3.9b)$$

$$\Omega_k \cap \Omega_j = \phi \quad \text{for} \quad k \neq j, \qquad (3.9c)$$

---

[1] Refer to Aravind, Civanlar and Reibman (1996) and Li (2001) for detailed discussions of scalable coding.
[2] See Section 6.2.2 for detailed discussion of this formulation.

$$p_{k,n} \geq 0, \tag{3.9d}$$

$$\sum_{k=1}^{K} \sum_{n=1}^{N} p_{k,n} \leq P, \tag{3.9e}$$

$$R_1 : R_2 : \ldots : R_K = \rho_1 : \rho_2 : \ldots : \rho_K. \tag{3.9f}$$

In this formulation, $g_{k,n}$ and $\sigma_{k,n}^2$ are the channel gain and the noise power of user k at channel n, respectively; and Ω_k and $p_{k,n}$ are the set of channels allocated to user k and the transmission power allocated to user k at channel n, respectively. The objective in (3.9a) indicates the maximization of the overall capacity. The constraints in (3.9b) and (3.9c) are on the bandwidth: They require that the whole set of channels should be given by $\{1, 2, \ldots, N\}$ and that no channel can be shared by more than two users. The constraints in (3.9d) and (3.9e) are on the transmission power: The total power consumption is limited to P. Finally, the constraint in (3.9f) is a performance-related constraint. In this example, the QoS policy is to maintain the relative ratio of the capacity achieved by each user to follow the predetermined criterion $\rho_1: \rho_2: \ldots: \rho_K$.

3.1.2.2 Centralized and Distributed Resource Management

Wireless resources may be managed in a centralized or distributed manner. In *centralized management* (see Figure 3.5), a resource manager at the central location collects the global information such as traffic load and channel state from each transmitter–receiver pair and allocates wireless resources based on the collected information. The main advantage of centralized management is that it can provide stable and consistent resource allocations as only a single operator manages the wireless resources. This point also simplifies the formulation of the resource management problem. In addition, centralized management is likely to achieve the global optimal solution to the formulated problem if the collected global information is properly taken into consideration. However, central management may possibly have serious disadvantage in *scalability*: the formulated problem may become too complicated to solve in reasonable time as the number of transmitter–receiver pairs increases to a large value. In addition, signaling for delivery of the local observation (e.g., channel gain of each user) and the resultant resource allocation can be a severe burden on the system. Further, the delay caused by the signaling procedure may impair the accuracy and timeliness of information, resulting in performance degradation in resource management. The scheduling problem in Chapter 5 and the transmission power allocation among parallel channels in Section 6.2 are two examples of centralized resource management.

The above-mentioned disadvantages may be resolved by employing a *distributed management* method (see Figure 3.5(b)). Each transmitter–receiver pair determines the amount and the usage of resources by itself based on local information. The signaling for information exchange among transmitter–receiver pairs may be allowed for improved resource management. Distributed management enables resolution of the scalability problem and enables timely adaptation of resource management with a light signaling overhead. The main distinction of distributed management is that multiple resource managers exist in the system. As the communication environment – such as the interference level or the amount of available resources – is affected by the resource allocation of other transmitter–receiver pairs, the optimal

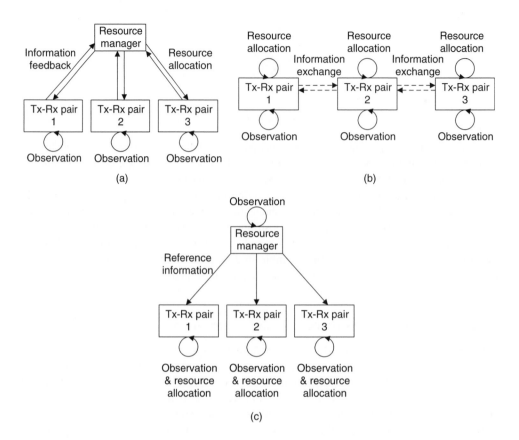

Figure 3.5 Resource management structures: (a) centralized type, (b) distributed type, and (c) hybrid type

resource allocation of a transmitter–receiver pair depends on the operation of the other pairs. As a result, distributed management has the weak point that the overall resource allocation may become unstable, since the resource allocation of each pair does not necessarily converge to a stable state. Even if it converges, the equilibrium point may be only a local optimality, and global optimality is not ensured in general. The distributed power control in Section 6.1.1 is a typical example of distributed resource management.

A hybridization of centralized and distributed managements can exploit the advantages of each method (see Figure 3.5(c)). In hybrid management, the central resource manager does not determine the specific resource allocation of each user every time but only distributes some reference information helpful for stable and efficient operation. This reference information is chosen in such a way that each local resource management can cooperate in a harmonized manner by using that information. The network load and the cost of using each resource are two possible examples of the reference information. Each transmitter–receiver pair then determines the usage of resources based on both the distributed reference information and its own collected local reference information. In this case, the overall resource management problem is divided into two sub-problems: one is determination of the reference information at the central resource manager, and the other is determination of the local resource allocation at each transmitter–receiver pair. The utility-based power control in Section 6.1.2 and the power-control-based inter-cell interference management in Section 8.1.4 are two examples of the hybrid approach.

In the resource management problem, the applied management type affects the available information used in determining resource allocations. All the information is available to the resource manager in centralized management if a perfect feedback channel is assumed for the transmitter–receiver pairs. In the distributed or hybrid case, however, only partial information is available in determining resource usage. In these cases, the information available to each resource manager equipped in each transmitter–receiver pair should be carefully considered according to the assumed signaling method.

3.2 Multiple-access Methods

Wireless communications inherently have the nature of broadcasting; that is, transmission signals of users are seen by all the receivers within the coverage. This causes interference among different transmissions in multiuser systems. Therefore, it is a fundamental requirement to establish a method that can separate out the signals. *Multiple access* refers to a mechanism that brings in orthogonality to multiuser communications so that multiple users can share the given medium without, or with a sustainable level of, interferences.

There are two basic scenarios in multiuser communications – the *downlink* (or the forward link) and the *uplink* (or the reverse link), as illustrated in Figure 3.6. In the case of the downlink, or the broadcast channel, one single transmitter sends signals to multiple receivers. A typical example of downlink is the transmission of base stations in cellular mobile systems. The transmitting signal, $s(t)$, is the sum of the signals transmitted to all the users. For user k, the desired signal $s_k(t)$ and the interfering signals $s_j(t), j \neq k$, propagate through the same channel

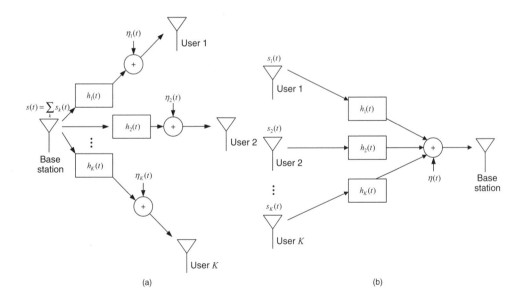

Figure 3.6   Illustration of (a) downlink, and (b) uplink channels

$h_k(t)$. This makes synchronization among different users relatively easy, since the signals of all the users arrive at the same time. In addition, it becomes much simpler to manage wireless resources in the downlink case, as the resource management function is generally located at the transmitter, the base station.

In the case of uplink, or the multiple-access channel, there are multiple transmitters sending their own signals to a common single receiver. Transmissions from user handsets in the cellular mobile system are one example of this. One important characteristic of the uplink channel is that different signals experience different channels and their propagation delays are different. So, synchronization becomes a difficult issue in the uplink case. Moreover, since the resource management function is separated from the transmitters, resource management becomes that much more difficult and limited: More signaling is required between the transmitters and the receiver, or wireless resources are required to be managed in a distributed manner. That is, each transmitter autonomously manages the resources used by itself with the limited information received from the other users or the receiver.

The multiple-access method divides the total resources into multiple dimensions, and different dimensions are allocated to different users. This means that it determines the type and characteristics of the wireless resources to be allocated to each user by resource management schemes. It determines the set of available resources and the interactions among the constituent users. Therefore, it is very important to reflect the type of employed multiple-access method in formulating the resource management problem. Depending on the type of dimension in which multiuser signals are separated, there are several different types of multiple-access methods.

Figure 3.7 pictorially compares the three widely used techniques – *frequency-division multiple-access* (FDMA), *time-division multiple-access* (TDMA), and *code-division multiple-access* (CDMA) – on the two-dimensional basis of frequency and time. FDMA allows multiple access among multiple users by allocating different frequency bands to different users, but TDMA allocates different time slots to different users. In contrast, CDMA provides multiple access not by allocating frequency or time but by allocating different codes to different users. *Orthogonal frequency-division multiple-access* (OFDMA) is similar in concept to FDMA but it divides the frequency band into orthogonal frequency subcarriers.

3.2.1 Frequency-division Multiple Access

In FDMA, the frequency axis is divided into multiple dimensions, with each different frequency band allocated to a different user. All the channels provided in FDMA are orthogonal to each other. Due to its simplicity, this FDMA based channelization has been widely employed in various systems: In the past it was used for radio and TV broadcasts and recently for analog cellular systems such as Advanced Mobile Phone System (AMPS).

To implement FDMA systems, each transmitter or receiver has to be equipped with radio circuitry that can operate in multiple frequency bands and can tune to the carrier associated with the allocated frequency band. Since transmission signals are modulated into, and demodulated from, each frequency band by using a band-pass filter, some guard band is necessary between every two adjacent bands to block interferences. As the guard band is not used for signal transmission, it decreases the usable frequency spectrum to an extent.

The main issue in resource management for FDMA is to determine whether a frequency band should be used or not. Inter-cell interference plays the key role in this resource management because the use of a frequency band at a cell appears as interference to the other cells that use the

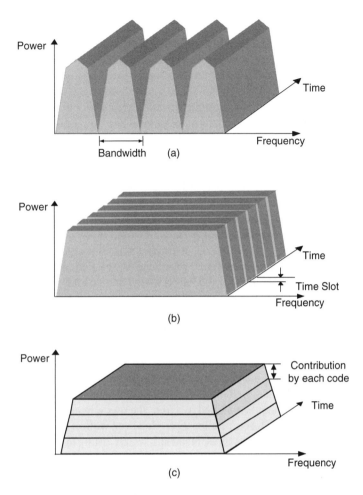

Figure 3.7 Pictorial comparison of multiple-access technologies: (a) FDMA, (b) TDMA, (c) CDMA (Reproduced with permission from B.G. Lee and S. Choi, *Broadband Wireless Access and Local Networks: Mobile WiMAX and WiFi*, Artech House, Norwood, MA. © 2008 Artech House, Inc.)

same frequency band. In other words, the users near the cell boundary may suffer from severe interference from neighboring cells if a frequency band is simultaneously used by two cells close to each other. The issue of inter-cell interference management is covered in detail in Section 8.1.

3.2.2 Time-division Multiple Access

In TDMA, the time axis is divided. Users are separated by transmitting signals at different time instances or in different time slots. An orthogonal channel is provided at every time slot, so no inter-user interference problem arises. TDMA has the advantage that it requires no additional signal processing operation for the channelization. TDMA makes the resource management problem quite simple: It only needs to control the assignment of transmission time slots, whereas power management should be considered as well in other multiple-access methods.

The main difficulty of TDMA technology is in maintaining the orthogonality of the channels. First, all transmissions in TDMA should be synchronized, otherwise signals from other users may be overlapped and severely corrupted. Synchronization becomes more important in the uplink because the propagation delay of the transmitted signals is different for different users as discussed above. To fix this problem, significant signaling overhead is required for communications between the users and the coordinator residing typically at the base station or the access point. Second, some compensation techniques are required to combat the time dispersion of wireless channels. When the delay spread of the multipath channel becomes a considerable fraction of the transmission time, transmission at each time instant may be interfered with by the previous transmission even if synchronization is perfectly achieved. Hence, TDMA systems often secure a guard time between two adjacent transmissions to compensate for the synchronization error and the delay spread.

In determining the time slot to transmit and the user to serve in TDMA, there are two different types of approach – *periodically repeated transmission* and *scheduled transmission*.

With periodically repeated transmission, a time slot is allocated to each user and the transmission epoch is repeated in every frame which consists of several time slots. An example of this approach is the *global system for mobile* (GSM) which was developed as the European second-generation digital cellular systems. Figure 3.8 depicts the GSM frame structure: A frame consists of eight slots and a slot is divided in time into the tail bits (T bits) that define the beginning and the end of a slot; 114 data bits; two 1-bit flag bits (S bits) which indicate the usage of the slot; 26-bit known training sequence for the channel estimation and time synchronization; and a guard time. The periodic transmission provides a "circuit" or a constant-rate transmission opportunity to each user. Resource management in this case is done simply by assigning time slots to each user. Signaling is needed only when initiating and terminating the connection and, once a time slot is assigned to a user, the user can initiate signal transmission without waiting for any further resource allocation. However, while a user does not have traffic to send, the corresponding time slots are sent vacant, which results in under-utilization of the

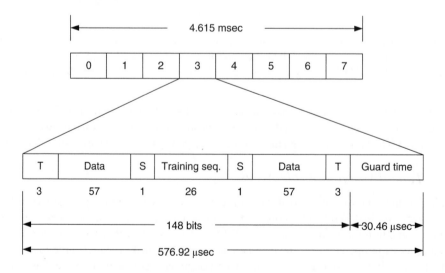

Figure 3.8   GSM frame structure

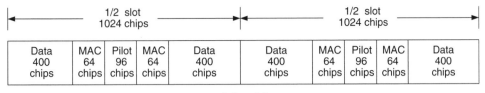

Figure 3.9   CDMA2000 1xEV-DO frame structure

resource. Therefore, periodically repeated transmission is more suitable for constant-rate services, such as voice telephony, which has to provide resources to ensure constant data rate and delay bound.

With scheduled transmission, time slots are shared by multiple users and, for every time slot, the scheduler determines the user to use that time slot. An example of this approach is the downlink of the CDMA2000 1xEV-DO system, developed for supporting high-speed data services while maintaining compatibility with the CDMA2000 system. Figure 3.9 depicts the time slot structure of the downlink of the 1xEV-DO system. Each slot is 1.66 ms and consists of the *medium access control* (MAC) field for managing the parameters related to the uplink transmissions; the pilot field for channel estimation; and a data portion to carry the traffic or control signal. For the user indication, different preambles are applied to different users in the first time slot of the physical layer packet.[3] In contrast to the GSM case, time slots are not allocated repeatedly and the user who uses the time slots may change at every transmission. Scheduled transmission inherently requires a well-designed scheduler that can effectively select the user to use the current time slot for the transmission of channel or traffic information as appropriate. This complicates the resource management substantially when compared with periodically repeated transmission, but it can enhance the resource efficiency as long as the employed scheduler performs well. The performance can improve significantly for a file transfer service, for which traffic generation is bursty and latency is of no great interest.

3.2.3 Code-division Multiple Access

In CDMA, users are separated by applying different codes. The overall CDMA process is depicted in Figure 3.10. The information signal of user k with rate R bits/sec, s_k, is modulated by the spreading code $\mathbf{c}_k$ and the spread signals, $s_k\mathbf{c}_k$, are simultaneously transmitted over the same time and frequency domain. A code $\mathbf{c}_k$ is a specified sequence of digital bits $\{c_{ik}\}$ and each bit in a spreading code is called a *chip*. Modulation of the information signal means that, when sending each information symbol in s_k, the spreading code $\mathbf{c}_k$ multiplied by the sending bit is transmitted. In order to separate multiuser signals, the chip rate, W chips/sec, is much larger than the information rate R, and the correlation between other codes

$$\mathbf{c}_k \cdot \mathbf{c}_j = \sum_i c_{ik} c_{jk} \quad \text{for } j \neq k$$

---

[3] According to the transmission rate, a single physical layer packet may occupy 1, 2, 4, 8, or 16 time slots. The preamble is located only in the first time slot and its length is determined by the transmission rate.

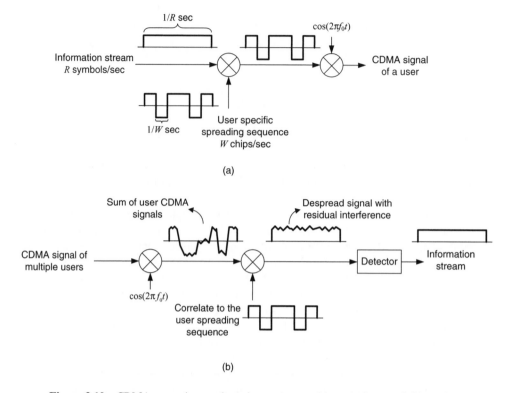

Figure 3.10 CDMA transmitter and receiver structure: (a) transmitter, and (b) receiver

should be low enough. The original information bits with rate R symbols/sec are transmitted over a much wider bandwidth of W chips/sec in CDMA and this is why the signal is said *spread*. The ratio of the signal spreading bandwidth to the information rate, W/R, is called the *spreading factor* (SF) or *processing gain* (PG) of the system. The receiver despreads the received signal by the applied spreading code and this renders the signal

$$s_k |\mathbf{c}_k|^2 + \sum_{j \neq k} s_j \mathbf{c}_j \cdot \mathbf{c}_k$$

Each information signal can be separated as long as the multiuser interference or the *multiple-access interference* (MAI)

$$\sum_{j \neq k} s_j \mathbf{c}_j \cdot \mathbf{c}_k$$

is sustainable.

The total resources are channelized in the code domain in CDMA, so the correlation property among the applied codes determines the overall characteristics of CDMA systems. The correlation property of a spreading code set is described by two factors. One is the autocorrelation, or the correlation of a spreading code with its time-shifted version. If a spreading code has low autocorrelation except for the case of zero time-shift, the receiver can easily find out the

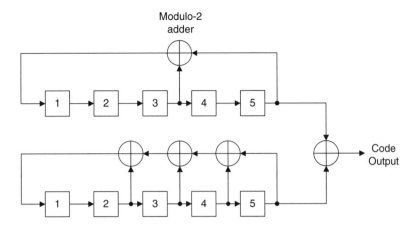

Figure 3.11   A Gold code generator with spreading factor 31

point where the spreading code begins and this enables a fast synchronization. In addition, low autocorrelation is desirable for rich multipath environments as the ISI is represented as a time-shifted version of each user's spreading code. The other factor is cross-correlation, or the correlation among different spreading codes in the code set. This property determines the multiuser interference characteristics, so it is of great importance in the capacity and the reliability of the CDMA communications.

The spreading codes may be determined to be orthogonal to each other; that is, $\mathbf{c}_k \cdot \mathbf{c}_j = 0$ for $j \neq k$. This implies that the cross-correlation becomes zero. An orthogonal spreading code set can be constructed like the Walsh–Hadamard code, for example

$$\mathbf{W}_2 = \begin{bmatrix} 1 & 1 \\ 1 & -1 \end{bmatrix} \quad \mathbf{W}_4 = \begin{bmatrix} \mathbf{W}_2 & \mathbf{W}_2 \\ \mathbf{W}_2 & -\mathbf{W}_2 \end{bmatrix} \quad \mathbf{W}_8 = \begin{bmatrix} \mathbf{W}_4 & \mathbf{W}_4 \\ \mathbf{W}_4 & -\mathbf{W}_4 \end{bmatrix} \quad \cdots \quad \mathbf{W}_{2N} = \begin{bmatrix} \mathbf{W}_N & \mathbf{W}_N \\ \mathbf{W}_N & -\mathbf{W}_N \end{bmatrix}.$$

Each row in the Hadamard matrix $\mathbf{W}_N$ is a binary orthogonal sequence with spreading factor N. In this orthogonal CDMA, all the users are provided with orthogonal channels and the operation is equivalent to the orthogonal TDMA case. However, since strict synchronization among different signals is also required to maintain the orthogonality, orthogonal code sets are usually used in the downlink.

When all the user signals are not perfectly synchronized, multiuser interferences are determined by the cross-correlation property of the spreading code set. In general, orthogonal code sets have poor correlation property in this case. As a method of applying the CDMA technique without strict synchronization, code sets with non-zero but low cross-correlation properties have been developed for the asynchronous case. Gold code (see Figure 3.11) is such an example that is obtained by adding two different pseudo-noise sequences generated by properly designed feedback shift registers. Different codes are obtained by changing the initial states of each shift registers.[4]

---

[4] Refer to Dixon (1994) and Proakis (1995) for more detailed discussion of the design of spreading codes and their properties.

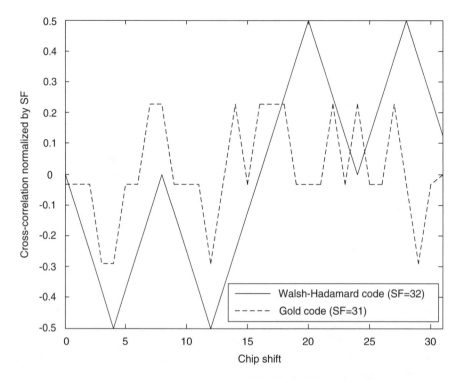

Figure 3.12 Cross-correlation property of Gold and Walsh–Hadamard codes

Figure 3.12 compares the cross-correlation property of different codes obtained from the Walsh–Hadamard and the Gold code sets. Walsh–Hadamard code exhibits that multiuser interference is completely removed with perfect synchronization, whereas the Gold code exhibits much lower cross-correlation with non-zero time shift. The non-orthogonal CDMA using well-designed spreading code sets has the advantage that no coordination is required among transmitters for maintenance of channel orthogonality. Each user signal may be transmitted at any time at any frequency as long as the cross-correlation of the used spreading codes is low. In this case, the system capacity is limited by the multiuser interference and no hard limit exists on the capacity, unlike with the orthogonal multiple-access methods. In general, each user signal has a sustainable level of interference and the interference above this level violates the QoS requirement of the user. Hence, to keep the interference below the sustainable level, it is required to employ a power control mechanism which manages the transmission power of each user signal. Non-orthogonal CDMA is usually applied to the uplink where coordination among different transmitters is difficult.

In addition to ease of operation, the non-orthogonal CDMA has the advantage that it can implicitly exploit the statistical multiplexing gain. By staying idle, a user reduces the multiuser interference to the others and thus helps to improve the service quality in CDMA, whereas the service quality remains constant in TDMA regardless of other users' activity. Since traffic is generated in a random manner even in voice telephony services, this statistical multiplexing property makes it possible to accommodate more users in the system. In addition, CDMA is more robust to multipath fading channels than TDMA systems. As CDMA, in principle, extracts the desired signal from interferences, the inter-symbol interference problem can be resolved by extracting (i.e., despreading) the signal several times at different chip times and then combining them into a more reliable signal. These properties can reduce the overheads such as guard time in TDMA which was required to maintain the channel orthogonality. Owing to these advantages, CDMA was used for the multiple-access method in the IS-95 digital cellular standard and the third-generation wireless cellular systems such as W-CDMA and CDMA2000.

3.2.4 Orthogonal Frequency-division Multiple Access

In OFDMA, users are separated by occupying different overlapping tones in the frequency domain. Basically, OFDMA falls within the category of FDMA in a wide sense but incorporates the orthogonality-without-guardband as its own characteristic. Conventional FDMA is inefficient in using spectrum because overlapping in spectrum bands is not allowed and a guard band is required to separate out adjacent bands. The main idea of OFDMA is to transform the data signal into the frequency domain by applying a *discrete Fourier transform* (DFT) and dividing the given frequency band into multiple subcarriers, each of which is equally spaced and is not affected by another subcarrier. As will be clear in the following discussions, OFDMA orthogonally parallelizes multiple user signals without requiring guard bands and can provide efficient communication in frequency-selective fading channels.

In the OFDMA transmitter, N data symbols $S[k]$,[5] $0 \leq k \leq N-1$, are sent over the subcarriers $\cos(2\pi(f_0 + k\Delta f)t)$, where f_0 and Δf respectively denote the frequency of the first subcarrier and the bandwidth of each subcarrier, to yield the expression

$$\hat{s}(t) = \sum_{k=0}^{N-1} S[k]e^{j2\pi(f_0 + k\Delta f)t} = e^{j2\pi f_0 t}\sum_{k=0}^{N-1} S[k]e^{j2\pi k\Delta f t}. \tag{3.10}$$

Since the N-point *inverse DFT* (IDFT) of $S[k]$ takes the expression (Oppenheim and Schafer, 1999)

$$s[n] = \text{IDFT}\{S[k]\} = \frac{1}{N}\sum_{k=0}^{N-1} S[k]e^{j2\pi kn/N}, \tag{3.11}$$

equation (3.10) is equivalent to the D/A converted version of the N-point IDFT of the data sequence $S[k]$. In the receiver the original sequences are reconstructed by taking the DFT on $s[n]$; that is

$$\text{DFT}\{s[n]\} = \sum_{n=0}^{N-1} s[n]e^{-j2\pi kn/N} = S[k]. \tag{3.12}$$

---

[5] Capital representation and usage of index k in this section implies that the signal is in the frequency domain. A data sequence of OFDMA is mapped to a frequency component in the frequency domain, and this is why the data symbols are written in the form of frequency-domain signals.

Since each symbol of $s[n]$ is sent in every $1/\Delta f N$ seconds, the D/A converted signal has the expression

$$\text{D/A}\{s[n]\} = \left[\frac{1}{N}\sum_{k=0}^{N-1} S[k]e^{j2\pi kn/N}\right]_{n=\Delta f N t} = \frac{1}{N}\sum_{k=0}^{N-1} S[k]e^{j2\pi k\Delta f t}. \qquad (3.13)$$

The two equations (3.13) and (3.10) reveal that the targeted transmission signal $\hat{s}(t)$ can be generated by first applying IDFT to the data symbols and then modulating the result with the frequency f_0. As both DFT and IDFT processes are easily implemented by the *fast Fourier transform* (FFT) and *inverse FFT* (IFFT) devices, the OFDMA process requires only an additional single carrier frequency modulator. Figure 3.13 illustrates two equivalent OFDMA implementations.

Figure 3.14 illustrates the OFDMA spectrum, which is divided into N narrow-band parallel subcarriers, with each subcarrier containing one data symbol. Unlike the conventional FDMA, adjacent subcarriers are not physically separated in the frequency spectrum but are rather overlapping. The figure demonstrates that the orthogonality among different subcarriers is maintained due to the nullified spectra of all other subcarrier components at the frequencies (indicated by dashed vertical bars) at which one particular subcarrier component reaches the peak. This indicates that transmitter–receiver synchronization is crucial in the OFDMA system, since any frequency offset, or the difference between the transmitter and the receiver carrier frequencies, would break the orthogonal property, thus incurring severe subcarrier-to-subcarrier interferences.

To maintain the orthogonality of data symbols in frequency-selective fading channels, OFDMA adopts *cyclic prefix* (CP), which means copying the last L symbols of the IDFT data at the front-most position, as illustrated in Figure 3.15. Then, a new transmission signal of length

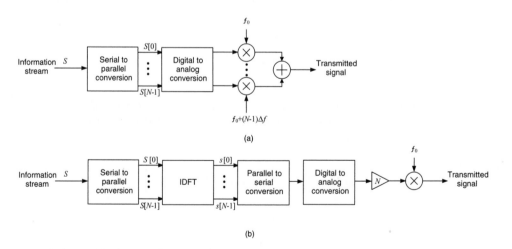

(a)

(b)

Figure 3.13 Two equivalent OFDMA implementations: (a) with N carrier frequency modulators, and (b) with a single carrier frequency modulator

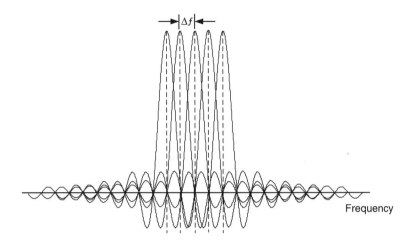

Figure 3.14 Illustration of OFDMA spectrum (Reproduced with permission from B.G. Lee and S. Choi, *Broadband Wireless Access and Local Networks: Mobile WiMAX and WiFi*, Artech House, Norwood, MA. © 2008 Artech House, Inc.)

$N + L$, $\tilde{s}[n]$, is generated such that

$$\tilde{s}[n] = \begin{cases} s[N-L+n] & \text{for } 0 \leq n \leq L-1, \\ s[n-L] & \text{for } L \leq n \leq N+L-1. \end{cases} \quad (3.14)$$

This new signal containing the N-point IDFT of the data sequence and the cyclic prefix is called an *OFDMA symbol*.

In a multipath environment, each path component may be represented by a delayed (by δ_i) and scaled (by α_i) copy of the original signal $\tilde{s}[n]$. So the received signal takes the expression

$$r[n] = \sum_{i=0}^{N_m-1} \alpha_i \tilde{s}[n-\delta_i], \quad (3.15)$$

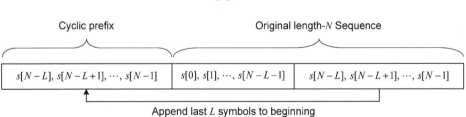

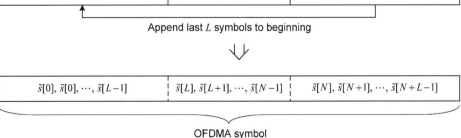

Figure 3.15 Illustration of (a) cyclic prefixing to obtain (b) OFDMA symbol

assuming that N_m multipath components exist. A delay in the time domain corresponds to a phase shift in the DFT domain and, in addition, the circular shift operation is valid to each received multipath component if its delay is smaller than the time duration corresponding to the cyclic prefix. Therefore the received signal in the multipath environment, $r[n]$ in (3.15), turns out to be a scaled version of the received signal in the single-path environment, $\tilde{s}[n]$, as long as the delay of each of the N_m multipath components does not exceed the time duration of the cyclic prefix (Lee and Choi, 2008).

Now consider the transmission of OFDMA in more detail. The received signal in (3.15) can be rewritten by the output signal of a discrete-time channel with a finite impulse response $h[n]$ by

$$r[n] = \sum_{l=0}^{L-1} h[l]\tilde{s}[n-l] \equiv h[n]*\tilde{s}[n], \tag{3.16}$$

where

$$h[n] = \begin{cases} \alpha_i, & \text{if } n = \delta_i, \\ 0, & \text{otherwise.} \end{cases} \tag{3.17}$$

In (3.16), the OFDMA symbol of length $N + L$ is convoluted by the channel response $h[n]$ to yield the received signal $r[n]$. To maintain the orthogonality of each subcarrier, we assume that the maximum multipath delay of $h[n]$, δ_{N_m-1}, is limited by the cyclic prefix duration L. The receiver takes only the last N symbols, $r[L], \ldots, r[N + L - 1]$, which may be equivalently represented as follows by adopting the circular convolution:

$$r[n] = \sum_{l=0}^{L-1} h[l]s[((n-L-l) \text{ modulo } N] \equiv h[n] \otimes s[n], \tag{3.18}$$

where $\otimes$ is the notation for the circular convolution (Oppenheim and Schafer, 1999). It is well known that the DFT of circularly convoluted signals is the product of the DFTs of the two signals. From this, we get

$$\text{DFT}\{r[n]\} = R[k] = \text{DFT}\{h[n] \otimes s[n]\} = \text{DFT}\{h[n]\} \times \text{DFT}\{s[n]\}. \tag{3.19}$$

Since $S[k]$ is the DFT of the original signal $s[n]$, we can recover the original signal $S[k]$ from the estimation of the frequency response of the channel, $H[k] = \text{DFT}\{h[n]\}$, by

$$S[k] = \frac{R[k]}{H[k]}. \tag{3.20}$$

Figure 3.16 depicts the functional structure of the OFDMA transmitter and receiver. It is noteworthy that each symbol is orthogonally separated in the OFDMA system. This is a major advantage of OFDMA since it removes the ISI in frequency-selective fading channels with marginal overhead. In contrast, in the TDMA case, some guard time is required at every slot to compensate for the delay spread. This property becomes more important in wireless

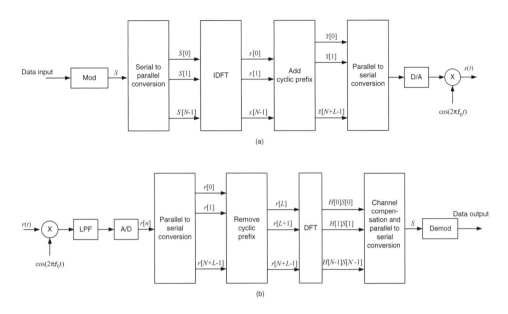

Figure 3.16 Functional structure of OFDM system: (a) transmitter, and (b) receiver

communications with wide bandwidth since an increase in transmission rate causes increased delay spread and decreased coherence bandwidth. Owing to this property, OFDMA is considered as the choice of multiple access for broadband wireless communications.

An OFDMA system may be considered as providing N parallel orthogonal channels with the channel gains $H[k]$, $k = 1, 2, \ldots, N$. This property makes it possible to employ an adaptive transmission technique in the frequency domain. The OFDMA transmitter can adjust its *modulation and coding scheme* (MCS) according to the channel condition of each subcarrier, and each subcarrier can be allocated to the user whose channel condition is relatively good at that frequency. Thus, OFDMA offers a more flexible resource management by bringing in the additional dimension – the frequency domain. A typical example of the OFDMA is the IEEE 802.16e standard system, called Mobile WiMAX, which was developed to provide mobile broadband access in the *wireless metropolitan area network* (WMAN). Figure 3.17 illustrates the frame structure of Mobile WiMAX, which contains the preamble for channel estimation, the frame header, the ranging channel, and the DL-MAP and UL-MAP for description of the downlink and uplink channels, and data bursts.[6] Data burst spaces are there to carry the user data in two-dimensional data space formed by frequency and time.[7]

3.3 Quality of Services

The rapid growth of multimedia applications has made *quality of service* (QoS) an important issue in today's communication networks. QoS is a rather broad concept considered as a metric

---

[6] Each MAP message contains information such as the range of transmission burst, the user index to whom each burst is allocated, and the employed modulation and coding scheme of each burst.

[7] Refer to Lee and Choi (2008) and IEEE (2006) for a detailed description of the Mobile WiMAX system.

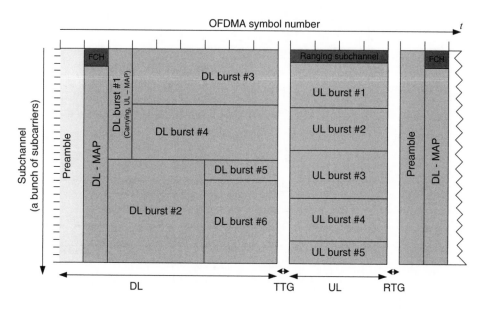

Figure 3.17   Example of the frame structure of the IEEE 802.16e system (IEEE, 2006)

of offering services in communication networks. It may be defined, in generic terms, as "the collective effect of service performance which determines the degree of satisfaction of a user of the service" (ITU, 1993) or as "the ability to deliver network services according to the parameters specified in a service level agreement" (Westerinen *et al.*, 2001). As such, QoS may mean the service quality itself or the ability to meet the requirement on service quality. Therefore, a QoS supportable communication system refers to the system that can satisfy each user's requirement on the provided service quality.

Supporting QoS implies an adaptation to various applications, since a communication system usually supports a large set of different services. This means that different policies are employed with different services. Each applied traffic has its own characteristics and the service requirement of each user is also unique. Thus, in order to support the QoS of each user, the system first has to figure out user requirements and then adjust the resource allocation depending on those requirements and the communication environment. The resources are shared by multiple users, so it may be difficult to satisfy all the users altogether: improvement of the QoS of one user may result in degrading the performance for other users. In order to enhance each user's performance in a balanced manner, it is necessary to approach the QoS issue at a system level by formulating pertinent resource management problems.

It is more challenging to support QoS in wireless communication systems than in wired networks. User mobility, the variance in noise and interference, and channel fading of wireless systems make quality unstable and unpredictable. It is not realistic to estimate the amount of resources required to satisfy each user's requirements in a deterministic way. Thus, supporting QoS in wireless communications is an adaptive process to be done in compliance with the applications as well as the channel states, which complicates the resource management problem.

3.3.1 QoS Classification

For easy and simple resource management, service traffic is categorized into several QoS classes in most communication systems. Each QoS class has its own specification to describe the traffic property and the QoS requirement. As an example of a well-defined QoS classification, consider the case of the Universal Mobile Telecommunications System (UMTS) described in 3GPP TS 23.107 (3GPP, 2006).

UMTS defines four QoS classes – namely, conversational, streaming, interactive, and background classes – as listed in Table 3.1. The main distinguishing factor among these classes is the delay sensitivity of the traffic.

First, the *conversational class* is the most sensitive class which delivers two-way interactive real-time traffic flows like video telephony. In the conversation class, the transfer time should be low enough and the time relation (or variation) between information entities should be preserved. The maximum transfer delay is given by the human perception of video and audio conversation, but it has long been recognized that real-time voice signals cannot be delayed more than 100 ms in end-to-end delivery, otherwise the human recipient finds difficulty in following the transmitted message. Most conversational services have a similar pattern – when one end-user starts his/her talk spurt, the other side becomes silent – and this pattern may be used in managing wireless resources. This class does not have a stringent requirement on reliability (i.e., error probability) as human recipients are error-resilient. Voice traffic is usually assumed to be still recognizable even when 5% of data are lost for the end-user.

Second, the *streaming class* is for delivery of real-time video and audio services. A distinctive difference from the conversational service is that it is a one-way transport and has no interactivity. This class also requires preservation of the time relation between information entities but has somewhat different characteristics. The receiving end of the streaming class is in general equipped with the capability of time alignment. The delivered information entities are buffered at the receiving end and played in sequence after time

Table 3.1 The UMTS QoS classes (© 2006. 3GPPTM TSs and TRs are the property of ARIB, ATIS, CCSA, ETSI, TTA and TTC who jointly own the copyright in them. They are subject to further modifications and are therefore provided to you "as is" for information purposes only. Further use is strictly prohibited.)

Traffic class	Conversational class Conversational RT	Streaming class Streaming RT	Interactive class Interactive best effort	Background Background best effort
Fundamental characteristics	Preserve time relation (variation) between information entities of the stream	Preserve time relation (variation) between information entities of the stream	Request response pattern	Destination not expecting the data within certain time
	Conversational pattern (stringent and low delay)		Preserve payload content	Preserve payload content
Example of the application	Voice	Streaming video	Web browsing	Background downloading of e-mails

alignment. Therefore, this class does not have a requirement of low transfer delay, as transfer delay affects only the initial access. In addition, the acceptable delay variation is much greater than the delay variation given by the limits of human perception, as delay jitter experienced in the channel can be smoothed by the time alignment capability.

Third, the *interactive class* is for applications in which the remote equipment sends a requesting message and the corresponding data is sent back as the response. Examples include web browsing, database retrieval, and server access. A distinguishing characteristic of this class is the request and response pattern of the end-user. First, a user who has sent a data request expects to receive the response within a certain time. Thus, round trip time is one of the key performance parameters; but, as the traffic is not in real time, the delay requirement is not as stringent as for the conversational and streaming classes. Second, the responding end sends the requested data traffic to the requesting end. This traffic pattern is bursty in nature: a relatively long time elapses between transmissions of requests, and the data amount of the response is generally much larger than that of the request message. The most important requirement of this class is the preservation of data integrity, which means that data should arrive at the recipient with very low probability of error. A partial error can make the whole data traffic useless in this class.

Fourth, the *background class* includes applications that are not time-sensitive. Examples are delivery of e-mails, the short message service, and downloading of databases. This class does not have any specific requirements as long as the payload content is preserved. It corresponds to the so-called best-effort service class.

3.3.2 Prioritization and Fairness

When a user application sets up a connection to a wireless communication system, its QoS requirement is described by the process of assigning the QoS class. The QoS requirement of each bit of traffic may be formulated within the resource management problem, but it complicates problem formulation if different QoS classes have to be considered. For example, when a common resource is shared by conversational traffic with a delay constraint and interactive traffic with a packet loss constraint, it is difficult to compare the urgency of each class and to decide which traffic to serve with how much resource; the performance metric is different for the two classes.

It is necessary to apply *traffic prioritization* to make QoS support more tractable. Each user is assigned a priority according to its QoS class and the stringency of its requirements. More resources can be allocated preferentially to higher priority traffic than to lower priority traffic. In this respect, traffic prioritization has the effect of transforming QoS requirements (e.g., delay, loss rate) into the resource allocation domain (e.g., order of bandwidth allocation). In general, a real-time service has a higher priority than a non-real-time service: For example, in the UMTS QoS classification in the previous subsection, the conversational class has the highest priority and the background has the lowest. Such a prioritized rearrangement of QoS requirements renders an easy means of comparing the QoS requirements of different classes and of making the resource management problem more tractable. However, the QoS provided by traffic prioritization is only *relative*; a higher priority does not ensure any absolute performance but only provides relatively better performance.

Traffic prioritization necessarily raises a fairness problem as it determines the order of access to resources. The fairness problem was originally dealt with in wired networks in managing a multiplexing processor and the associated buffer shared by multiple flows. The key issue is how to define the *fairness criterion*, or the state at which the resources may be claimed to be fairly shared by users. The fairness criterion in wired networks may be simply processor usage, as bandwidth is the only resource. In wireless communications, however, the fairness issue arises in a much more complicated form as the communication environment is different for different users: Equitable bandwidth sharing may be a good fairness criterion, but it results in unequal throughput among the users of the same QoS class, since a user with a strong channel gain achieves a higher throughput than a user with a weak channel. In this context, the equalized throughput may render another fairness criterion. The wireless communication environment has thus led to various types of fairness criteria being introduced in the literature. More discussion is provided in relation to bandwidth management in Chapter 5.

3.4 Resource Management in Protocol Layers

Since a communication system is composed of a large number of functionalities, it becomes overly complicated to try to design the system as a whole. So, practical design usually adopts the *layered approach*, in which the overall functionalities are arranged as a collection of hierarchical layers, with each layer containing a small manageable group of functionalities. A higher layer in the hierarchy treats the operation of a lower layer as a "black box" and takes only services provided by the adjacent lower layer. The processing module in each layer conducts a specific set of functions using its own *protocol*, or *a set of rules that govern how two communicating parties are to interact* (Bertsekas and Gallager, 1992).

Figure 3.18 depicts the structure and operation of this layered architecture. The processing module in the nth layer provides services to the upper layer by delivering the *service data unit* (SDU) of the nth layer. It generates the *protocol data unit* (PDU) of the layer by adding a header that contains the information required for the protocol operation. The PDU of the nth layer is delivered by the service of the lower layer (i.e., the $(n-1)$th layer), so it becomes the SDU of the $(n-1)$th layer. The processing modules at the same layer, respectively in the transmitter and receiver, exchange information, served by the lower layer processing modules.

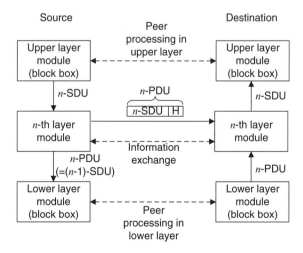

Figure 3.18 Protocol operation in layered architecture

The layering concept helps to divide the overall problem into multiple, independent sub-problems. Each sub-problem is easier to solve, as the designer of each layer may deal only with the internal details and operation of that particular layer. In addition, this approach secures modularity and flexibility in improving or modifying the communication network.

3.4.1 Classical Protocol Layering

The *open system interconnection* (OSI) model is a well-established classical framework for protocol layering in data networks (Leon-Garcia and Widjaja, 2000). It divides the basic communication functions into the seven layers shown in Figure 3.19. The *application layer* provides services specific to the particular application, such as file transfer and e-mail. The *presentation layer* is intended to provide the application layer with independence from the differences in representation of data. Its major functions are data compression, data encryption, and code conversion. The *session layer* handles interactions between the source and destination in setting up and tearing down a session. These three layers are in general concerned with the tasks of data processing for consistent and effective communications rather than tasks directly related to data transfer. Thus, most research for improving the efficiency of data transfer over wireless channels tends to focus on the lower four layers.

The *transport layer* is responsible for end-to-end transfer of messages from source to destination. It can perform error detection, recovery, and arrangement of the received messages for reliable and in-order delivery, and perform flow control to adjust the message sending rate to the network capacity.

The *network layer* provides data transfer in the form of packets across the communication network. This is performed by routing of the packets, or the procedure that is used to select a proper path from source to destination traversing a number of transmission links.

The *data link layer* is responsible for transfer of data across a transmission link that directly connects two communication nodes. Its function is similar to – but distinct from – the transport layer function in that it operates hop-by-hop, not end-to-end. In the environment where multiple communication nodes share a common communication medium, the data link layer assumes an additional function to control access of the constituent nodes to the medium with harmonized coexistence and sharing of multiple communication links. For this, the data link layer usually

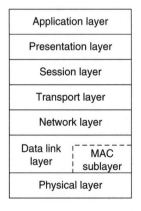

Figure 3.19   OSI seven-layered architecture

includes the *medium access control* (MAC) sublayer in it. The role of the MAC sublayer is to allocate the multiple-access channel so that each node can successfully transmit data without getting interference from the other nodes.

The *physical layer* deals with the transfer of bits over the communication medium. The key aspects of this layer is to define the relation which maps the information bits to the physical (i.e., electric) signal. This layer is also concerned with physical characteristics of the transmitted signal such as signal power and duration.

In view of the OSI seven-layer architecture, wireless resource management is part of the function of the data link layer, as its main functionality is to provide each connection with chances to access the wireless channel by allocating appropriate resources. However, as will become clearer in the next subsection, it is more beneficial to relax the strict layering so that some parameters of the upper and lower layers can be freely utilized in conducting wireless resource management.

3.4.2 Cross-layer Design in Wireless Resource Management

While the above classical layering strictly separates layers in order to fully exploit the advantage of protocol modularity, strict layering does not work well in supporting high-level QoS in wireless networks owing to its lack of adaptivity. In a layered structure, the benefits of joint optimization across multiple layers are precluded. This disadvantage becomes more salient in wireless networks where communication resources are scarce and communication environments vary more dynamically than in wired networks.

The lack of adaptivity in a strictly layered operation necessitates the development of new wireless resource management schemes that are able to determine the necessary and sufficient amount of resources based on dynamically changing environments, such as the channel gain, the interference level, the number of resource sharing users, the offered load, and so on. These resource management schemes necessarily require a *cross-layer design* in which information is exchanged between layers and each layer operates adaptively to this information. Figure 3.20 depicts the conceptual operation of the cross-layered protocol. In the figure, the upper and lower layers are not "black boxes" any longer and each layer cooperates while providing cross-layer adaptivity to others. To help understanding of this, we can consider two examples of cross-layer design.

3.4.2.1 TCP Over Wireless Links

The *transport control protocol* (TCP) has been developed as a transport layer protocol and has become the *de facto* standard in the current Internet. As a transport layer protocol, TCP provides reliable data transfer from source to destination. There exist several variations of TCP, but all have a common operating principle: TCP is equipped with an *automatic repeat request* (ARQ) process which retransmits data packets that are not acknowledged by the destination within a predetermined time. An unacknowledged packet is regarded as a lost packet due to network congestion, and TCP reduces the sending rate of packets to adjust it to the network capacity. Thus, the ARQ process of the TCP performs not only packet loss recovery but also flow control to the network capacity.

This operating principle of TCP may be detrimental in wireless networks in which the probability that transmission error occurs is much higher than in a wired network. Since TCP

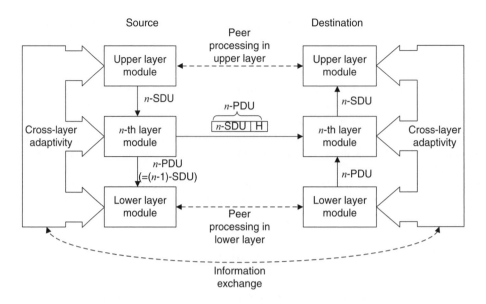

Figure 3.20   Protocol operation with cross-layer adaptivity

interprets all losses as being related to congestion, the source decreases the sending rate whenever a packet loss happens due to transmission errors occurring over the wireless channel, leading to reduced network throughput. One method to prevent this throughput loss is to hide any non-congestion-related loss from the TCP sender and solve transmission error problem locally at the data link layer which is responsible for the reliability of the wireless link. This can be considered as a cross-layer operation between the data link and the transport layer, as the wireless link is required to be aware of the operation and the current status of the TCP. Another solution to this problem is to provide a means of differentiating between losses due to congestion and those related to the wireless channel. This represents an interoperation between the physical and the transport layer as the physical layer information (e.g., the occurrence of a transmission error) is used for flow control at the transport layer.

Refer to Balakrishnan *et al.* (1996) for detailed comparisons of the schemes for improving TCP performance in wireless networks.

3.4.2.2 Channel-aware Scheduling

Scheduling is to determine when to access the channel to transmit data packets, so it is one of the major tasks of the MAC sublayer. In time-varying wireless channels, if each user can be scheduled when his/her channel quality becomes relatively better than usual conditions, the signal-to-noise ratio (SNR) provided to each user can be improved without additional use of transmission power. This can lead to an improvement in network throughput or the mitigation of transmission errors. *Channel-aware scheduling* refers to scheduling algorithms that utilize the channel condition of the constituent users. This is a cross-layer design between the physical and the data link layer, as the physical layer information (e.g., the channel condition) is used at the data link layer. Further discussion on channel-aware scheduling will be found in Chapter 5.

References

3GPP (2006) TS 23.107: *Quality of Service Concept and Architecture*, version 6.4.0.

Aravind, R., Civanlar, M.R. and Reibman, A.R. (1996) Packet loss resilience of MPEG-2 scalable video coding algorithms. *IEEE Transactions on Circuits and Systems for Video Technology*, **6**, 426–435.

Balakrishnan, H. *et al.* (1996) A comparison of mechanisms for improving TCP performance over wireless links. In: Proceedings ACM Sigcomm, Stanford, CA.

Bertsekas, D. and Gallager, R. (1992) *Data Networks*, Prentice-Hall.

Dixon, R.C. (1994) *Spread Spectrum Systems with Commercial Applications*, John Wiley & Sons.

FCC (2007) Federal Communication Commission Rules, Part 15, Section 15.247: *Radio Frequency Devices*.

Haykin, S. (2001) *Communication Systems*, 4th edn, John Wiley & Sons.

IEEE (2006) Standard 802.16e-2005 and Standard 802.16-2004/Corrigendum 1-2005, Part 16: *Air Interface for Fixed and Mobile Broadband Wireless Access Systems*.

ITU (1993) Recommendation E.800: *Terms and Definitions Related to Quality of Service and Network Performance Including Dependability*.

Lee, B.G. and Choi, S. (2008) *Broadband Wireless Access and Local Networks: Mobile WiMAX and WiFi*, Artech House.

Leon-Garcia, A. and Widjaja, I. (2000) *Communication Networks*, McGraw-Hill.

Li, W. (2001) Overview of fine granularity scalability in MPEG-4 video standard. *IEEE Transactions on Circuits and Systems for Video Technology*, **11**, 301–317.

Oppenheim, A.V. and Schafer, R.W. (1999) *Discrete-time Signal Processing*, Prentice-Hall.

Proakis, J.G. (1995) *Digital Communications*, McGraw-Hill.

Shenker, S. (1995) Fundamental design issues for the future internet. *IEEE Journal on Selected Areas in Communications*, **13**, 1176–1188.

Telatar, I.E. (1999) Capacity of multi-antenna Gaussian channels. *European Transactions on Telecommunications*, **10**, 585–595.

Westerinen *et al.* /IETF (2001) IETF Request for Comments 3198: *Terminology for Policy-based Management*.

4

Mathematical Tools for Resource Management

Engineering problems in practical applications are very diverse in form and number but, if modeled or formulated as mathematical problems, they converge to a limited number of problem patterns. Wireless communication engineers have set up mathematical models by taking account of physical phenomena such as propagation, fading, additive noise, and so on. The mathematical models obtained after simplifying, approximating, or ignoring certain aspects that have only negligible effects usually involve some simple and abstract mathematical expressions, but these are detailed enough to provide useful insights and interpretation of the physical phenomena. The formulated problems, in general, take the form of optimizing some functions under one or multiple constraints. In many cases, such *constrained optimization problems* may be cast into some well-defined categories for which numerical algorithms are readily available to make the problem-solving a straightforward matter. Therefore, mathematical modeling is the very first and crucial step in solving engineering problems.

The constrained optimization problem, in general, takes the form

$$\text{minimize } f(x)$$
$$\text{subject to } x \in C$$

where $f(x)$ is the objective (or cost) function, x the optimization variable, and C the constraint set. Optimization problems may be classified into different categories according to the properties of the objective function and the constraint set. For example, if the objective function is linear and the constraint set is a polyhedron, the optimization problem is classified as a linear programming problem, which can be solved by the well-known simplex method, interior point method, and other algorithms. So, in the past, it was believed that linearity determined the complexity of optimization problems. However, it is now recognized that the critical factor is not the linearity but convexity. Provided a problem can be formulated as a convex optimization problem, it can be solved in an efficient way by utilizing its convex structure.

In some cases, objective functions require multistage decision-making. Then the best decision is made at each stage based on all decisions previously made. The overall problem is split into multiple subproblems and then the optimal solutions of all subproblems are combined to determine the optimal solution of the overall problem. Dynamic programming techniques are well developed to solve such multistage optimization problems efficiently.

This chapter will examine the basic mathematical tools available. First it deals with convex optimization which is the foundation of all optimization problems. Then it addresses dynamic programming techniques, including the Markov decision process. Finally, it discusses the analogy between wireless resource management and economics theory in terms of basic mathematical formulations.

4.1 Convex Optimization

Convex optimization is a very fundamental technique that has many desirable properties. First, all local solutions are global solutions, so it is not necessary to check the validity of a local solution once it is determined. Second, a duality theory applies that helps to determine a solution with the support of an optimality condition. Third, efficient numerical algorithms already exist for determining the optimal solutions. Fourth, most importantly, as many optimization problems can be cast into convex optimization problems, the technique provides a basic tool for many engineering problems.[1]

4.1.1 Basic Concepts

The term "convex" is defined as follows. A set C is *convex* if, for any $x,y \in C$ and θ with $0 \leq \theta \leq 1$, the point $\theta x + (1-\theta)y$ also lies in C. In other words, every point on the line segment connecting two points x and y in a convex set C also belongs to the set. This implies that a convex set is connected. Figure 4.1 depicts some examples of a convex set and a non-convex set. An interval on a real line and any triangle on the two-dimensional space are typical examples of convex sets. A *convex combination* of $x_i \in C$ is a linear combination of the form $\theta_1 x_1 + \theta_2 x_2 + \cdots + \theta_n x_n$, where all coefficients θ_i are non-negative and sum to one. If C is a convex set, then any convex combinations of any elements in C also belong to C.

A set C is a *convex cone* if, for any $x,y \in C$ and $\theta_1 \geq 0, \theta_2 \geq 0$, the point $\theta_1 x + \theta_2 y$ also lies in C. By definition, a convex cone is also convex. A *conic combination* or *non-negative combination* of $x_i \in C$ is a linear combination of the form $\theta_1 x_1 + \theta_2 x_2 + \cdots + \theta_n x_n$, where all coefficients θ_i are non-negative. So, a convex cone contains any conic combinations of its elements.

A function $f : \mathbf{R}^n \to \mathbf{R}$ with the domain $D(\subseteq \mathbf{R}^n)$ is *convex* if, for any $x,y \in D$ and $0 \leq \theta \leq 1$, the following inequality holds:

$$f(\theta x + (1-\theta)y) \leq \theta f(x) + (1-\theta)f(y), \tag{4.1}$$

which is called *Jensen's inequality*. This implies that the line segment connecting $(x, f(x))$ and $(y, f(y))$ is above the graph of f. A function f is strictly convex if the inequality holds strictly in (4.1) when $x \neq y$ and $0 < \theta < 1$. If $-f(x)$ is convex or strictly convex, then $f(x)$ is *concave* or *strictly concave*, respectively. So, for the concave functions, the inequality in (4.1) is reversed. The α-*sublevel set* $\{x | f(x) \leq \alpha\}$ of a convex function f is convex for

---

[1] Refer to Boyd and Vandenberghe (2004) for more details.

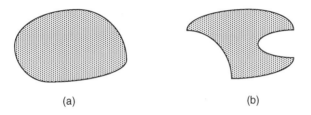

Figure 4.1 Examples of (a) a convex set, and (b) a non-convex set

any α. However, a function whose α-sublevel sets are convex sets is not always a convex function; such a function is called a *quasi-convex function*. For example, $f(x) = \log x$ is quasi-convex because the α-sublevel sets are convex and f is concave. There are some operations that preserve the convexity of functions. For example, the non-negative weighted sums of convex functions, $\sum_i w_i f_i(x)$, and the point-wise maximum of convex functions, $\max_i f_i(x)$, are convex.

The following are some examples of convex and concave functions. The *indicator function*

$$1_A(x) = \begin{cases} 1, & x \in A, \\ \infty, & \text{otherwise}, \end{cases}$$

is convex for a convex set A. Also:

- *Affine function* $f(x) = a^\mathsf{T} x + b$ is both convex and concave for $a, x \in \mathbf{R}^n$ and $b \in R$.
- *Power function* $f(x) = x^a$ is convex on $x \in \mathbf{R}_{++}$ if $a \geq 1$ or $a \leq 0$, and concave for $0 \leq a \leq 1$.
- *Exponential function* $f(x) = e^{ax}$ is convex for any $a \in \mathbf{R}$.
- *Logarithmic function* $f(x) = \log x$ is concave.
- *Least-square function* $f(x) = |Ax - b|^2$ is convex for any $\mathbf{A}$.
- *Entropy function* $H(x) = -\sum_i x_i \log x_i$ is concave.
- *Log determinant function* $f(X) = \log \det \mathbf{X}$ is concave for positive definite $\mathbf{X} \in \mathbf{R}^{n \times n}$.

Any convex function satisfies the Jensen's inequality in (4.1). Since many functions are convex or concave, Jensen's inequality is widely used to derive useful bounds associated with those functions. For example, the concavity of the logarithmic function $f(x) = \log x$ helps to derive the inequality of arithmetic and geometric means

$$\frac{1}{N} \sum_{k=1}^{N} x_k \geq \left(\prod_{k=1}^{N} x_k \right)^{1/N}, \tag{4.2}$$

since

$$\log \left(\frac{1}{N} \sum_{k=1}^{N} x_k \right) = f\left(\sum_{k=1}^{N} \frac{1}{N} x_k \right) \geq \frac{1}{N} \sum_{k=1}^{N} f(x_k) = \frac{1}{N} \log \left(\prod_{k=1}^{N} x_k \right). \tag{4.3}$$

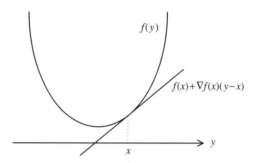

Figure 4.2 Tangent line of a convex function

As another example, Gibbs' inequality is an important inequality in information theory:

$$-\sum_{i=1}^{N} p_i \log p_i \leq -\sum_{i=1}^{N} p_i \log q_i, \qquad (4.4)$$

where p_i and q_i are probability distributions with $\sum_{i=1}^{N} p_i = 1$ and $\sum_{i=1}^{N} q_i = 1$. This may be easily proved by using Jensen's inequality on the concave function $f(x) = \log x$.

When a function $f\colon \mathbf{R}^n \to \mathbf{R}$ is continuously differentiable, f is convex if and only if

$$f(y) \geq f(x) + \nabla f(x)^T (y - x) \qquad (4.5)$$

holds for all $x, y \in \mathbf{R}^n$. This implies that the first-order Taylor expansion is a *global underestimator* of f. Geographically, the tangent line of f lies below the function f as depicted in Figure 4.2. When a function $f\colon \mathbf{R}^n \to \mathbf{R}$ is twice continuously differentiable, f is convex if and only if $\nabla^2 f(x) \geq 0$, where the inequality $\mathbf{A} \geq \mathbf{B}$ indicates that $\mathbf{A} - \mathbf{B}$ is positive semi-definite. Therefore, the quadratic form $f(x) = x^T \mathbf{A} x + b^T x + c$ is convex if and only if $\mathbf{A} \in \mathbf{R}^{n \times n}$ is positive semi-definite.

4.1.2 Constrained Optimization

Consider the standard minimization problem

$$\begin{aligned}
\text{minimize} \quad & f_0(x) \\
\text{subject to} \quad & f_i(x) \leq 0, \quad i = 1, 2, \ldots, m, \\
& h_i(x) = 0, \quad i = 1, 2, \ldots, p.
\end{aligned} \qquad (4.6)$$

The function $f_0(x)$ is called an *objective function* or *cost function*, and the variable $x \in \mathbf{R}^n$ is called an *optimization variable*. The inequalities $f_i(x) \leq 0$ and the equality $h_i(x) = 0$ are called *inequality constraints* and *equality constraints*, respectively. This problem is a constrained optimization, whereas the problem without any constraints is called unconstrained optimization. The optimization variable x is *feasible* if it satisfies the constraints $f_i(x) \leq 0$ and $h_i(x) = 0$, and it is *infeasible* otherwise. The set of all feasible variables is called a *feasible set*. A feasible solution x^* is (*globally*) *optimal* if $f_0(x^*) \leq f_0(x)$ for all feasible x. Or, equivalently, a feasible solution x^* is globally optimal if $f_0(x^*) = p^*$ for the global minimal value of (4.6). The set of all globally optimal points is called an *optimal set*. In contrast, a feasible point $\tilde{x}$ is *locally optimal* if there exist $\varepsilon > 0$ such that $f_0(\tilde{x}) \leq f_0(x)$ for feasible x with $|\hat{x} - x| \leq \varepsilon$. If $f_i(x) = 0$ for a feasible point x, then the inequality constraint $f_i(x) \leq 0$ is *active* at x. Otherwise, the inequality constraint $f_i(x) \leq 0$ is *inactive* at x.

A type of constrained optimization problem related to convexity is the *convex optimization problem*

$$
\begin{aligned}
\text{minimize} \quad & f_0(x) \\
\text{subject to} \quad & f_i(x) \leq 0, \quad i = 1, 2, \ldots, m, \\
& a_i x - b_i = 0, \quad i = 1, 2, \ldots, p,
\end{aligned}
\tag{4.7}
$$

where $f_0(x), f_1(x), \ldots, f_m(x)$ are convex functions. So, a minimization problem is a convex optimization problem if 1. the objective function is *convex*, 2. the inequality constraint functions are *convex*, and 3. the equality constraint functions are *affine*. The following maximization problem can also be a convex optimization problem:

$$
\begin{aligned}
\text{maximize} \quad & f_0(x) \\
\text{subject to} \quad & f_i(x) \leq 0, \quad i = 1, 2, \ldots, m, \\
& a_i x - b_i = 0, \quad i = 1, 2, \ldots, p,
\end{aligned}
\tag{4.8}
$$

where $f_0(x)$ is a concave function and $f_1(x), \ldots, f_m(x)$ are convex functions, because $-f_0(x)$ is convex and (4.8) can be transformed into a form of (4.7).

An important property of convex optimization problems is that any local optimal solution is also a globally optimal solution. In addition, from (4.5), a feasible point x is optimal if

$$
\nabla f_0(x)^{\mathrm{T}}(y - x) \geq 0
\tag{4.9}
$$

for all feasible y when the objective function is differentiable. In the case of the unconstrained problem, the necessary and sufficient optimality condition for optimal x is given by

$$
\nabla f_0(x) = 0.
\tag{4.10}
$$

4.1.3 Lagrange Dual Function

Consider the standard optimization problem

$$
\begin{aligned}
\text{minimize} \quad & f_0(x) \\
\text{subject to} \quad & f_i(x) \leq 0, \quad i = 1, 2, \ldots, m, \\
& h_i(x) = 0, \quad i = 1, 2, \ldots, p,
\end{aligned}
\tag{4.11}
$$

which is not necessarily a convex optimization problem. The *Lagrangian function* L:$\mathbf{R}^n \times$ $\mathbf{R}^m \times \mathbf{R}^p \rightarrow \mathbf{R}$ associated with (4.11) is defined by

$$L(x, \lambda, \nu) = f_0(x) + \sum_{i=1}^{m} \lambda_i f_i(x) + \sum_{i=1}^{p} \nu_i h_i(x). \qquad (4.12)$$

The vectors λ and ν are called *dual vectors* or *Lagrange multiplier vectors* associated with the inequality constraints and the equality constraints, respectively. The optimization variable x is called the *primal variable vector*, and the problem (4.11) is called the *primal optimization problem*. The Lagrangian function is an augmentation of the objective function with all the inequality and equality constraint functions incorporated by using Lagrange multipliers.

The *Lagrangian dual function* g:$\mathbf{R}^m \times \mathbf{R}^p \rightarrow \mathbf{R}$ associated with the primal optimization problem (4.11) is the minimum value of the Lagrangian function over x; that is

$$g(\lambda, \nu) = \min_{x} L(x, \lambda, \nu). \qquad (4.13)$$

The Lagrangian dual function in (4.13) is always concave irrespective of the convexity of the primal function, because the point-wise minimum of the affine functions is concave. The Lagrangian multiplier set (λ, ν) is *dual feasible* if both $\lambda \geq 0$ and $g(\lambda, \nu)$ is finite.

Let p^* denote the optimal value of the primal problem. Then, for $\lambda \geq 0$ and any ν, the Lagrangian dual function is a lower bound of the optimal value p^*

$$g(\lambda, \nu) \leq p^*. \qquad (4.14)$$

As this lower bound depends on λ and ν, the best lower bound can be found by maximizing the lower bound; that is

$$\begin{aligned} \text{maximize} \quad & g(\lambda, \nu) \\ \text{subject to} \quad & \lambda \geq 0. \end{aligned} \qquad (4.15)$$

This maximization problem is called the *dual optimization problem*. Optimal solutions of the dual problem are the Lagrange multipliers for the primal problem. The dual problem is always a convex optimization because the objective $g(\lambda, \nu)$ is concave and the constraints are convex.

If the optimal value of the dual problem is denoted by d^*, then it is the best lower bound on the primal value; that is, $p^* \geq d^*$. This important property is called *weak duality* and the difference $p^* - d^*$ is called a *duality gap*. The duality gap is always non-negative and it can be zero under some mild condition. *Strong duality* refers to this case of the zero duality gap. Even though strong duality does not hold in general, it holds under some constraint qualification conditions, such as the existence of strict interior points in the feasible set. For the following form of the convex optimization problem

$$\begin{aligned} \text{minimize} \quad & f_0(x) \\ \text{subject to} \quad & f_i(x) \leq 0, \qquad i = 1, 2, \ldots, m, \\ & Ax + b = 0, \end{aligned} \qquad (4.16)$$

Slater's point is a vector x satisfying

$$\begin{aligned}
f_i(x) &< 0, \quad \text{for all } i \text{ if } f_i(x) \text{ is nonlinear,} \\
f_i(x) &\leq 0, \quad \text{for all } i \text{ if } f_i(x) \text{ is linear,} \\
Ax + b &= 0.
\end{aligned} \tag{4.17}$$

If a Slater's point exists, then strong duality holds for the convex optimization.

In order to set up a graphical interpretation of the primal and dual optimization problems, consider a case having only one inequality constraint. The primal problem is given by

$$\begin{aligned}
&\text{minimize} \quad f_0(x) \\
&\text{subject to} \quad f_1(x) \leq 0.
\end{aligned} \tag{4.18}$$

Figure 4.3 depicts the optimal values of the primal and dual problems over the set $A = \{[f_1(x), f_0(x)] | x \in D\}$ in a two-dimensional space. Here, the domain of the problem is denoted by D. In this figure, the primal optimum p^* is located at the point where the boundary of A crosses the vertical axis. The Lagrangian function $L(x, \lambda_1) = f_0(x) + \lambda_1 f_1(x)$ corresponds to the hyperplane associated with the normal vector $[\lambda_1, 1]$. Then, the minimization of $L(x, \lambda_1)$ with respect to $x \in D$ implies finding the supporting hyperplane of the set A, and the Lagrangian dual function $g(\lambda_1) = \min_x L(x, \lambda_1)$ equals the vertical-axis intercept of the supporting hyperplane with the normal vector $[\lambda_1, 1]$. We first observe that, for any $\lambda_1 \geq 0$, this intercept is always below the primal optimum and this implies the weak duality $g(\lambda_1) \leq p^*$. The dual optimization problem is equivalent to finding the maximum of the achievable vertical-axis intercept of the supporting hyperplane for various different $\lambda_1 \geq 0$. We can observe in this figure that the optimized supporting hyperplane is tangential to the set A at the very point that the primal

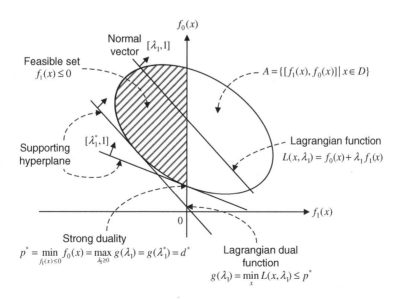

Figure 4.3   Graphical interpretation of strong duality

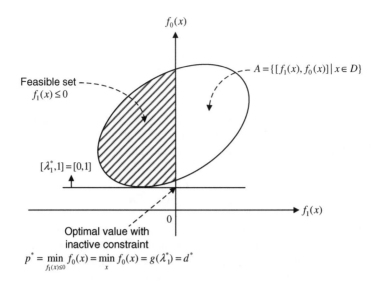

Figure 4.4   Strong duality for an inactive constraint

optimum p^* is attained. This implies that the strong duality $p^* = g(\lambda_1^*) = d^*$ holds for the optimized normal vector $[\lambda_1^*, 1]$ in the case plotted in Figure 4.3.

Figure 4.4 plots a special case where the constraint $f_1(x) \leq 0$ is inactive; that is, the solution of the constrained problem is equal to that of the unconstrained problem $\min_x f_0(x)$. As observed in the figure, the solution of the unconstrained problem is already located in the feasible set in this case, thereby making the constraint meaningless. Then, the normal vector of the optimized supporting hyperplane is given by $[\lambda_1^*, 1] = [0, 1]$ and this implies that the inequality-constraint part of the Lagrangian function is inactive; that is, $L(x, \lambda_1^*) = f_0(x)$. Therefore, the Lagangian dual function becomes the solution of the unconstrained problem, and strong duality also holds in this case.

Figures 4.3 and 4.4 illustrate cases where strong duality holds. Note that this strong duality holds if the minimal elements of set A form the boundary of a convex set.[2] This means that a segment connecting two different minimal elements of A is still included in A. If the primal problem is a convex one, this condition is satisfied as follows. Suppose that two different vectors x_1 and x_2 are related with two different minimal elements of A, $[f_1(x_1), f_0(x_1)]$ and $[f_1(x_2), f_0(x_2)]$. Then, a point on the segment connecting x_1 and x_2 is represented by $[\theta f_1(x_1) + (1 - \theta)f_1(x_2), \theta f_0(x_1) + (1 - \theta)f_0(x_2)]$ for $0 \leq \theta \leq 1$. This point cannot be located out of set A because the point $[f_1(x_3), f_0(x_3)]$ associated with $x_3 = \theta x_1 + (1 - \theta)x_2$ is closer to the boundary of set A. This fact comes from Jensen's inequality

$$f_i(x_3) \leq \theta f_i(x_1) + (1 - \theta)f_i(x_2) \tag{4.19}$$

for $i = 0, 1$ and $x_3 \in D$ by the convexity of the domain D.

---

[2] A point $x \in A$ is a minimal element of a set A if there exists no point $y \neq x$ in A such that $y \leq x$.

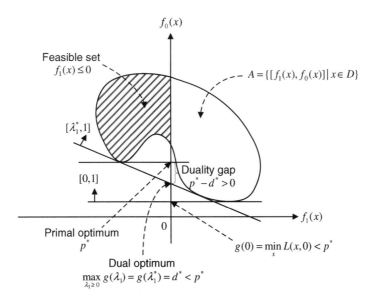

Figure 4.5 Illustration of non-convex minimal points

If the minimal elements of set A do not form the boundary of a convex set, strong duality is no longer guaranteed. Figure 4.5 depicts such an example. In this figure, the supporting hyperplane at the primal optimum has the normal vector $[0, 1]$. However, the Lagrangian dual function with $\lambda_1 = 0$ does not coincide with this supporting hyperplane as the Lagrangian function is minimized at an infeasible point. The solution of the dual optimization problem $d^* = g(\lambda_1^*)$, the optimized supporting hyperplane, renders a vertical-axis intercept lower than the primal optimum, and as a result the duality gap is larger than zero in this case.

Example 4.1. Consider the following convex optimization problem

$$\text{minimize } |\mathbf{A}x|^2 \\ \text{subject to } \mathbf{C}x \leq d. \tag{4.20}$$

The Lagrangian function associated with (4.20) is

$$L(x, \lambda) = x^T \mathbf{A}^T \mathbf{A} x + \lambda^T (\mathbf{C}x - d) \tag{4.21}$$

and the Lagrangian dual function is given by $g(\lambda) = \min_x L(x, \lambda)$. As (4.21) is a convex function, we can find the optimal x by using the optimal condition in (4.10); that is

$$\nabla_x L(x, \lambda) = 2\mathbf{A}^T \mathbf{A} x + \mathbf{C}^T \lambda = 0. \tag{4.22}$$

If we insert $x = -1/2(\mathbf{A}^\mathrm{T}\mathbf{A})^{-1}\mathbf{C}^\mathrm{T}\lambda$ into (4.21), we get the dual function

$$g(\lambda) = -\frac{1}{4}\lambda^T\mathbf{C}(\mathbf{A}^T\mathbf{A})^{-1}\mathbf{C}^T\lambda - d^T\lambda. \tag{4.23}$$

For this problem, strong duality holds.   □

4.1.4 Karush–Kuhn–Tucker Optimality Condition

We assume that the objective and constraint functions are all differentiable and the strong duality holds. Let x^* and (λ^*, ν^*) denote the primal optimal and the dual optimal variables. Then the optimal solution for the optimization problem is given by

$$f_i(x^*) \leq 0 \tag{4.24a}$$

$$h_i(x^*) = 0 \tag{4.24b}$$

$$\lambda_i^* \geq 0 \tag{4.24c}$$

$$\lambda_i^* f_i(x^*) = 0 \tag{4.24d}$$

$$\nabla f_0(x^*) + \sum_{i=1}^m \lambda_i^* \nabla f_i(x^*) + \sum_{i=1}^p \nu_i^* \nabla h_i(x^*) = 0 \tag{4.24e}$$

which are called the *Karush–Kuhn–Tucker* (KKT) conditions. Equations (4.24a) and (4.24b) come from the primal feasibility and (4.24c) from the dual feasibility. If the constraint $f_i(x) \leq 0$ is inactive, this constraint becomes meaningless and the associated Lagrange multiplier λ_i becomes 0 as shown in Figure 4.4. If the constraint $f_i(x) \leq 0$ is active, then we have $f_i(x^*) = 0$. From these two relations, we have equation (4.24d), which is called the *complementary slackness*. Equation (4.24e) is obtained by applying the optimality condition in (4.10) to the unconstrained Lagrangian function. When strong duality holds, the primal optimal x^* and the dual optimal (λ^*, ν^*) hold the KKT conditions. Conversely, if the primal problem is convex and (x', λ', ν') satisfies the KKT conditions, then x' is the primal optimal and (λ', ν') is the dual optimal with zero duality gap. So, the KKT conditions play a very important role in determining the optimal solution in constrained optimization problems.

Example 4.2. Consider the minimization problem

$$\begin{aligned} &\text{minimize } x^2 + y^2 + z^2 \\ &\text{subject to } x + 2y + 3z - 14 = 0. \end{aligned} \tag{4.25}$$

The Lagrangian function associated with (4.25) is

$$L = x^2 + y^2 + z^2 + v(x + 2y + 3z - 14). \qquad (4.26)$$

With the KKT conditions applied to this optimization problem, we get

$$x + 2y + 3z - 14 = 0, \qquad (4.27a)$$

$$\begin{bmatrix} 2x \\ 2y \\ 2z \end{bmatrix} + v \begin{bmatrix} 1 \\ 2 \\ 3 \end{bmatrix} = 0. \qquad (4.27b)$$

On solving the linear equations in (4.27a,b), we get the optimal solution

$$\begin{bmatrix} x \\ y \\ z \end{bmatrix} = \begin{bmatrix} 1 \\ 2 \\ 3 \end{bmatrix}, \qquad (4.28)$$

and the minimum value of the problem, which is 14.   □

4.1.5 Application of Convex Optimization

When allocating a limited resource X to n users for maximization of the total benefit (e.g., throughput), we usually get the following type of constrained optimization problem

$$\text{maximize} \sum_{i=1}^{n} \varphi_i(x_i)$$

$$\text{subject to} \sum_{i=1}^{n} x_i = X, \qquad (4.29)$$

$$x_i \geq 0,$$

where $\varphi_i(x)$ denotes a mapping from the allocated resource x_i to the benefit of user i. We assume the objective functions $\varphi_i(x)$ are strictly concave, differentiable, and strictly increasing functions. If we let $\psi_i(x) = -\varphi_i(x)$ and substitute the objective by "minimize $\sum_{i=1}^{n} \psi_i(x)$," this optimization problem is convex and the Lagrangian function is given by

$$L(x, \lambda, v) = \sum_{i=1}^{n} \psi_i(x_i) - \sum_{i=1}^{n} \lambda_i x_i + v \left(\sum_{i=1}^{n} x_i - X \right) \qquad (4.30)$$

On applying the KKT conditions for this optimization problem, we get

$$x_i \geq 0 \qquad (4.31a)$$

$$\sum_{i=1}^{n} x_i = X \tag{4.31b}$$

$$\lambda_i \geq 0 \tag{4.31c}$$

$$\lambda_i x_i = 0 \tag{4.31d}$$

$$\psi_i'(x_i) - \lambda_i + \nu = 0. \tag{4.31e}$$

The optimal resource allocation can be obtained by determining the x_i that satisfies the KKT conditions. From (4.31c) and (4.31e), $\lambda_i = \psi_i'(x_i) + \nu \geq 0$. Since $\psi_i(x) = -\varphi_i(x)$, we have $\lambda_i = -\varphi_i'(x_i) + \nu \geq 0$. In addition, the complementary slackness in (4.31d) implies that λ_i is zero if $x_i > 0$ and λ_i is positive if $x_i = 0$. In other words, we have

$$\varphi_i'(x_i) = \nu, \quad \text{if} \quad x_i > 0, \tag{4.32a}$$

$$\varphi_i'(x_i) < \nu, \quad \text{if} \quad x_i = 0. \tag{4.32b}$$

Since $\varphi_i(x)$ is concave increasing, the first derivative $\varphi_i'(x)$ is a positive and strictly decreasing function. Thus, the optimal point x_i satisfying $\varphi_i'(x_i) = \nu$ is unique and the inverse function of $\varphi_i'(x)$ exists. By (4.32a, b) we get

$$x_i = \max\{(\varphi_i')^{(-1)}(\nu), 0\}. \tag{4.33}$$

On substituting (4.33) into (4.39) we get

$$\sum_{i=1}^{n} \max\{(\varphi_i')^{(-1)}(\nu), 0\} = X. \tag{4.34}$$

As the left-hand side of (4.34) is a strictly decreasing function of ν, it is possible to determine the unique dual optimal variable satisfying (4.34).

The optimal condition shows two different ways to obtain the optimal solution. First, we can adjust the dual variable ν such that the optimal solutions x_i are attained at the point where the marginal increment is identical with respect to ν and their sum is equal to X. If $\varphi_i'(0) < \nu$, then there exists no $x_i \geq 0$ satisfying $\varphi_i'(x_i) = \nu$ because $\varphi_i'(x)$ is strictly decreasing. Second, we can adjust the primal variables that sum to X. We initially choose vector x such that $\sum_{i=1}^{n} x_i = X$ and then determine $x_{i_{max}}$ and $x_{i_{min}}$ such that $\varphi_i'(x_{i_{max}})$ is the largest and $\varphi_i'(x_{i_{min}})$ is the smallest. If we decrease $x_{i_{min}}$ by a small value δ and increase $x_{i_{max}}$ by δ, then summation of the x_i is still equal to X and the overall objective will increase. Finally, this adjustment converges to the point where all the marginal increment is the same as shown in (4.32a, b).

One typical example of the optimization problem in (4.29) is the channel capacity over parallel channels. This case corresponds to $\varphi_i(x_i) = \log_2(1 + x_i/\sigma_i^2)$, where σ_i^2 is the channel

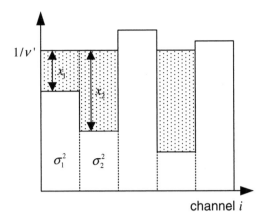

Figure 4.6   Water-filling for parallel channels

noise variance of the ith channel and x_i is the power allocated to the ith channel. Then, the optimal power allocation is derived by using (4.33) such that

$$x_i = \max\{1/\nu' - \sigma_i^2, 0\}, \tag{4.35}$$

with $\nu' = \nu \ln 2$. This optimal solution is called *water-filling*, as illustrated in Figure 4.6. The vertical levels indicate the noise level of the channels. The power is allocated to the low-noise channels so that the power-plus-noise level remains the same.

Consider now what would happen if $\varphi_i(x)$ in (4.29) changes from concave to convex. For simplicity, we consider the maximization problem:

$$\text{maximize} \quad \sum_{i=1}^{N} \varphi(x_i)$$

$$\text{subject to} \quad \sum_{i=1}^{N} x_i = X, \tag{4.36}$$

$$0 \le x_i \le a_i,$$

$$a_1 \ge a_2 \ge \cdots \ge a_N \ge 0.$$

Here, $\varphi(x)$ is a convex increasing function. Note that this is the maximization of a convex function, which is not a convex optimization problem. So, we cannot apply the KKT optimality condition to this problem. Instead, we have to obtain the solution in different ways. In this case, there exists at most one optimal solution x^* in the range $0 < x_i^* < a_i$. This implies that almost all elements take the value 0 or a_i. So, the maximizer would take the form $x^* = (a_1, a_2, \ldots, a_{k-1}, x_k^*, 0, \ldots, 0)$, where $x_k^* = X - \sum_{i=1}^{k-1} a_i$. Now we consider the "greedy algorithm"[3]

---

[3] A *greedy algorithm* is referred to as any algorithm that makes the locally optimum choice at each stage with the hope of finding the global optimum.

$$x_1^G = \min(a_1, X),$$

$$x_{k+1}^G = \min\left(a_{k+1}, X - \sum_{i=1}^{k} a_i\right). \tag{4.37}$$

We can prove that the solution to this greedy algorithm is the optimal solution to the problem (4.36) by contradiction. Let $g(x) \equiv \sum_{i=1}^{N} \varphi(x_i)$ and suppose that $x^* \neq x^G$. If we take any permutation π such that $x_{\pi(1)}^* \geq x_{\pi(2)}^* \geq \cdots \geq x_{\pi(N)}^*$, then $x_\pi^* = (x_{\pi(1)}^*, x_{\pi(2)}^*, \ldots, x_{\pi(N)}^*)$ is also feasible and maximizes $g(x)$. So, without loss of optimality, we assume that $x_1^* \geq x_2^* \geq \cdots \geq x_n^*$. As $x^* \neq x^G$, we can find (i, j) such that $0 < x_i^* < x_i^G$ and $x_j^* > x_j^G$ $(i < j)$. Define x^{**} by $x^{**} = (x_1^*, \ldots, x_i^* + \Delta, \ldots, x_j^* - \Delta, \ldots, x_n^*)$, where Δ is chosen such that $\Delta = \min(x_i^G - x_i^*, x_j^*)$. Then, x^{**} is also feasible. Now we evaluate the difference

$$g(x^{**}) - g(x^*) = \{\varphi(x_i^* + \Delta) - \varphi(x_i^*)\} - \{\varphi(x_j^*) - \varphi(x_j^* - \Delta)\} > 0 \tag{4.38}$$

where the inequality holds due to the convexity of $\varphi(x)$. This implies that the feasible x^{**} is chosen such that $g(x^{**}) > g(x^*)$, which is a contradiction. So, the greedy solution of (4.37) is the optimal solution to the problem (4.36). The above demonstrates that convexity is important in obtaining the optimal solution even in the case of a non-convex optimization problem.

4.2 Dynamic Programming

Some optimization problems may be split into subproblems. A typical way of solving these is to construct the overall solution by combining each solution to each subproblem. *Dynamic programming* is a method that first decomposes a given problem into a series of smaller problems and then applies a divide-and-conquer type sequential optimization. For example, it converts one N-variable problem into N single-variable problems, which usually requires less computational complexity. The term "dynamic" represents *multistage* or *time-varying*, and "programming" represents *decision-making* or *plan-finding*. Even though named "dynamic programming," it is irrelevant to computer programming.[4]

4.2.1 Sequential Optimization

Sequential optimization problems are classified into finite-horizon and infinite-horizon dynamic programming problems according to the number of decision stages. Besides that, they belong to stochastic or deterministic dynamic programming depending on the randomness in problems.

4.2.1.1 Finite-horizon Deterministic Dynamic Programming

Consider first a general optimization problem that is hard to solve and then introduce some assumptions that make the problem solvable in an iterative way. In general, the optimization problem has the form

---

[4] Refer to Bertsekas (2001) for more details.

$$\underset{\{u_k\}}{\text{minimize}} \quad g(x_0, x_1, \ldots, x_N, u_0, u_1, \ldots, u_{N-1})$$

$$\text{subject to} \quad f(x_0, x_1, \ldots, x_N, u_0, u_1, \ldots, u_{N-1}) \leq 0, \qquad (4.39)$$

$$u_k \in U_k, \ k = 0, 1, \ldots, N-1,$$

$$x_0 \text{ is given.}$$

Here x_k denotes a state vector that describes the state of the system, and u_k a vector of decision variables that can be chosen to minimize the objective. The objective function $g(\bullet)$ is a function of both the states and the decision variables, and the constraint function $f(\bullet)$ describes the system dynamics. Basically, we can treat it as a conventional optimization problem and solve it by using the KKT conditions for the associated Lagrangian. However, if N is very large, the problem is difficult to solve, especially when the constraint function $f(\bullet)$ is nonlinear.

In order to make the problem tractable, a time-separable assumption is introduced on the objective and constraint functions. We assume the objective takes the expression

$$g(x_0, x_1, \ldots, x_N, u_0, u_1, \ldots, u_{N-1}) = g_0(x_0, u_0) + g_1(x_1, u_1) + \cdots + g_{N-1}(x_{N-1}, u_{N-1})$$
$$+ g_N(x_N). \qquad (4.40)$$

Also, we assume that the state transition follows the *Markovian property*; that is, the next state $x_{k+1} \in S_{k+1}$ is determined by the current state $x_k \in S_k$ and the current decision $u_k \in U_k(x_k)$; that is

$$
\begin{aligned}
x_1 &= f_0(x_0, u_0) \\
x_2 &= f_1(x_1, u_1) \\
&\vdots \\
x_N &= f_{N-1}(x_{N-1}, u_{N-1}).
\end{aligned}
\qquad (4.41)
$$

This implies that the state transition is memoryless in the sense that the next state depends only on the current state and the current decision, and is independent of all the past states and decisions that have led to the current state. Now the problem becomes

$$\underset{\{u_k\}}{\text{minimize}} \quad \sum_{k=0}^{N-1} g_k(x_k, u_k) + g_N(x_N)$$

$$\text{subject to} \quad x_{k+1} = f_k(x_k, u_k), \ k = 0, 1, \ldots, N-1, \qquad (4.42)$$

$$u_k \in U_k, \ k = 0, 1, \ldots, N-1,$$

$$x_0 \text{ is given.}$$

This problem is called *deterministic dynamic programming*. The states and the decisions are ordered by the stages associated with the index variable $k = 0, 1 \ldots, N-1$. At stage k, the system is described by the state variable x_k and the decision is made according to the state and the stage. After a decision is made, the system makes a transition to the next state depending on the current state and the decision variables through $x_{k+1} = f_k(x_k, u_k)$ and receives the immediate cost

$g_k(x_k, u_k)$ after state transition. The ultimate goal is to find the optimal decisions at each stage to minimize the overall objective under the state transition rules.

The problem in (4.42) may be solved also by applying the KKT conditions, but the complexity would still be high. However, it can be solved more efficiently by using the underlying special structure of the problems – that is, the time-separable and Markovian properties. *Bellman's principle of optimality* states that an optimal policy has the property that, no matter what the initial state x_0 and the initial decision u_0 may be, the remaining decisions u_1, $u_2, \ldots, u_{N-1}$ must constitute an optimal policy with regard to the state resulting from the first decision u_0 (Bellman, 1953). Let the admissible policies be $\pi = \{\mu_0, \mu_1, \ldots, \mu_{N-1}\} \in \Pi$, where $u_k = \mu_k(x_k) \in U_k(x_k)$. Then, the expected cost of π starting at x_0 is given by

$$V_\pi(x_0) = \sum_{k=0}^{N-1} g_k(x_k, \mu_k(x_k)) + g_N(x_N). \tag{4.43}$$

The optimal policy π^* that minimizes V among all admissible policies satisfies

$$V^*(x_0) = V_{\pi^*}(x_0) = \min_\pi V_\pi(x_0). \tag{4.44}$$

Let $\pi^* = (\mu_0^*, \mu_1^*, \ldots, \mu_{N-1}^*)$ be the optimal policy. We consider the *truncated subproblem* where the overall cost is minimized from time i to N:

$$
\begin{aligned}
\underset{\{u_k\}}{\text{minimize}} \quad & \sum_{k=i}^{N-1} g_k(x_k, u_k) + g_N(x_N) \\
\text{subject to} \quad & x_{k+1} = f_k(x_k, u_k), \; k = i, i+1, \ldots, N-1, \\
& u_k \in U_k, \; k = i, i+1, \ldots, N-1, \\
& x_i \text{ is given.}
\end{aligned}
\tag{4.45}
$$

Then, the truncated optimal policy $(\mu_i^*, \mu_{i+1}^*, \ldots, \mu_{N-1}^*)$ is the optimal policy for the truncated subproblem due to the principle of optimality.

For example, consider the shortest path-finding problem illustrated in Figure 4.7. Let highway 1 be the shortest path from city A to city C. Since highway 1 passes through city B, we can show by contradiction that the shortest path from B to C will also be highway 1. Suppose that there exists another way, highway 2, from city B to city C, which is shorter than highway 1 from B to C. Then, the shortest path from city A to city C is not highway 1 any longer, because the shortest path will be a combination of highway 1 from A to B and

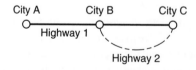

Figure 4.7   Graph for the shortest-path finding problem

highway 2 from B to C. This is a contraction and it shows that the shortest path from city B to city C should be highway 1. So, the optimal solution consists only of the optimal solutions to the subproblems.

The principle of optimality is the key to solving the sequential control problem in dynamic programming. It allows us to solve the problem by sequentially solving the smaller problem. The optimal cost $V^*(x_0)$ equals $V_0(x_0)$ by using the backward iteration

$$V_N(x_N) = g_N(x_N),$$
$$V_k(x_k) = \min_{u_k \in U_k(x_k)} \{g_k(x_k, u_k) + V_{k+1}(f_k(x_k, u_k))\}, \quad k = 0, 1, \ldots, N-1. \tag{4.46}$$

This is called the *Bellman equation*. The policy $\pi^* = (\mu_0^*, \mu_1^*, \ldots, \mu_{N-1}^*)$ consisting of the minimizer of (4.46), $\mu_k^*(x_k) = u_k^*$, is the optimal decision for the above sequential control problem.

Now, consider the following deterministic dynamic programming with $N = 2$:

$$
\begin{aligned}
&\underset{\{u_k\}}{\text{minimize}} && \sum_{k=0}^{N-1}(x_k + u_k)^2 + x_N^2 \\
&\text{subject to} && x_{k+1} = 2x_k + u_k, \quad k = 0, 1, \ldots, N-1, \\
& && u_k \in U_k, \quad k = 0, 1, \ldots, N-1, \\
& && x_0 \text{ is given.}
\end{aligned}
\tag{4.47}
$$

By using the principle of optimality, we can apply the backward iteration to (4.47). First, for $k = 2$, we have

$$V_2(x_2) = x_2^2. \tag{4.48}$$

For $k = 1$, the problem is now expressed as

$$
\begin{aligned}
V_1(x_1) &= \min_{u_1}[(x_1 + u_1)^2 + V_2(x_2)] \\
&= \min_{u_1}[(x_1 + u_1)^2 + x_2^2] \\
&= \min_{u_1}[(x_1 + u_1)^2 + (2x_1 + u_1)^2].
\end{aligned}
\tag{4.49}
$$

The optimal solution u_1 to the problem (4.49) is simply obtained by taking its first-order derivative with respect to u_1 and is given by

$$u_1 = \mu_1^*(x_1) = -\frac{3}{2}x_1. \tag{4.50}$$

With this optimal decision, (4.49) becomes

$$V_1(x_1) = \frac{1}{2}x_1^2. \tag{4.51}$$

For $k = 0$, we have the subproblem

$$V_0(x_0) = \min_{u_0}[(x_0 + u_0)^2 + V_1(x_1)]$$

$$= \min_{u_0}\left[(x_0 + u_0)^2 + \frac{1}{2}x_1^2\right] \qquad (4.52)$$

$$= \min_{u_0}\left[(x_0 + u_0)^2 + \frac{1}{2}(2x_0 + u_0)^2\right].$$

The optimal decision u_0 to the problem (4.52) is given by

$$u_0 = \mu_0^*(x_0) = -\frac{4}{3}x_0, \qquad (4.53)$$

and this optimal decision turns (4.52) into

$$V_0(x_0) = \frac{1}{3}x_0^2. \qquad (4.54)$$

If the initial state is given by $x_0 = 1$, then the optimal decisions at stage 0 and stage 1 are $u_0 = -4/3$ and $u_1 = -1$ by using (4.53) and (4.50), respectively. In addition, the optimal reward from state transitions is $V_0(1) = 1/3$ from (4.54). The principle of optimality helps to solve the multistage decision-making problem without carrying Lagrangian multipliers.

4.2.1.2 Finite-horizon Stochastic Dynamic Programming

Some optimization problems may involve randomness in objective functions and state transitions. For example, the state transition is affected by random parameter w_k such that

$$x_{k+1} = f_k(x_k, u_k, w_k), \quad k = 0, 1, \ldots, N-1. \qquad (4.55)$$

For the Markov property of the state transition, we assume that the probability distribution of w_k is independent of $w_0, w_1, \ldots, w_{k-1}$. Then the stochastic dynamic programming problem can be formulated as

$$\begin{aligned}
\text{minimize}_{\{u_k\}} \quad & \mathbb{E}\left[\sum_{k=0}^{N-1} g_k(x_k, u_k, w_k) + g_N(x_N)\right] \\
\text{subject to} \quad & x_{k+1} = f_k(x_k, u_k, w_k), \ k = 0, 1, \ldots, N-1, \\
& u_k \in U_k, \ k = 0, 1, \ldots, N-1, \\
& x_0 \text{ is given.}
\end{aligned} \qquad (4.56)$$

The principle of optimality can still be applied to this stochastic problem, and we get

$$\begin{aligned}
V_N(x_N) &= g_N(x_N), \\
V_k(x_k) &= \min_{u_k \in U_k(x_k)} \mathbb{E}[g_k(x_k, u_k, w_k) + V_{k+1}(f_k(x_k, u_k, w_k))], \quad k = 0, 1, \ldots, N-1.
\end{aligned}$$

$$(4.57)$$

For deterministic dynamic programming, the forward-iteration Bellman equation can be derived by applying the principle of optimality either to the head parts of the stages or to the tail parts. In contrast, for stochastic dynamic programming, only the backward recursion is possible because no states are guaranteed to arrive due to the uncertainty in w_k.

As an example of stochastic dynamic programming, consider a power control problem of maximizing the delay-constrained capacity (Negi and Cioffi, 2002). The considered channel is a frequency-flat block fading channel with i.i.d. channel gain w_k. The transmitter is allowed to distribute the total power NP within N blocks in order to maximize the expected capacity. If the channel gains are fully known at the transmitter, then the water-filling in (4.35) becomes the optimal decision. Now we assume the causal channel information such that, at stage k, only the channel gains $w_0, w_1, \ldots, w_k$ are known at the transmitter. Let state x_k be the power remaining for stages k to $N - 1$. Since the total power is NP, the initial state is $x_0 = NP$. If u_k denotes the power allocated to stage k, then the remaining power at stage $k + 1$ becomes

$$x_{k+1} = NP - \sum_{j=0}^{k} u_j = x_k - u_k, \tag{4.58}$$

which characterizes the state transition. Now, the problem can be formulated as

$$
\begin{aligned}
\underset{\{u_k\}}{\text{maximize}} \quad & \mathbb{E}\left[\sum_{k=0}^{N-1} \log_2(1 + w_k u_k)\right] \\
\text{subject to} \quad & x_{k+1} = f_k(x_k, u_k) = x_k - u_k, \quad k = 0, 1, \ldots, N-1, \\
& 0 \le u_k \le x_k, \quad k = 0, 1, \ldots, N-1, \\
& x_0 = NP.
\end{aligned} \tag{4.59}
$$

We apply the relation $\max f(x) = -\min(-f(x))$ to modify the objective to

$$\underset{\{u_k\}}{\text{minimize}} \ \mathbb{E}\left[-\sum_{k=0}^{N-1} \log_2(1 + w_k u_k)\right] \tag{4.60}$$

and apply the principle of optimality in (4.46) to obtain

$$
\begin{aligned}
V_N(x_N) &= 0, \\
V_k(x_k) &= \underset{u_k \in U_k(x_k)}{\max} \ \mathbb{E}[\log_2(1 + w_k u_k(x_k)) + V_{k+1}(f_k(x_k, u_k(x_k)))], \\
&= \underset{0 \le u_k(x) \le x_k}{\max} \ \mathbb{E}[\log_2(1 + w_k u_k(x_k)) + V_{k+1}(x_k - u_k(x_k))], \\
u_k(x_k) &= \arg \underset{0 \le u_k(x_k) \le x_k}{\max} \ \{\log_2(1 + w_k u_k(x_k)) + V_{k+1}(x_k - u_k(x_k))\},
\end{aligned} \tag{4.61}
$$

for $k = 0, 1, \ldots, N-1$. Note that the decision at each stage written in (4.61) depends on the statistical distribution of w_k. By using this backward recursion, the optimal decision for stage k is given by

$$u_k^* = u_k(x_k^*), \tag{4.62}$$

where

$$x_k^* = NP - \sum_{j=0}^{k-1} u_j^*(x_j^*),$$

$$x_0^* = NP.$$

(4.63)

Finally, the maximum achievable rate is given by $V_0(NP)$. Notice that the principle of optimality has helped to solve the stochastic dynamic programming.

4.2.1.3 Infinite-horizon Dynamic Programming

So far, we have considered only cases where N is finite, and the objective function $g_k(x_k, u_k)$ and the state transition function $f_k(x_k, u_k)$ are allowed to be time-varying; that is, they may depend on the time index k as in (4.42). In most cases of infinite-horizon problems, however, it is assumed that the problems are stationary, so that the objective function and the state transition function are assumed independent of the time index for simplicity. Extending the finite-horizon problem in (4.42) to the stationary infinite-horizon dynamic programming problem, we get

$$\begin{aligned} \underset{\{u_k\}}{\text{minimize}} \quad & \sum_{k=0}^{\infty} g(x_k, u_k) \\ \text{subject to} \quad & x_{k+1} = f(x_k, u_k), \\ & u_k \in U, \\ & x_0 \text{ is given.} \end{aligned}$$

(4.64)

However, in this case, the objective may diverge for some optimization problems, as it is an infinite sum. To prevent this irritating divergence, we introduce a discount factor λ so that the objective function at the kth stage is discounted by λ^k for the discounted rate $0 < \lambda < 1$. Then the problem is modified to

$$\begin{aligned} \underset{\{u_k\}}{\text{minimize}} \quad & \sum_{k=0}^{\infty} \lambda^k g(x_k, u_k) \\ \text{subject to} \quad & x_{k+1} = f(x_k, u_k), \\ & u_k \in U, \\ & x_0 \text{ is given.} \end{aligned}$$

(4.65)

An infinite-horizon problem with a discount factor can be considered as a finite-horizon problem with an uncertain number of stages. If the system stage transition stops at any stage k with a probability $1 - \lambda$, then the probability of arriving at the kth stage is λ^k. So, $\sum_{k=0}^{\infty} \lambda^k g(x_k, u_k)$ denotes the average return when the stopping probability at each stage is $1 - \lambda$. In the monetary example, the discount means evaluating the present value of the future money taking account of an interest rate and uncertainty. The current 100 dollars have a value of $100(1 + i)^2$ dollars two years later when the annual compound interest rate is i. Conversely, 100 dollars after 3 years has the current value of $100/(1 + i)^3$ dollars. Similarly, if we set $\lambda = 1/(1 + i)$, the cost $g(x_k, u_k)$ at stage k is equivalent to $\lambda^k g(x_k, u_k)$ at stage 0.

As given in (4.46), the principle of optimality yields

$$V_k(x_k) = \min_{u_k \in U_k(x_k)} \{\lambda^k g(x_k, u_k) + V_{k+1}(f_k(x_k, u_k))\}. \tag{4.66}$$

Since $V_k(x_k)$ is the cost in terms of the value at stage 0, it is equivalent to the cost $V_k(x_k)/\lambda^k$ at stage k. If we divide (4.66) by λ^k, we obtain the Bellman equation

$$W_k(x_k) = \min_{u_k \in U_k(x_k)} \{g(x_k, u_k) + \lambda W_{k+1}(f_k(x_k, u_k))\}, \tag{4.67}$$

where $W_k(x_k) \equiv V_k(x_k)/\lambda^k$ is the current cost. Iterations on (4.67) starting from any initial bounded continuous W_0 converge as the number of iteration becomes large. Using the converged W, the optimal decision for (4.65) can be determined.

Note that the optimal solution of (4.64) depends on the discount factor λ used in the formulation; that is, different discounting factors may result in different optimal policies. A policy is called *Blackwell optimal* if there exists a $\lambda_0 \in (0, 1)$ such that the policy is the optimal solution of (4.65) for all $\lambda \in [\lambda_0, 1]$. This implies that the Blackwell optimal policy is optimal for all large discounting factors. Then, it is shown that there exists a Blackwell optimal policy and it is also optimal for the average cost problem formulated (Bertsekas, 2001) as

$$\begin{aligned} \text{minimize} \quad & \lim_{N \to \infty} \frac{1}{N} \sum_{k=0}^{N-1} g(x_k, u_k) \\ \{u_k\} \\ \text{subject to} \quad & x_{k+1} = f(x_k, u_k), \\ & u_k \in U, \\ & x_0 \text{ is given.} \end{aligned} \tag{4.68}$$

As a result, it is sufficient to find an optimal policy for a discounted cost problem with a large discounting factor.

4.2.2 *Markov Decision Process*

The Markov chain is a discrete stochastic process with the Markov property that the conditional probability of the future state, given the present and past states, depends only on the current state; that is

$$\begin{aligned} \Pr(x_{k+1} = b_{k+1} | x_k = b_k, x_{k-1} = b_{k-1}, \ldots, x_0 = b_0) \\ = \Pr(x_{k+1} = b_{k+1} | x_k = b_k) \end{aligned} \tag{4.69}$$

The Markov chain is time-invariant if the state transition probability is independent of time index k; that is, $\Pr(x_{k+1} = j | x_k = i) = P_{ij}$. When the number of states, m, is finite, we can form the $m \times m$ transition matrix $\mathbf{P} = [P_{ij}]$ and the stationary distribution row vector $\phi = (\phi_1, \phi_2, \ldots, \phi_m)$ that satisfy

$$\phi = \phi\mathbf{P}, \quad \sum_{i=1}^{m} \phi_i = 1. \qquad (4.70)$$

If the finite state Markov chain is irreducible and aperiodic, then the stationary distribution is unique and the distribution of the states tends to be a stationary distribution as $k \to \infty$ irrespective of the initial state. If there is a reward $R(x_k)$ for each state x_k, then the expected reward becomes

$$R = \sum_{i=1}^{m} \phi_i R(i). \qquad (4.71)$$

For example, consider a two-state business model of a company. The company experiences different business environments each month and makes more money when it is in a good state than in a bad state. Let state 0 denote the bad state and state 1 the good state. If the state transition depends only on the current state, then it forms a Markov chain. Figure 4.8 depicts the state diagram for the two-state Markov chain for this business model. The transition probabilities are given by

$$\mathbf{P} = [P_{ij}] = \begin{bmatrix} 0.5 & 0.5 \\ 0.3 & 0.7 \end{bmatrix}, \qquad (4.72)$$

and the company earns 5 units of money at state 0 and 10 at state 1. So, $R(0) = 5$ and $R(1) = 10$. From (4.70), the stationary probability for this Markov chain is given by

$$\phi = \left(\frac{3}{8}, \frac{5}{8} \right). \qquad (4.73)$$

So, the expected earning for this company in the steady state is

$$R = \frac{3}{8} \times 5 + \frac{5}{8} \times 10 = \frac{65}{8}. \qquad (4.74)$$

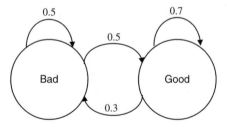

Figure 4.8   Markov chain for a two-state business model

Next we consider the *Markov decision process* (MDP) that is an extension of the Markov chain. MDP is a Markov chain in which the state transition probabilities are controlled to get the maximum reward. MDP is characterized by

- a set of states: $S = \{s_1, s_2, \ldots, s_m\}$
- a set of actions associated with each element $x \in S$: $A_x = \{a_1, a_2, \ldots, a_n\}$
- a transition probability for a given action a: $\Pr(x_{k+1} = j | x_k = i, a) = p_{ij}(a)$
- reward function: $R : \{(i, a) | i \in S, a \in A_s\} \rightarrow \Re$.

Transition probabilities are time-invariant for the stationary MDP and time-varying for the non-stationary MDP. The goal of the MDP is to maximize the total rewards earned at each state; that is[5]

$$\underset{\{a_k\}}{\text{maximize}} \sum_{k=0}^{N-1} R(x_k, a_k) + R(x_N), \tag{4.75}$$

where the initial state x_0 is given. For the infinite-horizon MDP, the reward at state k is discounted by λ^k and the goal is to maximize the total discounted rewards

$$\underset{\{a_k\}}{\text{maximize}} \sum_{k=0}^{\infty} \lambda^k R(x_k, a_k). \tag{4.76}$$

We can apply the dynamic programming approach to the MDP because the objective functions are time-separable, the state transition follows the Markov property, and the actions for each state play the role of decision variables.

The policy of the finite-horizon MDP is a mapping $\pi : \{0, 1, \ldots, N-1\} \times S \rightarrow A$. The value of each policy π starting at x_0 is given by

$$V_\pi(x_0) = \mathbb{E}\left[\sum_{k=0}^{N-1} R(x_k, \pi(k, x_k)) + R(x_N) \right]. \tag{4.77}$$

The optimal policy π^* that maximizes V among all admissible policies satisfies

$$V^*(x_0) = V_{\pi^*}(x_0) = \max_\pi V_\pi(x_0). \tag{4.78}$$

If we apply the principle of optimality to this MDP problem, then the Bellman equation is obtained by using the backward recursion

$$
\begin{aligned}
V_N(x_N) &= R_N(x_N), \\
V_k(x_k) &= \max_{a_k \in A} \mathbb{E}[R_k(x_k, a_k) + V_{k+1}(x_{k+1})] \\
&= \max_{a_k \in A} \sum_{x_{k+1}} p_{x_k, x_{k+1}}(a_k) \{ R_k(x_k, a_k) + V_{k+1}(x_{k+1}) \},
\end{aligned}
\tag{4.79}
$$

---

[5] The objective of the MDP may be minimization of the cost, but one could equivalently state it as the maximization of the reward.

and the optimal policy $\pi(k, x)$ is the maximizer of (4.79). For every possible initial state, this policy is optimal for the problem in (4.76).

For the infinite-horizon MDP, consider the stationary state transition probabilities. The policy is stationary if the policy is time-invariant; that is, $\pi(k, x) = \pi(x)$. The value of policy π in state x is given by

$$V_\pi(x) = \mathbb{E}\left[\sum_{k=0}^{\infty} \lambda^k R(x, \pi(x))\right].$$
(4.80)

As given in (4.66), the principle of optimality also yields the Bellman equation

$$V(x) = \max_{a \in A_x} \sum_{x' \in S} p_{x,x'}(a)(R(x, a) + \lambda V(x')).$$
(4.81)

The optimal values $V^*(x)$ are the unique solutions of (4.81). The optimal policy π can be determined such that it satisfies (4.81) for given optimal values; that is

$$\pi(x) = \arg\max_{a \in A_x} \sum_{x' \in S} p_{x,x'}(a)(R(x, a) + \lambda V(x')).$$
(4.82)

Conversely, if the optimal policy is given, it is easy to obtain the values of the states. So, there are dual methods to solve the MDP problem.

Figure 4.9 shows the value-iteration algorithm that yields the optimal value $V^*(x)$ after convergence. The value iteration repeats computing new values $V_{new}(x)$ by using the previous values $V_{old}(x)$ until their difference becomes smaller than a prespecified small constant ε. Note that the optimal policy π is computed only after the optimal values are obtained.

The second method to solve the Bellman equation is the policy iteration as shown in Figure 4.10. The policy iteration consists of two loops. The inner loop computes the values when the policy given by the outer loop is followed. In the outer loop, the policy is updated based on the values computed in the inner loop until the updated policy becomes the same as the old policy. There are some variants of the policy-iteration algorithm to expedite the inner loop

---

Given a discount factor λ and a required tolerance $\varepsilon > 0$

Initialize $V_{old}(x) = 0$ and $V_{new}(x) = 0$

Repeat

   for all $x \in S$

   $V_{old}(x) := V_{new}(x)$

   $V_{new}(x) := \max_{a \in A_x} \sum_{x' \in S} p_{x,x'}(a)(R(x, a) + \lambda V_{old}(x'))$

Until $|V_{old}(x) - V_{new}(x)| < \varepsilon$ for all $x \in S$

for all $x \in S$

   $\pi(x) := \arg\max_{a \in A_x} \sum_{x' \in S} p_{x,x'}(a)(R(x, a) + \lambda V_{old}(x'))$

Figure 4.9   Algorithm for value iteration

Given a discount factor λ and a required tolerance $\varepsilon > 0$

Initialize arbitrary π_{new}

Repeat

$\qquad \pi_{old}(x) := \pi_{new}(x)$

$\qquad V_{old}(x) := 0 \quad V_{new}(x) := 0$

$\qquad$ **Repeat**

$\qquad\qquad$ for all $\quad x \in S$

$\qquad\qquad\qquad V_{old}(x) := V_{new}(x)$

$\qquad\qquad\qquad V_{new}(x) := \sum_{x' \in S} p_{x,x'}(\pi_{old}(x)) \left(R(x, \pi_{old}(x)) + \lambda V_{old}(x') \right)$

$\qquad\qquad$ **Until** $\left| V_{old}(x) - V_{new}(x) \right| < \varepsilon$ for all $\quad x \in S$

$\qquad\qquad$ for all $\quad x \in S$

$\qquad\qquad\qquad \pi_{new}(x) := \arg\max_{a \in A_x} \sum_{x' \in S} p_{x,x'}(a) \left(R(x, a) + \lambda V_{new}(x') \right)$

Until $\pi_{new}(x) = \pi_{old}(x)$

for all $\quad x \in S$

$\qquad \pi(x) := \pi_{new}(x)$

Figure 4.10 Algorithm for policy iteration

computation. We can directly compute the values by solving the linear equations under the policy given by the outer loop. Or, we can iterate the inner loop several times rather than iterating until convergence.

The policy iteration updates the policy iteratively to find the optimal policy while the value iteration updates the values iteratively to find the optimal value. If we already know a good policy (but not necessarily an optimal one), then it is better to use the policy iteration to find the optimal policy with this initial policy. Also, the policy iteration is a good choice when there are a lot of actions associated with each state. If there are few actions, the value iteration will be better.

We shall now revisit the example of the Markov chain for a two-state business model. Recalling that more money is made in the good state than in the bad state, we can advertise the business to prevent the business state from falling into the bad state and thus make the business more profitable. Action 0 denotes no advertisement and action 1 making an advertisement, respectively. With action 0, the state transition probabilities and the rewards are the same as for the previous business model. With action 1, the reward is reduced by 1 which is the cost of the advertisement. So:

$$R(0,0) = 5 \quad R(0,1) = 4$$
$$R(1,0) = 10 \quad R(1,1) = 9. \tag{4.83}$$

The transition probabilities with action 0 is the same as the previous one in (4.72), and those with action 1 are given by

$$p_{00}(1) = 0.3 \quad p_{01}(1) = 0.7$$
$$p_{10}(1) = 0.2 \quad p_{11}(1) = 0.8. \tag{4.84}$$

It is not clear whether the advertisement is beneficial because it changes the transition probabilities to increase the probability of staying in state 1 at the cost of reduced reward. So, we formulate the Markov decision process to obtain the optimal decision as in (4.75). By applying the value iteration for $\lambda = 0.99$, the optimal policy is $\pi(0) = 1$ and $\pi(1) = 0$. So, in order to maximize the expected rewards, we have to advertise at state 0 only. With this optimal policy, the stationary probabilities are $(\phi_0, \phi_1) = (0.3, 0.7)$ and the expected earning for this company in the steady state is

$$R = \frac{3}{10} \times 4 + \frac{7}{10} \times 10 = \frac{82}{10}, \tag{4.85}$$

which is strictly higher than (4.74). This example demonstrates that the principle of optimality works well in the case of a Markov decision process.

4.3 Analogy of Economics and Wireless Resource Management

Since wireless resources are scarce, they are worth managing most efficiently by applying all possible means. Since economists, too, study how to make decisions to manage scarce resources, some fundamental methodologies in economics may possibly become useful tools for wireless resource management.

4.3.1 Economics Model

There are many principles governing economic systems. One basic principle is that the resources are scarce and demand exceeds availability. Therefore, there exists a fundamental trade-off: to obtain something one has to give up something else. When there are two or more uses for a resource, the *opportunity cost* of using resources in a particular way is the value of the next-best alternative. In other words, the opportunity cost is the benefit we might have received by taking an alternative action. For example, if we decide to grow roses in the garden, then the opportunity cost is the alternative crops (e.g., lilies) that we have to give up.

To understand the opportunity cost in detail, consider the *production possibility frontier* (PPF), which refers to all the combinations of outputs that can be produced with a fixed amount of resources. Figure 4.11 illustrates an example of the PPF for the case of producing products X and Y with a limited resource. If all resources are allocated to product Y, then we can produce 30 units of Y and 0 units of X (point "a"). At the other extreme, all resources are allocated to product X, then we can produce 10 units of X and 0 units of Y (point "f"). There is also an option to produce, for example, 3 units of X and 25 units of Y (point "b"). Any combination of X and Y that corresponds to points on or inside the PPF can be produced by allocating resources appropriately.

One point in the PPF is better than another point if it produces more than that other point. Even though a point inside the PPF is a possible choice, it is not an efficient choice as it does not use all available resources. So, the most efficient use of the limited resources appears at points *on* the PPF curve, and a trade-off exists among different points on the PPF in terms of the products X and Y. A point on the PPF is said to be *Pareto-optimal* if there is no better point on the PPF. In Figure 4.11, all the points on the PPF are Pareto-optimal, so no point on the PPF is better than any other point on the PPF. Among the many Pareto-optimal points, we need to

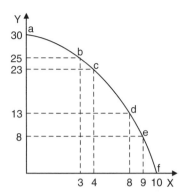

Figure 4.11   Illustration of the production possibility frontier

decide which point is the most desirable. Sometimes this decision is not made in a scientific way but by using some normative analysis. It may depend on what is more desirable or what is valued more. For example, consider that we allocate the wireless resources to two users in a communication system. Option 1 is to allocate 2 Mbps to each user and option 2 is to allocate 4 Mbps to user 1 and 1 Mbps to user 2. One may prefer the first option because both users get the same data rates but someone else may prefer option 2 because the sum of rates is higher. Therefore, the value judgment plays an important role in determining the best point on the PPF.

The slope of the tangential line at any point on the PPF curve represents the opportunity cost of producing one more good in terms of the others. Suppose that we produce 3 units of X and 25 units of Y (i.e., point "b" in Figure 4.11). If we want to produce an additional unit of X, then we need to give up 2 units of Y (point "c"). The opportunity cost of an additional production of X is the production of Y we give up. In this case, the opportunity cost of producing an additional unit of X is 2 units of Y. Now, suppose we produce 8 units of X and 13 units of Y (point "d"). If we want to produce an additional unit of X, then the opportunity cost is 5 units of Y (point "e"). Consequently, the opportunity cost depends on the location on the PPF, as the opportunity cost is nothing but the slope at that particular point on the PPF. If one product is being produced a lot, then the opportunity cost increases; in other words, it becomes harder to produce more of that product, since the productivity of the resources may be different for each product. If it is desired to produce more of one particular product, it requires that much more resources, which otherwise might be efficiently used to produce other products. In other words, it causes inefficient use of resources. In the PPF in Figure 4.11, the curve is concave with respect of X, so the slope of the PPF is decreasing (or the magnitude of the slope is increasing.). Basically, this explains why the opportunity cost increases when product X is readily being produced.

Some particular goods might be preferred to others. For example, we may prefer three apples to one watermelon. In order to model this preference, we introduce a utility function $U(x)$ with respect to a vector of goods, x. Then we may express that we prefer x to y by $U(x) > U(y)$. Since the utility function represents the order of preference, the utility representation is not necessarily unique: there may be different utility functions that can represent the same preference. In the situation that (x_1, x_2) is preferred to (y_1, y_2) if and only if $x_1 + x_2 > y_1 + y_2$, the utility functions may take the expressions $U(x_1, x_2) = x_1 + x_2$, $U(x_1, x_2) = 2(x_1 + x_2)$, and so on. More specifically, if $U(x)$ is a utility representation for some preference, then $f(U(x))$ is

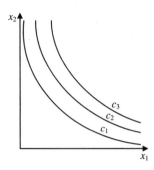

Figure 4.12   Illustration of indifference curves

also a utility representation for the same preference as long as $f : \Re \to \Re$ is a strictly increasing function. In addition, a utility representation is ordinal, not cardinal. The utility $U(x) = 100$ itself has no meaning but x is preferred to y if $U(x) > U(y)$.

An *indifference curve* (IC) gives all combinations of goods for which the customer is indifferent. In terms of the utility function, the indifference curve is given by $U(x_1, x_2, \ldots, x_N) = c$, where c is a constant. In other words, the indifference curve is the level curve of the utility function. Any two indifference curves do not intersect each other because they are different level curves of the utility functions. Figure 4.12 illustrates some typical indifference curves.

The *marginal utility* (MU) is a change of utility that occurs when we consume a little more of a good. Mathematically, a marginal utility MU_i is given by the partial derivative of $U(x_1, x_2, \ldots, x_N)$ with respect to x_i. From the total differential of a utility function $U(x_1, x_2)$, we get

$$dU(x_1, x_2) = \frac{\partial U}{\partial x_1} dx_1 + \frac{\partial U}{\partial x_2} dx_2. \tag{4.86}$$

When the utility stays constant – that is, $dU(x_1, x_2) = 0$ – then we get the *marginal rate of substitution* (MRS)

$$\text{MRS} = -\frac{dx_2}{dx_1} = \frac{\partial U}{\partial x_1} \Big/ \frac{\partial U}{\partial x_2}. \tag{4.87}$$

So, MRS is the slope of the indifference curve. It is also the ratio of marginal utilities; that is

$$\text{MRS} = \frac{MU_1}{MU_2}. \tag{4.88}$$

The MRS may be interpreted as the amount of good x_2 that we are willing to give up to get one unit of good x_1. Note that it is quite similar to the opportunity cost in the PPF. It is often assumed that the MRS is decreasing. If we have a small number of x_1 and a lot of x_2, the MRS is high and thus we are willing to give up many of x_2 for one additional unit of x_1. On the other hand, if we have a large number of x_1 and a little x_2, then the MRS is low and thus we need to give up a little of x_2 to get one additional unit of x_1. This diminishing property of MRS stems from the fact that the indifference curves are convex as illustrated in Figure 4.12. One example of a utility function is the *Cobb–Douglas function* $U(x, y) = x^\alpha y^\beta$, for which the MRS is given by

$$\text{MRS} = \frac{\alpha y}{\beta x}. \tag{4.89}$$

Consider now the utility maximization problem for the case of two goods. We assume that the utility function $U(x, y)$ is twice continuously differentiable and, in addition, that $U(x, y)$ is a concave function. Denote by I the total income of the customer and by p_x and p_y the prices of goods x and y. Then, the budget constraint is given by $p_x x + p_y y \leq I$. However, since it is better if more goods are consumed, we take the budget constraint with equality. The main problem is to determine what combination of goods to take to maximize the total utility. In a constrained optimization problem, we have

$$\begin{array}{c} \underset{x,y}{\text{maximize}} \ \ U(x,y) \\ \text{subject to} \ \ p_x x + p_y y = I. \end{array} \tag{4.90}$$

The Lagrangian function is taken to (4.90) to get

$$L = U(x,y) - \lambda(p_x x + p_y y - I). \tag{4.91}$$

From the KKT condition, we have

$$\frac{\partial}{\partial x} U(x, y) - \lambda p_x = 0,$$

$$\frac{\partial}{\partial y} U(x, y) - \lambda p_y = 0, \tag{4.92}$$

$$p_x x + p_y y = I.$$

Equation (4.92) is solved to determine the optimal combination of goods (x^*, y^*). On rearranging (4.92), the MRS is obtained as the ratio of the prices

$$\text{MRS} = \frac{p_x}{p_y}. \tag{4.93}$$

So, as shown in Figure 4.13, the optimum point is achieved where the slope of the budget curve is the same as the MRS of an indifference curve. In the case of Cobb–Douglas utility, if we use (4.89) and (4.93) and the budget constraint, we get

$$x^*(p_x, p_y, I) = \frac{\alpha}{\alpha + \beta} \frac{I}{p_x},$$

$$y^*(p_x, p_y, I) = \frac{\beta}{\alpha + \beta} \frac{I}{p_y}. \tag{4.94}$$

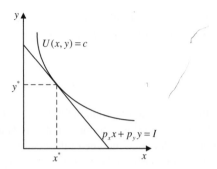

Figure 4.13   Utility maximization under constrained income

where $x^*(p_x, p_y, I)$ and $y^*(p_x, p_y, I)$ are called the *demand functions*. Rearranging (4.92) also gives the relation

$$\frac{MU_x}{p_x} = \frac{MU_y}{p_y} = \lambda, \tag{4.95}$$

which implies that, at the optimum, the marginal utilities per price are the same. To express the maximal utility in terms of prices and income, define $U^*(p_x, p_y, I) \equiv U(x^*(p_x, p_y, I), y^*(p_x, p_y, I))$. On taking the first derivative of $U^*(p_x, p_y, I)$ with respect to I, we get

$$\frac{\partial}{\partial I} U^*(p_x, p_y, I) = \lambda. \tag{4.96}$$

This indicates that if income increases by one unit, the optimal utility increases by λ. So, λ is called the *shadow price*.

4.3.2 Examples of Wireless Resource Allocation

The economic model finds good applications in wireless resource management. The term "utility" is broad enough in meaning to be interpreted in different ways in resource management. It could be the throughput, the quality of a service, a combination of the two, or any others. Two typical examples of applying the economic model to wireless resource management are now considered.

Example 4.3. *Throughput Maximization.* Consider an example related to packet scheduling. Suppose that there are two users, user 1 and user 2, in a cell. The system operates in time-division manner such that only one user is served at each time slot. Assume that the supportable transmission rate is time-varying and takes the rates 400 kbps or 1200 kbps with equal probability; and transmission is distributed independently and identically across users. Now we consider two extreme transmission modes of allocating all the time slots either to user 1 or to user 2. If all time slots are allocated to user 1, then the throughput of user 1 is

$$T_1 = 400 \times 0.5 + 1200 \times 0.5 = 800 \tag{4.97}$$

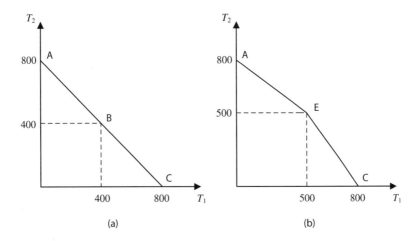

Figure 4.14 Achievable regions for (a) conventional TDMA scheme, and (b) intelligent TDMA scheme

and the throughput of user 2, T_2, becomes 0. If all the time slots are allocated to user 2, the achievable throughputs become $T_1 = 0$ and $T_2 = 800$ kbps. Avoiding these two extreme cases, we consider allocating time slots alternatively, odd-numbered time slots to user 1 and even-numbered time slots to user 2. Then we can achieve $T_1 = 400$ kbps and $T_2 = 400$ kbps. In general, if we allocate a portion θ of the time slots to user 1 and the remaining portion, $1 - \theta$, to user 2 (for $0 \leq \theta \leq 1$), then we achieve

$$\begin{aligned} T_1 &= 800\theta \\ T_2 &= 800(1-\theta). \end{aligned} \tag{4.98}$$

This scheme corresponds to a conventional *time-division multiple access* (TDMA) scheme. The PPF for this conventional TDMA scheme is shown in Figure 4.14(a). Due to the naivety of the time-sharing, the PPF curve is a straight line connecting the two extreme points A and C as shown.

Now consider a more intelligent system. Assume the channel state information is available at the base station (BS). Then the BS can select the user to allocate the current time slot based on the current channel information. Suppose, at a time slot, that the supportable transmission rate for user 1 is 400 kbps and the rate for user 2 is 1200 kbps. If the BS selects user 1 for transmission, then the opportunity cost is 1200 kbps, since transmission to user 2 has to be given up. On the other hand, if the BS selects user 2 for transmission, the opportunity cost is 400 kbps. In general, if we want to minimize the opportunity cost at each time slot, we should select the user whose rate is higher. When a tie occurs (i.e., two user rates are the same), either user can be chosen randomly. This system may be called an "intelligent TDMA scheme". Since the transmission rate distributions are i.i.d. for the two users in this problem, the average throughput of each user is a half of the sum throughput of user 1 and user 2. If the BS selects the user whose rate is higher than the other, the probability of selecting 400 kbps is 1/4 because this happens if and only if all user rates are 400 kbps. On the contrary, the probability of selecting 1200 kbps is then 3/4, which is the probability that one of the user rates is 1200 kbps. So, the sum throughput becomes

$$T_1 + T_2 = 400 \times \frac{1}{4} + 1200 \times \frac{3}{4} = 1000. \tag{4.99}$$

Due to the user symmetry, $T_1 = T_2 = 500$ kbps. Figure 4.14(b) depicts the achievable throughput of this intelligent TDMA scheme with a combination of two extreme points A and C. Note that any point on the line connecting A and E is obtained when all time slots are allocated to user 2 in some portion of time slots, while time slots are allocated by the intelligent TDMA scheme in the other portion. Likewise, we can also achieve any point on the line connecting E and B by allocating time slots in a manner symmetrical to the above.

Note that the PPF for the conventional TDMA is a proper subset of that for the intelligent TDMA in Figure 4.14. More specifically, the user throughput at point E for both user 1 and user 2 is 100 kbps higher than the throughput at point B. This gain comes from the fact that the intelligent TDMA selects the higher rate user to minimize the opportunity cost. So, the wireless resource is allocated to the user who can take more at each time slot. However, the conventional TDMA selects the user in a deterministic way without considering the channel condition, so only the average of the individually supportable gains is obtained. The throughput gain of the intelligent TDMA over the conventional TDMA is called *multiuser diversity gain*. This is discussed in more detail in Chapter 5.   ☐

Example 4.4. *Utility Maximization.* Consider an example of wireless resource allocation in the downlink channel in a single-cell wireless network. The system operates in a time-division manner in which the base station transmits to a single user at any given time. There are M users and the channel states of all users are known at the BS. Let R_k denote the rate in packets per second at which the user can receive data in a time slot. During a frame, the BS allocates a time $t_k \in [0, 1]$ to transmit to user k at rate R_k, which are related by (Marbach and Berry, 2002)

$$\sum_{k=1}^{M} t_k = \sum_{k=1}^{M} \frac{x_k}{R_k} \le T, \tag{4.100}$$

where x_k denotes the throughput of user k in terms of the number of packets. The traffic is assumed to be elastic such that the utility function $U_k(x_k)$ associated with user k is increasing, strictly concave, and continuously differentiable. So, the quality of service is measured solely as a function of throughput. Define the demand function $D_k(u)$ to be the amount of packets that user k would request if the cost per unit packet were u; that is

$$D_k(u) \equiv \arg \max_{x \ge 0} \{U_k(x) - ux\}. \tag{4.101}$$

The goal of this optimization is to find the user's optimal throughput that maximizes the net benefit given by the utility less the cost. As $U_k(x_k)$ is strictly concave and continuously differentiable, the inverse function of the first derivative exists; that is

$$D_k(u) = (U'_k)^{-1}(u). \tag{4.102}$$

So, the demand goes down when price goes up because $D_k(u)$ is a decreasing function. For example, if $U_k(x) = m_k \ln x$, then $D_k(u) = (m_k/u)$.

Assume that the BS wants to maximize the total users' utility. The optimization problem may be stated as

$$
\begin{aligned}
\text{maximize} \quad & \sum_{k=1}^{M} U_k(x_k) \\
\text{subject to} \quad & \sum_{k=1}^{M} \frac{x_k}{R_k} = T, \\
& x_k \geq 0, \quad k = 1, 2, \ldots, M.
\end{aligned}
\tag{4.103}
$$

When compared with (4.90), the utility is now additive-separable and the $1/R_k$ plays the role of the price of user k's throughput. On taking the Lagrangian, we get

$$
L = \sum_{k=1}^{M} U_k(x_k) - \lambda \left(\sum_{k=1}^{M} \frac{x_k}{R_k} - T \right).
\tag{4.104}
$$

The KKT condition gives

$$
U'_k(x_k) - \frac{\lambda}{R_k} = 0, \quad k = 1, 2, \ldots, M.
\tag{4.105}
$$

So, the optimal throughput of user k is

$$
x_k^* = D_k(u_k),
\tag{4.106}
$$

where $u_k = \lambda/R_k$. As shown in (4.96), λ is the shadow price where each user is charged the same price per time. The price of a unit packet for user k is given by u_k. So, economics concepts can also be applied to this utility maximization problem. □

References

Bertsekas, D.P. (2001) *Dynamic Programming and Optimal Control*, 2nd edn, Athena Scientific.

Bellman, R. (1953) The theory of dynamic programming. *Bulletin of the American Mathematical Society*, **60**, 503–515.

Boyd, S. and Vandenberghe, L. (2004) *Convex Optimization*, Cambridge University Press.

Marbach, P. and Berry, R. (2002) Downlink resource allocation and pricing for wireless networks. In: *Proceedings IEEE Infocomm*, New York, 1470–1479.

Negi, R. and Cioffi, J.M. (2002) Delay-constrained capacity with causal feedback. *IEEE Transactions on Information Theory*, **48**, 2478–2494.

Further Reading

Kelly, F.P. (1997) Charging and rate control for elastic traffic. *European Transactions on Telecommunications*, **8**, 33–37.

Kelly, F.P., Maullo, A.K., Tan, D.K.H. (1998) Rate control for communication networks: Shadow price, proportional fairness and stability. *Journal of the Operational Research Society*, **49**, 237–252.

King, I. (2002) *A Simple Introduction to Dynamic Programming in Macroeconomic Models*, University of Auckland.

Luo, Z.-Q. and Yu, W. (2006) An introduction to convex optimization for communication and signal processing. *IEEE Journal on Selected Areas in Communications*, **24**, 1426–1438.

Mankiw, N.G. (2006) *Principles of Microeconomics*, Cengage Learning.

Nicholson, W. (2004) *Microeconomics Theory: Basic Principles and Extensions*, 9th edn, South-Western College Publications.

Part Two

Wireless Resource Management Technologies

5

Bandwidth Management

Bandwidth is a key wireless resource which refers to the range of frequencies (denoted by Hz) occupied by a transmitted signal. Since bandwidth determines the maximum symbol transmission rate, it puts a fundamental limit on the channel access rate. In a wide sense, bandwidth may be considered as the transmission opportunity to access the wireless medium; that is, how often a user can transmit symbols over the wireless medium. It may be shared among multiple users if they are contending each other to access the same bandwidth. For example, in the case of the *Aloha system*, each user tries packet transmission whenever packets are generated and, if collision occurs (i.e., more than two users transmit over the same channel simultaneously), then each user retries to access the channel later. This contention-based access scheme is inefficient because a large amount of available bandwidth is wasted due to unavoidable collisions. So, in order to use wireless resources efficiently, a certain type of scheduled transmission is needed.

Scheduling refers to a method of allocating resources to users over time by following some predetermined procedures. Since resources for wireless communications are scarce and constrained, efforts are exerted to allocate resources in an optimal way based on some criteria. There are different types of scheduling algorithms, with each achieving its own optimality. A simple form of scheduling policy is the round-robin policy: this selects users according to a fixed cyclic order, thereby yielding a fair allocation of transmission opportunity. However, when there are urgent packets to deliver within a certain time duration, the round-robin policy may fail to serve them as they have to wait until their turns come. So, scheduling priority is often considered to make a scheduling decision take account of various constraints. The scheduler may select users based on their priority, which can be static or dynamic. In general, *static priority* is assigned at the time of connection creation, while *dynamic priority* is made based on the system behavior.

This chapter discusses the concept of bandwidth management. It first examines the essential differences between wired and wireless bandwidth management. Then it discusses the scheduling policies evolved from wired bandwidth management and introduces various wireless packet scheduling algorithms that can maximize the total throughput. Next, it investigates the performances of bandwidth management schemes in terms of throughput and delay. Finally, it discusses the admission control scheme in conjunction with wireless packet scheduling and QoS provision.

Wireless Communications Resource Management   Byeong Gi Lee, Daeyoung Park, and Hanbyul Seo
© 2009 John Wiley & Sons (Asia) Pte Ltd

5.1 Differences between Wired and Wireless Communications

In wired packet networks, a fixed amount of bandwidth is shared by multiple packet flows in time-division manner and a packet scheduler selects one of the flows for transmission. In general, the arrival packet flows are bursty; that is, the arrival packet sizes vary over time. However, the maximum size of the packets that can be transmitted into the output link is not time-varying. In contrast, since the wireless channel is basically time-varying in nature, the maximum size of packets is also time-varying. So, the wireless network and the wired network have a dual relation if buffers are infinitely backlogged in the wireless network.

5.1.1 Statistical Multiplexing in a Wired Network

Consider a wired link shared among multiple users. Each user wants to send B Mbps through the output link and the user traffic is very bursty, so that user traffic is idle with a probability p. Assume that the output link can handle C Mbps. When the output link to users is allocated statically, the maximum number of users is C/B. However, the probability that all C/B users are active is $(1-p)^{C/B}$. When C/B is high, this wired link is not fully utilized all the time.

Now increase the number of users to N. Then the overflow probability – that is, the probability that the active users are more than the link can handle – is given by

$$\Pr(\text{more than } C/B \text{ active users}) = \sum_{k=C/B+1}^{N} \binom{N}{k} p^{N-k}(1-p)^{k}. \tag{5.1}$$

For example, take $B=1$, $C=5$, and $p=0.9$. In this case, the overflow probabilities for the cases $N=10$, $N=15$, and $N=20$ are 1.4×10^{-4}, 2.2×10^{-3}, and 0.011, respectively. In the case of static allocation, the maximum number of allowable users is $C/B=5$. However, if the output link is shared statistically, then we can accommodate more users at the cost of reasonably low overflow probability. This gain is called *statistically multiplexing gain*.

5.1.2 Multiuser Diversity in a Wireless Network

Consider the uplink block fading channel with K users. Then the received signal is given by

$$r[m] = \sum_{k=1}^{K} h_k[m]s_k[m] + \eta[m], \tag{5.2}$$

where $h_k[m]$ is a fading channel that user k experiences and $\eta[m]$ is a zero-mean additive white Gaussian noise with variance σ^2. In the block fading channel the channel gains remain the same during the coherence time slot and vary independently from slot to slot. In addition, each user channel gain is assumed to be independent across users and known to that user. Each user adapts its power in order to maximize the sum rate capacity under the average power constraint

$\mathbb{E}P_k[m] = \mathbb{E}|s_k[m]|^2 \leq P$. When the channel gains h_k are given, the sum-rate capacity of this uplink channel is

$$C = \log_2\left(1 + \frac{\sum_{k=1}^{K} g_k[m]P_k[m]}{\sigma^2}\right), \tag{5.3}$$

where $g_k[m] = |h_k[m]|^2$. Since users adapt their powers based on the current channel gain to maximize the sum-rate capacity, this optimization problem can be formulated as follows:

$$\text{maximize } \frac{1}{M}\sum_{m=1}^{M}\log_2\left(1 + \frac{\sum_{k=1}^{K} g_k[m]P_k[m]}{\sigma^2}\right) \tag{5.4}$$

$$\text{subject to } \frac{1}{M}\sum_{m=1}^{M}P_k[m] \leq P, k = 1, 2, \cdots, K,$$

where M is the number of slots, which we can increase to infinity. If we apply the Karush–Kuhn–Tucker (KKT) condition to this problem, the optimal solution is given by (Knopp and Humblet, 1995)

$$P_k[m] = \begin{cases} \dfrac{1}{\lambda_k} - \dfrac{\sigma^2}{g_k[m]}, & \text{if } \dfrac{g_k[m]}{\lambda_k} = \max_j \dfrac{g_j[m]}{\lambda_j} \text{ and } g_k[m] > \lambda_k\sigma^2, \\ 0, & \text{otherwise,} \end{cases} \tag{5.5}$$

where the Lagrange multiplier λ_k is a constant satisfying the average power constraint in (5.4). Note that the Lagrange multipliers play the role of cut-off thresholds as well as scaling factors to determine the strongest channel gains. Figure 5.1 depicts this optimal transmission-power allocation policy in a two-user case.

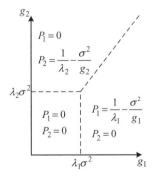

Figure 5.1   Transmission power allocation for a two-user case

Now consider a case where the channel gains are identically distributed. The Lagrange multipliers become identical. So the one user whose channel gain is the largest transmits packets with a power allocation according to the water-filling policy, while the other users remain silent until their channel gains become the largest. This transmission scheme is called *opportunistic transmission*, which implies that each user transmits only at the time that its channel is in the best state. If the number of coherent slots, M, goes to infinity, then we get the sum-rate capacity

$$C = \mathbb{E}\left[\log_2\left(1 + \frac{g_{k^*}(\mathbf{g})P_{k^*}(\mathbf{g})}{\sigma^2}\right)\right], \tag{5.6}$$

where $k^*(\mathbf{g})$ is the user index whose channel gain is the largest and $P_{k^*}(\mathbf{g})$ is the transmit power of that user according to the water-filling policy.

For a downlink fading channel with channel state information available at the transmitter, we get the same capacity result as in (5.6). Therefore, in the uplink and downlink fading channels, opportunistic transmission is the best transmission policy. In fact, this is a kind of TDMA system in which only one user is allowed to transmit. However, the conventional TDMA system usually serves each user in a round-robin manner; that is, it serves one user at a time in a deterministic order, irrespective of the current channel states. As discussed in Section 4.3.2, the opportunistic transmission scheme has a larger achievable region than has the conventional TDMA system.

The multiuser opportunistic transmission system has a capacity gain over the single-user system in the sense that its effective channel gain is $\max_k g_k$ while the channel gain of the single-user system is g_1. For the Rayleigh distributed channel, the square of the channel magnitude is exponentially distributed with mean 1; that is

$$F_{g_1}(x) = 1 - e^{-x}, \tag{5.7}$$

and the cumulative distribution function (CDF) of the maximum of K i.i.d. exponential distribution is

$$F_{\max_k g_k}(x) = (1 - e^{-x})^K. \tag{5.8}$$

In the extreme-value theory, $\max_{k=1,\ldots,K} g_k - \log K$ converges in distribution to $\exp(-e^{-x})$ as $K \to \infty$. This implies that $\max_{k=1,\ldots,K} g_k$ grows like $\log K$ for large K. So, the effective signal-to-noise ratio (SNR) in multiuser opportunistic systems becomes $\log K$ times larger than in the single-user system, which is called *multiuser diversity gain*.

This multiuser diversity originates from the fact that each user transmits only when its channel performance is near its own peak. This is a very different concept compared with the usual diversity. In conventional diversity, multiple independently faded signals were combined to reduce the fading fluctuation for reliable communications based on the property that the probability that all the independent signals drop below a certain threshold is small. However, the opportunistic transmission system intelligently utilizes the independent fading variation by serving only the user with the strongest channel because it is highly probable that one of the user channels lies in its own peak state among a large number of users.

The sum-rate capacity of the multiuser round-robin system remains the same provided each user channel is *independent and identically distributed* (i.i.d.). So, the per-user capacity is $O(1/K)$, which tends to approach zero. In contrast, the multiuser opportunistic transmission system has an effective SNR gain of $\log K$ and its capacity is $O(\log \log K)$. So, as the number of users grows to infinity, the sum-rate capacity also goes to infinity. However, the capacity per user goes to zero because it is

$$O\left(\frac{\log \log K}{K}\right).$$

So, adding one more user in a multiuser opportunistic system increases the sum-rate capacity while it reduces per-user capacity, as the round-robin system does so.

5.2 Schedulers based on Generalized Processor Sharing

Originally, scheduling was an important issue in wired packet networks. It was an essential component of a packet multiplexed link where an output link is shared, in time, by multiple input packet flows. In this packet multiplexer, scheduling refers to a policy that determines the packet to transmit next, upon the completion of each packet transmission. Since any incoming packet should wait for a certain period of time unless it is served right away, each packet flow has a queue, which buffers each incoming packet until it is transmitted.

The main focus of the scheduling in wired networks was to provide each packet flow a "fair" chance to occupy the output link. Each flow generates packets of different size at different times, and the packet arrival patterns are also different for different flows. So, for a fair scheduling, the arrival pattern and each packet size should be carefully taken into consideration. If they are not properly considered, as is the case with a simple *first-in first-out* (FIFO) scheduler, a flow with a high generation rate of large packets is likely to occupy the output link longer than others. For that reason, research on packet scheduling in wired networks has mainly concentrated on developing fair scheduling methods, while achieving the given objectives.

The scheduling algorithms developed for use in wired networks do not necessarily perform efficiently in wireless networks. The main reason is that the transmission rate, or the capacity of the output link, is not static but varies in time in wireless networks. Thus, the channel condition should be taken into account for efficient wireless network scheduling. In order to overcome this limitation, scheduling algorithms developed for wired networks have been modified for efficient operation in wireless networks.

The following subsections discuss some wireless scheduling algorithms that are modifications of scheduling policies developed for wired networks. Since the main focus of wired schedulers was on guaranteeing fairness, these wireless schedulers focused more on the fairness issue than on channel adaptation. We first review the *generalized processor-sharing* scheme which is an ideal example of fair schedulers in wired networks, and then introduce some wireless schedulers that are developed based on this.

5.2.1 Generalized Processor Sharing

Generalized processor sharing (GPS) (Parekh and Gallager, 1993), in its operation, assumes that each incoming packet can be divided into informational elements whose size is infinitesimal and the transmission can be done element-wise. Under this assumption, multiple packets

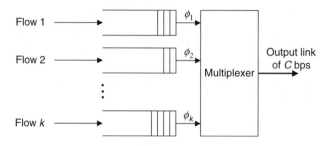

Figure 5.2   System model of GPS

can share the output link by transmitting their information elements in a round-robin manner with very fast switching. In essence, each packet is regarded as fluid and infinitely divisible, so this modeling is usually called the *fluid-flow model*. It is obvious that this fluid-flow model is an unrealistic assumption but it gives a guideline on achieving fairness. A realistic version of it is the packetized GPS, which will be discussed later.

The system considered by GPS is as shown in Figure 5.2. There are K packet flows sharing the server of capacity C bits per second. A flow k is associated with a weight ϕ_k which is a positive real number. This weight indicates how much priority is granted on the flow by the scheduler, as GPS performs like a fluid version of the weighted round-robin scheduling as follows. At each scheduling round, GPS serves ϕ_k bits from the non-empty queue k. As a result, at time t, the flow k is served at the rate of

$$R_k(t) = \frac{\phi_k 1_k(t)}{\sum_{i=1}^{K} \phi_i 1_i(t)} C \quad \text{(bits/sec)}, \tag{5.9}$$

where $1_k(t)$ is an indicator which becomes 1 when the queue k is non-empty at time t, and 0 otherwise. Thus, the GPS provides more service opportunity to a user with a higher weight.

Figure 5.3 illustrates the scheduling operation of the GPS for the case of two flows, $\phi_1 = 1$ and $\phi_2 = 1$, with the output link capacity C set to 1. The upper two plots show the arrival times and the sizes of packets in the two flows. For example, a packet with a size of 3 arrives at flow 2 at time 0, and a packet with a size of 1 arrives at flow 1 at time 1. The lower two plots describe how the GPS multiplexes the two flows. The dotted line represents the accumulated packet arrival for each flow, while the solid line depicts the accumulated transmission – that is, the amount of services provided by the multiplexer up to that time. The difference of the heights of the two lines indicates the amount of data buffered in the queue. In the interval $(0, 1]$, flow 1 has nothing to transmit, so flow 2 wholly occupies the output link. Thus, service is provided at the rate 1 in this interval, which means that the accumulated transmission increases with the slope of 1. In the interval $(1, 4]$, a packet arrives at flow 1 and the output link is shared by the two flows. According to their weights, the service rates are 1/3 and 2/3 for flows 1 and 2, respectively. Note that two packets are served simultaneously in this interval. The first packet of flow 2 departs the system at time 4 and its buffer is emptied. So, flow 1 is served at rate 1 until another packet arrives at flow 2. (Note that the second packet of flow 2 arrives at time 5.)

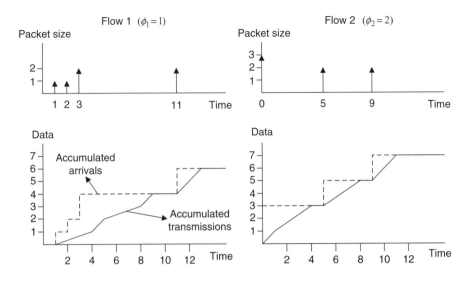

Figure 5.3   Scheduling operation of the GPS

The above operation of the GPS results in an interesting consequence that the duration of a scheduling round depends on the number of non-empty queues. In other words, when more flows have packets in their queues, the duration of a scheduling round becomes longer and each flow should wait longer to send ϕ_k bits. This means that the service rate of each flow varies in time but, if a static service rate is assumed for each flow, this effect may be regarded as making time elapse more slowly when more non-empty queues exist.

This discussion leads us to introduce the concept of the *virtual time* as follows. At time u, $\sum_{i=1}^{K} \phi_i 1_i(u)$ bits are transmitted within one round and thus there are $C / \sum_{i=1}^{K} \phi_i 1_i(u)$ rounds per second. For example, if there are two non-empty queues with $\phi_1 = 1$ and $\phi_2 = 2$, the server makes $C/3$ rounds per second. The virtual time at time t is defined by the accumulated round number of the server at that time, which is given by

$$V(t) = \int_0^t \frac{C}{\sum_{i=1}^{K} \phi_i 1_i(u)} 1(u)du, \tag{5.10}$$

where the indicator function $1(t)$ is 1 when at least one queue is non-empty at time t, and 0 otherwise. The virtual time is non-decreasing with t and piecewise linear with the slope determined by the set of non-empty queues. The slope becomes zero when all the queues are empty. Figure 5.4 illustrates the virtual time $V(t)$ for the arrival process in Figure 5.3. Note that $V(t)$ has a slope of $1/2$ in the interval $(0, 1]$ where the flow 2 is the only backlogged flow, and has a slope of $1/3$ in the interval $(1, 4]$ where both flows have packets to transmit. After time 13 where all the buffers are empty, $V(t)$ is kept constant. Note that the virtual time is a global function which applies to the entire queues in the multiplexer.

Now consider how the arrival and departure of each packet are represented in terms of the virtual time. For the nth packet arriving in the queue k, we denote by $a_k^{(n)}$ the arrival time, $L_k^{(n)}$ the amount of data, $s_k^{(n)}$ the starting time of transmission, and $d_k^{(n)}$ the departure time from the

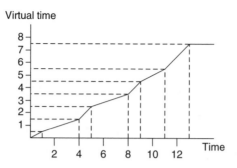

Figure 5.4 Virtual time for the arrival process in Figure 5.3

queue. Then, the virtual time when the $(n + 1)$th packet of the queue k departs the multiplexer is given by

$$V(d_k^{(n+1)}) = \int_0^{s_k^{(n+1)}} \frac{C}{\sum_{i=1}^K \phi_i 1_i(u)} 1(u)du + \int_{s_k^{(n+1)}}^{d_k^{(n+1)}} \frac{C}{\sum_{i=1}^K \phi_i 1_i(u)} 1(u)du. \qquad (5.11)$$

The first term on the right-hand side of (5.11) is equal to $V(s_k^{(n+1)})$ by the definition of the virtual time. On the other hand, accumulation of the instantaneous service rate from the starting time to the departure time of a packet becomes equal to the amount of data of that packet. Since the instantaneous service rate is given by (5.9), we get

$$\int_{s_k^{(n+1)}}^{d_k^{(n+1)}} \frac{\phi_k C}{\sum_{i=1}^K \phi_i 1_i(u)} du = L_k^{(n+1)}. \qquad (5.12)$$

Hence, the relation in (5.11) may be rewritten as

$$V(d_k^{(n+1)}) = V(s_k^{(n+1)}) + \frac{L_k^{(n+1)}}{\phi_k}. \qquad (5.13)$$

When the queue is empty, an arriving packet starts its transmission as soon as it arrives; otherwise, its transmission begins just after the previously arrived packet departs the queue. Thus we have $s_k^{(n+1)} = \max\{d_k^{(n)}, a_k^{(n+1)}\}$, and hence (5.13) is finally rearranged to

$$V(d_k^{(n+1)}) = \max\left\{V(d_k^{(n)}), V(a_k^{(n+1)})\right\} + \frac{L_k^{(n+1)}}{\phi_k}. \qquad (5.14)$$

This recursive equation rules the departing processes of the entire queues scheduled by the GPS.

The main advantage of the GPS is that it guarantees a minimum service rate for each flow. According to (5.9), a flow is served at a lower rate as more users have non-empty queues. Thus, the minimum service rate of flow k is guaranteed by

$$R_k(t) \geq \frac{\phi_k}{\sum\limits_{i=1}^{K} \phi_i} C \quad (\text{bits/sec}), \tag{5.15}$$

independently of other flow's traffic. As a result, each flow is guaranteed to be provided with a throughput in (5.15) and the delay experienced at the queue can be bounded.

5.2.1.1 Weighted Fair Queueing: a Packetized Version of GPS

The GPS scheduling policy is a "fair scheduler" in the sense that it always distributes the output link capacity to each backlogged flow in proportion to its own weight. However, as remarked above, this GPS scheduling based on the fluid-flow model cannot be used in real packet networks because packets must be served in their entirety, not in bits. So we now consider a packetized version of GPS, called *weighted fair queueing* (WFQ). The WFQ operates as follows:

1. The virtual time process, $V(t)$, is simulated as if a GPS scheduler were employed.
2. Each arriving packet is marked with its virtual finish time by (5.14).
3. When a packet is selected for transmission, it is transmitted completely.
4. After a packet departs the queue, the scheduler selects, for the next transmission, the packet that has the smallest virtual finish time among all the packets in the system.

WFQ approximates the GPS by transmitting the packet first which would depart the system earliest under the GPS. The operation of WFQ for the arrival process in Figure 5.3 is as illustrated in Figure 5.5. In contrast to the GPS, the slope of the accumulated transmission is one when the flow is selected for transmission, and zero otherwise.

It is obvious that GPS and WFQ have the same order of departures for the packets present together at the start of a scheduling instance. In other words, if two packets are present in the multiplexer at a scheduling instant, a packet is selected to be served by WFQ if and only if it departs the system earlier than the other packet in the GPS. This is because the WFQ selects a packet for transmission which would depart earliest under the GPS. However, this property does not hold for a packet which arrives after the scheduling instant. Suppose that a packet with size 11 (say, packet 1) arrives at flow 1 at time 0, and another packet with size 1 (say, packet 2) arrives at flow 2 at time 1. Further, assume that both queues are empty at time 0. Under the GPS, the departure times of packets 1 and 2 are given by 12 and 3, respectively. However, the WFQ begins transmission of packet 1 at time 0 and continues this transmission until packet 1 departs the system and, as a result, the departure times under the WFQ become 11 and 12, respectively. As in this example, the order of departures may differ in the GPS and the WFQ if a shorter packet arrives while a longer packet is served.

We can consider the WFQ as a good approximation of the GPS, the ideal reference of fair scheduling. According to Parekh and Gallager (1993), the departure times of a packet p under the GPS and the WFQ, d_p and $\hat{d}_p$, respectively, are related by

$$\hat{d}_p - d_p \leq \frac{L_{\max}}{C}, \tag{5.16}$$

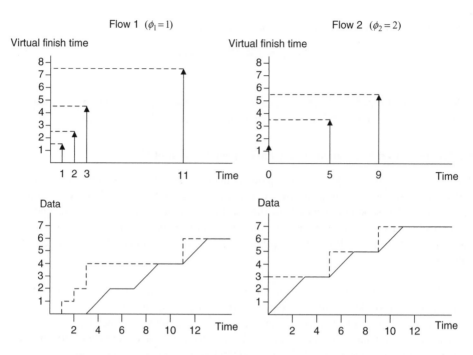

Figure 5.5   Operation of WFQ for the arrival process in Figure 5.3

where $L_{\max}$ is the maximum packet length. This implies that the difference between the departure times of the two schedulers is bounded as long as the packet length is bounded. In addition, the amounts of flow k served in the interval $[\tau, t]$ under the GPS and the WFQ, $S_k(\tau, t)$ and $\hat{S}_k(\tau, t)$, respectively, are related by

$$S_k(0, \tau) - \hat{S}_k(0, \tau) \leq L_{\max} \tag{5.17}$$

for any τ. This implies that the deviation of the throughput of each user rendered by the WFQ from that by the GPS is also bounded. In summary, the WFQ operates scheduling with a bounded error from the ideal reference and thus may be a good approximation of the GPS.

5.2.2 Modifications of GPS for Wireless Channels

One of the biggest differences between wired and wireless networks is the variability of transmission links. Due to the fading effect, the quality of a wireless channel varies dynamically in time when compared with a wired channel. As a result, it can happen that the quality of a wireless link becomes too poor to make any successful transmission. This time-dependent link variability may appear to any user but in an independent manner. At a particular instance, some users may experience a preferable channel condition while the others suffer from hostile link quality. Since transmitting to a user with a poor channel condition may be almost meaningless owing to the very high error probability, it is demanding to develop a scheduling policy that can deal with this time-dependent link variability most effectively.

Since the issue of scheduling originates from wired networks, a series of researches in the past concentrated on modifying schedulers developed for wired networks in such a way that they can perform effectively in wireless networks as well. Since "fairness" was the main objective to achieve in wired networks, the modified approaches tend to emphasize fairness guarantee on some simplified wireless channel model. More specifically, most of those approaches employ the two-state Markov channel model, in which the wireless link is in either a *good* state or a *bad* state, and is error-free in the good state and erred in the bad state. Transitions between the two states occur randomly, and each transition probability is determined by the communication environment such as the mobility of the user.

Under the above two-state channel model, one simple way to take the channel condition into account is not to serve a user in a bad state. This method is called *channel-state-dependent packet scheduling* (CSDPS) (Bhagwat *et al.,*1996). The CSDPS improves the scheduling performance substantially when compared with a scheduler that does not consider the channel state. However, it has a serious drawback in view of fairness: A user yields its transmission opportunity to other users when its channel state is bad, but the CSDPS does not provide any mechanism to compensate for this loss of transmission opportunity. As a result, a user may be served less than its share which would be granted in wired cases, and this makes the CSDPS seriously unfair in channel sharing.

"Modification" approaches tried to solve this unfairness by introducing fair schedulers developed in wired networks. Some tried to use the GPS as a reference in scheduling. The following discusses two different GPS-based wireless schedulers.[1]

5.2.2.1 Idealized Wireless Fair Queueing

In order to resolve the unfairness problem of CSDPS, *idealized wireless fair queueing* (IWFQ) was proposed (Lu *et al.*, 1999). In IWFQ, an *error-free* service is defined as the GPS service with identical arrival processes and completely error-free channels. IWFQ uses this error-free service as the baseline of scheduling. It selects the packet that has the smallest virtual finish time for transmission as long as the selected user is in a good state. If the selected user is in a bad state, IWFQ chooses an alternative packet in a good state. Since the transmission is blocked for users in a bad state and is allowed only for users in a good state, a user whose channel just transits to a good state is much more likely to backlog packets with small virtual finish time. Thus, the flows recovered from a bad state have higher priority in scheduling and this method inherently has a compensation mechanism.

However, it can happen that the scheduler serves only the recovered flow exclusively for an extended period, while causing serious starvation for the other users. Thus, in order to provide a fair compensation mechanism for the flows recovered from a bad state, IWFQ defines the following three states by using the error-free service as reference. A flow is defined to be *lagging*, *leading*, or *in-synch*, if its queue length is longer than, shorter than, or equal to the queue length of its error-free service at the same time instant. A lagging flow implies that the flow has suffered or is currently suffering from a bad channel state and it has lost its transmission opportunity which

[1] There are schedulers that employ the two-state channel model but solve the fairness problem in different ways without referencing to GPS. Interested readers may refer to Cao and Li (2001) for a survey of wireless schedulers employing the two-state channel model.

would be granted in the error-free system. A leading flow has served more than its grant by using the opportunities abandoned by the lagging flows.

The key feature of the compensation mechanism of IWFQ is to allow lagging flows to make up their lag by causing leading flows to give up their lead. Since an unbounded compensation may result in a long starvation of the leading flows, which can cause serious delay problems, IWFQ defines a bound on the lag and the lead for each flow k. The lagging bound b_k operates such that each lagging flow k is allowed to retain at most b_k bits with a virtual finish time less than the current virtual time. The leading bound l_k means that a leading flow k will only surrender up to l_k bits of service share to other flows later on.

Under the above definitions, IWFQ operates as follows. The error-free service with GPS scheduling is simulated for each flow. Each arriving packet is marked with its virtual start and finish times by (5.14). The marking of each flow is readjusted as follows. For each lagging flow k, if the aggregated length of packets with finish tags less than the current virtual time is greater than b_k, then only the first N_k packets are retained and other lagging packets are deleted from the queue. Here, N_k is the largest integer such that $\sum_{i=0}^{N_k-1} L_k^{(HOL_k+i)} \leq b_k$, where HOL_k denotes the index of the packet at the *head-of-line* (HOL) of the flow k. For each leading flow k, if the start tag of the HOL packet, $V(s_k^{(HOL_k)})$, is greater than the virtual time $V(t)$ by more than l_k/ϕ_k, then the tags are replaced by

$$V(s_k^{(HOL_k)}) = V(t) + l_k/\phi_k, V(d_k^{(HOL_k)}) = V(s_k^{(HOL_k)}) + L_k^{(HOL_k)}/\phi_k. \qquad (5.18)$$

The scheduler picks the packet with the smallest finish tag among the flows in the good state. After completing the transmission of the selected packet, the above procedure is repeated to determine the next transmission.

Readjustment of the lagging flows implies that lagging more than b_k bits is not allowed and the excessive amount of data is deleted from the queue. Thus, user k recovered from the bad state is served by up to b_k bits as the compensation for transmission opportunity loss during the bad state. This operates as an upper bound on the compensation of lagging flows. The deleted packets are expected to be recovered by an upper layer protocol such as *transport control protocol* (TCP). The readjustment of the leading flows prevents a leading flow from having too large a virtual finish time. By this, the finish tag of any leading flow is maintained within a limited range from the current virtual time, and thus, the leading users become comparable with the users recovered from the bad state. As a result, leading users do not have to give up too much transmission opportunity for the compensation of the lagging flows. Owing to this mechanism, the amount of compensation for the lagging flows and the penalty for the leading flows are bounded. Consequently, throughput and delay guarantees are achievable for IWFQ. Analytical results of the throughput and delay bounds of IWFQ are derived in Lu *et al.* (1999). It is noteworthy that the delay bounds are derived only for the packets not deleted from the queue by the scheduler to enforce the artificial lagging bound.

5.2.2.2 Packet Fair Queueing Independent of Channel Condition

Even though IWFQ has desirable properties in fairness and QoS guarantees, it also has some limitations (Cao and Li, 2001; Ng *et al.*, 1998): First, IWFQ gives absolute priority to the flow with the smallest virtual finish time. Consequently, all the other flows cannot receive services at all as long as a lagging flow exists in the system. Moreover, compensation for all the lagging flows will take the same amount of time regardless of their guaranteed rate, violating the semantics that a larger guaranteed rate implies better quality of service. Second, there is a conflict between the delay and fairness properties. To achieve long-term fairness, a lagging session should be allowed to catch up as much as possible, which requires a large enough lagging bound. However, a large lagging bound causes other error-free flows to be unserved for a long period and thus experience a large delay. Thus, the IWFQ algorithm cannot ensure perfect fairness while achieving a low delay bound for an error-free flow, at the same time.

In order to overcome the above limitations, an algorithm called *channel-condition-independent packet fair queueing* (CIF-Q) was proposed (Ng *et al.*, 1998). CIF-Q is similar to IWFQ in the sense that it also uses an error-free fair queueing reference system and tries to approximate the real service to an ideal system. CIF-Q defines a reference error-free system S^r and associates it with a real error-prone system S. For a flow k, we define by lag_k the difference between the service amounts received by S and S^r. Similar to the IWFQ case, flow k is classified as *leading*, *lagging*, or *satisfied*, if lag_k is positive, negative, or zero, respectively. Since lag of a flow is caused by serving another flow more than those in S^r, we get $\sum_{k=1}^{K} lag_k = 0$.

Each packet arrival is put into queues in both S and S^r. Service order is determined by the scheduling in S^r.[2] If the selected user is in a good state, the chosen packet is transmitted in both S and S^r. However, if the selected user is in a bad state, the real packet in S is kept but the virtual packet in S^r is still served and the virtual time of the corresponding flow is updated. It is noteworthy that, in CIF-Q, the virtual time is only kept and updated in S^r, not in S.

The biggest advantage of CIF-Q is that it achieves graceful degradation in service for the leading flows while not giving an absolute priority on the lagging flows. To support this property, CIF-Q defines a parameter $\alpha(0 \le \alpha \le 1)$ to control the minimal fraction of service retained by a leading flow. This means that a leading flow has to give up at most $(1 - \alpha)$ amount of its service to compensate for the lagging flows. If the selected flow in S^r is in a good state but is a leading flow, it is served only if the normalized service it has received since it became leading is no larger than a portion α of the normalized service it should have received based on its share. That is, a leading flow is served only when $a_k \le \alpha v_k$ for a_k updated by $a_k = a_k + L/\phi_k$ for each service of a length-L packet in S after being leading and v_k updated by $v_k = v_k + L/\phi_k$ for each service in S^r during the same time. Since it became leading, a leading flow k has received service of v_k bits in the error-free system while a_k bits are transmitted actually. Therefore, this mechanism withdraws the transmission opportunity only from a flow which is regarded as leading too much, and any leading flow is guaranteed to be served up to a faction α of its share in S^r. A larger value of α yields more graceful degradation experienced by the leading flows. At the limit, no compensation is given to the lagging sessions when α is set to one.

As discussed above, there are two cases when CIF-Q does not transmit the packet chosen in S^r. One is when the channel state is bad and the other is when the selected flow is currently leading too much. In these cases, CIF-Q tries to allocate the transmission opportunity to the lagging flows in proportion to their weight. Each flow accounts the normalized amount of

---

[2] In Ng *et al.* (1998), CIF-Q is developed based on start-time fair queueing (Goyal *et al.*, 1996) which selects the flow with the smallest start time. The reason for this choice is that it is much easier to consider the channel state based on the start time than on the finish time. Nevertheless, CIF-Q can be implemented by using other fair queueing schemes developed for wired networks, such as WFQ.

compensation by $c_k = c_k + L/\phi_k$ for each service of a length-L packet since it became lagging. When the selected user in S^r does not take the opportunity, the flow with the smallest c_k is selected to take the transmission opportunity. This means that the lagging flow that has received the least compensation takes the opportunity. If no lagging flow can take this compensation due to the channel state, the additional service is distributed to non-lagging flows. If the originally selected user in S^r is in a good state, this flow is selected again; otherwise, the additional service is distributed in proportion to the users' weights in a way similar to the compensation mechanism for the lagging flows.

CIF-Q is advantageous in that it provides both long-term and short-term fairness guarantees. In addition, two different QoS parameters, packet delay and throughput, can also be guaranteed for each flow. Analysis of fairness and packet delay is given in (Ng *et al.*, 1998). These desirable properties are achieved by virtue of the graceful degradation of the leading flows and the proportional distribution of the compensation opportunities to the lagging flows. Compared with IWFQ, CIF-Q improves the scheduling fairness by associating the compensation rate and penalty rate with the weight of the flow.

5.3 Schedulers for Throughput Maximization

In Section 5.1.2, the opportunistic transmission system utilized multiuser diversity in such a way that it serves the best user channel at each time. If all the user channels are independent and identically distributed, each user gets the same number of chances to get service as long as it can wait for enough time. However, all user signals experience different signal propagation paths, so they are not necessarily identical in general even though they are independent. The users in more favorable channel conditions may get more chances to be served than the other users in poor channel conditions, which results in poor fairness among users. So, it is necessary to utilize the multiuser diversity gain with a fair resource allocation among users whose channel conditions are not identically distributed. This fairness issue was considered in Section 5.2, but the schedulers in that section were developed under a simple two-state channel model. As a result, those schedulers could not fully exploit the potential of channel adaptation via AMC (adaptive modulation and coding), thereby incurring some performance loss in terms of system throughput.

This section considers a downlink system in which the base station (BS) selects one user to serve at each time slot such that the throughput is maximized under some fairness constraints with a generalized channel model. Figure 5.6 depicts a block diagram of this wireless packet scheduling system. Each user measures the downlink channel, computes the maximum possible rate that the downlink can support if the next time slot is assigned to it, and feeds this rate information back to the BS.[3] Based on the rate information, the BS determines which user (i.e., mobile station) to serve at the next time slot and transmits a packet to that user.

Let $R_k(n)$ denote the maximum available transmission rate of user k, for $k = 1, 2, \ldots, K$, at time slot n; $k^*(n)$ is the user selected for data transmission at time slot n. Assume that each $R_k(n)$ is an independent and stationary random sequence, and denote by $F_{R_k}(r)$ the cumulative density

---

[3] In the case of CDMA2000 1xEV-DO, such a feedback channel is called the *data rate control* (DRC) channel (Bender *et al.*, 2000).

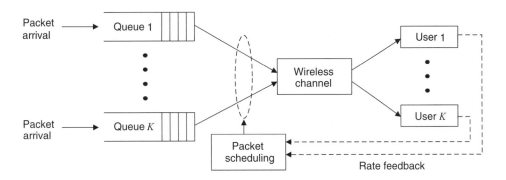

Figure 5.6   Block diagram of wireless packet scheduling

function of $R_k(n)$. The memoryless wireless packet scheduling algorithm can be expressed in the general form

$$k^*(n) = \arg \max_k h_k(R_k(n)),\tag{5.19}$$

where $h_k(x)$ is a non-decreasing function that does not depend on the history of the user selection. When a tie occurs (i.e., when two or more users yield the same maximum value), one of the users may be selected in a random fashion. So, a memoryless packet scheduler may be characterized by the set of the functions $\{h_k(x)|k = 1, 2, \ldots, K\}$. In some cases, there may exist two or more memoryless scheduling algorithms that yield the same scheduling results but are not necessarily to be distinguished. In order to handle such algorithms collectively, we define the term "equivalent" as follows. Two memoryless scheduling algorithms characterized by $\{h_k(x)|k = 1, 2, \ldots, K\}$ and $\{\tilde{h}_k(x)|k = 1, 2, \ldots, K\}$, respectively, are said to be *equivalent* to each other if $\arg \max_k h_k(R_k) = \arg \max_k \tilde{h}_k(R_k)$ for all possible R_k other than those that lead to a tie. If we apply a random-selection rule whenever a tie occurs, a scheduler can possibly yield many different scheduling results for some R_k. So, we exclude the cases of those R_k that lead to a tie.

The above equivalence corresponds to the equivalence relation applied on the set of memoryless scheduling algorithms. By applying the equivalence concept, we can put together all the scheduling algorithms that are equivalent to a particular scheduling algorithm S to form an equivalence class determined by S. This enables us to partition the whole memoryless scheduling algorithms into disjoint subsets, with each subset consisting of equivalent scheduling algorithms.

This equivalence relation has an interesting property. Two memoryless scheduling algorithms $\{h_k(x)|k = 1, 2, \ldots, K\}$ and $\{\tilde{h}_k(x)|k = 1, 2, \ldots, K\}$ are equivalent to each other if and only if there exists a monotonically increasing function $m(x)$ such that $m(h_k(x)) = \tilde{h}_k(x)$ for all k (Park *et al.*, 2005). So, monotonically increasing functions play an important role in deriving equivalent scheduling algorithms in a simple manner.

5.3.1 Maximal-rate Scheduling

First consider the simplest packet scheduling algorithm to obtain multiuser diversity without bearing fairness in mind. A pioneering work on packet scheduling has been reported in Knopp and Humblet (1995). As discussed in Section 5.1, in order to maximize the total information-theoretic capacity, the optimal strategy is to schedule at any time instant only the user with the best channel. Diversity gain arises from the fact that, in a system with many users whose channels vary independently, it is likely that a user exists whose channel is near its peak state at any time instant. Overall system throughput can be maximized by allocating the common channel resource to the user who can best exploit it. It can also be considered as a form of selection diversity. Maximal-rate scheduling selects the user for the next data transmission, $k^*(n)$, to be the user that has the largest value of the maximum available transmission rate; that is

$$k^*(n) = \arg\max_k R_k(n). \tag{5.20}$$

This maximal-rate scheduling corresponds to $h_k(x) = x$ in the memoryless scheduling in (5.19). This maximal-rate scheduling maximizes the instantaneous throughput at each slot as well as the total throughput. However, it may yield unfair resource allocation in terms of the number of allocated time slots, with the more favorable channel user occupying more time slots, when the channel gain distributions are not identical.

Now assume that each user channel is *independent and identically distributed* (i.i.d.). Then, the probability that user k is selected for transmission, given $R_k(n) = r_k$, is

$$\Pr(k^*(n) = k|R_k(n) = r_k) = F_{R_k(n)}(r_k)^{K-1}. \tag{5.21}$$

So, the average sum throughput of all users is

$$\begin{aligned} T = \sum_{k=1}^{K} T_k &= \sum_{k=1}^{K} \int r_k F_{R_k}(r_k)^{K-1} dF_{R_k}(r_k) \\ &= K \int r_1 F_{R_1}(r_1)^{K-1} dF_{R_1}(r_1). \end{aligned} \tag{5.22}$$

Consider the case where K users share the same wireless link, with the user rates being i.i.d. so that the SNR is the same. Figure 5.7 compares the sum throughput of the maximal-rate scheduling with that of the conventional TDMA. In the conventional TDMA scheme, the sum throughput is the same as the throughput of one user and adding more users does not affect the sum throughput. However, maximal-rate scheduling exploits the multiuser diversity gain. As the number of users increases, the sum throughput also increases.

5.3.2 Proportional Fairness Scheduling

One popular fairness criterion is *proportional fairness* (PF), which was first introduced in wired networks to fairly share fixed bandwidth. A feasible vector of rates $\mathbf{x} = \{x_1, x_2, \ldots, x_K\}$ is said to be proportionally fair if, for any other feasible vector $\mathbf{x}^*$, the aggregate of the proportional changes is zero or negative; that is

$$\sum_{k=1}^{K} \frac{x_k^* - x_k}{x_k} \le 0. \tag{5.23}$$

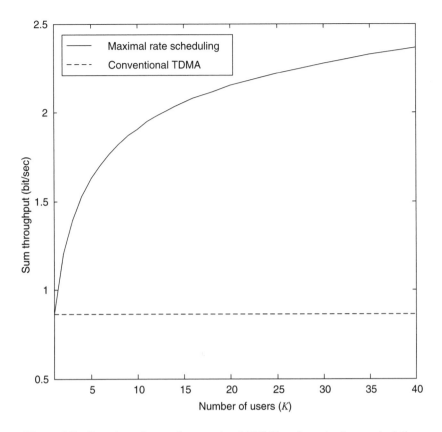

Figure 5.7   Sum throughputs of conventional TDMA and maximal rate scheduling

This implies that increasing the throughput of one user by 1% yields more than 1% cumulative decrease in all the other users. If $\mathbf{x}$ is a proportional fair vector, for any infinitesimal perturbation vector $\delta\mathbf{x} = \{\delta x_1, \delta x_2, \ldots, \delta x_K\}$, we have

$$
\begin{aligned}
\sum_{k=1}^{K} \ln(x_k + \delta x_k) &= \sum_{k=1}^{K} \ln x_k + \sum_{k=1}^{K} \ln\left(1 + \frac{\delta x_k}{x_k}\right) \\
&= \sum_{k=1}^{K} \ln x_k + \sum_{k=1}^{K} \frac{\delta x_k}{x_k} + O((\delta x_k)^2) \\
&\leq \sum_{k=1}^{K} \ln x_k,
\end{aligned}
\tag{5.24}
$$

where the last inequality holds by the definition of proportional fairness in (5.23). So, the proportionally fair vector is a feasible vector that maximizes $\sum_{k=1}^{K} \ln x_k$.

   To achieve proportional fairness in wireless packet scheduling, proportional fairness scheduling has been proposed (Jalali *et al.*, 2000). Let $T_k(n)$ be the average throughput of user k at time slot n:

$$T_k(n+1) = \left(1-\frac{1}{t_c}\right)T_k(n) + \frac{1}{t_c}R_k(n)1_{k=k^*(n)}, \qquad (5.25)$$

where $1_{k=k^*(n)}$ becomes 1 if user k is selected for transmission, and 0 otherwise. Equation 5.25 corresponds to a first-order *infinite-impulse response* (IIR) filter with the parameter t_c indicating the effective window size during which duration the transmission rate is averaged. So t_c is chosen to be much larger than the channel coherence time. In order to maximize the logarithm of the throughput, the current time slot is assigned to the user who maximizes the logarithm difference of the throughput; that is

$$
\begin{aligned}
\max_k[\ln T_k(n+1)-\ln T_k(n)] &= \max_k \ln\left(1-\frac{1}{t_c}+\frac{1}{t_c}\frac{R_k(n)1_{k=k^*(n)}}{T_k(n)}\right) \\
&= \ln\left(1-\frac{1}{t_c}+\frac{1}{t_c}\max_k\frac{R_k(n)1_{k=k^*(n)}}{T_k(n)}\right).
\end{aligned}
\qquad (5.26)
$$

So, in order to maximize the PF, the scheduling algorithm selects user $k^*(n)$ for which the ratio $R_k(n)$ over $T_k(n)$ is the largest; that is

$$k^*(n) = \arg\max_k \frac{R_k(n)}{T_k(n)}. \qquad (5.27)$$

This proportional fair scheduling algorithm maximizes the sum of the logarithm of the long-term average throughput, or the product of the long-term average throughputs.

   Round-robin scheduling refers to the scheduling algorithm that allocates the same number of time slots to all the users in a round-robin fashion. It performs no better than the single-user system as it schedules users regardless of their channel conditions. *Maximal-rate scheduling* allocates the time slot to the strongest channel user without considering user channel distribution, which may yield an unfair resource allocation for a user with poor channel condition. These are two extreme scheduling algorithms: one is dedicated to resource fairness and the other to throughput performance. PF scheduling is a compromise between those two extremes: it allocates each time slot to a user whose channel gain is relatively-strong compared with its average. Thus it obtains multiuser diversity and fair resource sharing at the same time.

   Now consider the case where the transmission rate $R_k(n)$ is determined by the capacity/SNR relation

$$R_k(n) = \log_2(1+\rho s_k f_k(n)), \qquad (5.28)$$

where ρ denotes the SNR, s_k the path loss of user k, and $f_k(n)$ the fast-fading channel gain of user k at time slot n. The signal propagation in this model is a combination of the path loss determined by the distance between the base station and the user and the fast-fading component determined by the local scattering environment. If the fast-fading component is i.i.d., the long-term average sum-rate throughput achieved by PF scheduling, for a fixed s_k, is given by (Caire *et al.*, 2007)

$$\sum_{k=1}^{K} T_k = \frac{1}{K} \sum_{k=1}^{K} \int \log_2(1 + \rho s_k x) dF_{\max\{f_1,\ldots,f_K\}}(x), \tag{5.29}$$

where $F_{\max\{f_1,\ldots,f_K\}}(x)$ is the CDF (cumulative distribution function) of $\max\{f_1,\ldots,f_K\}$. So, PF scheduling works as if it performed scheduling based on the i.i.d. fast-fading component disregarding the fixed path loss. This implies that each user is served at its own peak in PF scheduling.

5.3.3 Temporal Fairness Scheduling

Even though PF scheduling provides multiuser diversity with fair resource allocation, it still schedules more time slots to the users in a favored channel condition than to other users with poor channel condition. So, it would be desirable to allocate an equal number of time slots to each user while maximizing the total sum-rate throughput. This is *temporal fairness scheduling* that provides equal opportunity to each user rather than equal outcome.

Consider a pre-assigned, not necessarily equal, time fraction for each user. Let f_k denote the time fraction assigned to user k, where

$$\sum_{k=1}^{K} f_k = 1.$$

A fraction f_k of the whole time should be scheduled to user k on average. The goal of temporal fairness scheduling is to develop a scheme that exploits the time-varying channel conditions to maximize system performance, under the time-fraction constraint f_k. To achieve such fairness, the *optimal temporal fairness scheduling algorithm* has been proposed (Liu *et al.*, 2001). This algorithm determines to which user the current time slot should be assigned to achieve optimal temporal fairness. The optimal temporal fairness scheduler selects the user for which the sum of the rate $R_k(n)$ and an offset v_k is the largest; that is

$$k^*(n) = \arg\max_{k} R_k(n) + v_k, \tag{5.30}$$

where the offset v_k is determined to satisfy a given time-fraction assignment requirement by using some stochastic approximation techniques. The temporal scheduling algorithm in (5.30) corresponds to $h_k(x) = x + v_k$ in the memoryless scheduling in (5.19). With the time-fraction requirement met, this scheduler maximizes the average total throughput. To prove this, denote by $\tilde{k}$ the temporal fairness scheduling that allocates the time slot according to the time-fraction assignment requirement – that is, $\Pr(\tilde{k} = k) = f_k$ – and denote by k^* the scheduling as shown in (5.30). Then by (5.30) we have

$$\mathbb{E}\sum_{k=1}^{K}(R_k + v_k)1_{\tilde{k}=k} \leq \mathbb{E}\sum_{k=1}^{K}(R_k + v_k)1_{k^*=k}, \tag{5.31}$$

and hence the throughput is upper-bounded by

$$\mathbb{E}\sum_{k=1}^{K}R_k 1_{\tilde{k}=k} \leq \mathbb{E}\sum_{k=1}^{K}(R_k + v_k)1_{k^*=k} - \sum_{k=1}^{K}v_k f_k$$

$$= \mathbb{E}\sum_{k=1}^{K}R_k 1_{k^*=k},$$

(5.32)

where we use the relation $\mathbb{E}1_{k^*=k} = f_k$. So, the temporal fairness scheduling given in (5.30) is the optimal one that maximizes the total throughput while satisfying the time-fraction constraints.

Even though temporal fairness scheduling guarantees the pre-assigned time fraction to each user, weak users may not enjoy the multiuser diversity gains as the scheduler prefers strong users. For example, consider two users with the associated time fractions f_1 and $f_2 = 1 - f_1$. The transmission rates of these users are given by R_1 and R_2, where $R_2 = \alpha R'_2$. As the scaling factor α increases, the transmission rate of user 2 increases. In this case we investigate the effect of α on the scheduling performance. The time fraction of user 1 is given by

$$f_1 = \int \Pr(k^* = 1|R_1 = r_1)dF_{R_1}(r_1)$$

$$= \int \Pr(r_1 + v_1 \geq \alpha R'_2 + v_2|R_1 = r_1)dF_{R_1}(r_1)$$

$$= \int F_{R'_2}\left(\frac{r_1 + v}{\alpha}\right)dF_{R_1}(r_1),$$

(5.33)

where $v = v_1 - v_2$ should be appropriately chosen such that the selection probability of user 1 becomes f_1. If $v \to v_0 < \infty$ as $\alpha \to \infty$, then the temporal fairness condition cannot be met because $F_{R'_2}(r_1 + v)/\alpha$ converges to 0 and consequently f_1 converges to 0. To avoid this case, we need the relation $v \to \infty$ as $\alpha \to \infty$. In addition, in order to satisfy (5.33), v/α should converge to a fixed constant such that

$$F_{R'_2}\left(\frac{r_1 + v}{\alpha}\right) \to f_1.$$

(5.34)

So, the selection probability of user 1 for given $R_1 = r_1$, $\Pr(k^* = 1|R_1 = r_1)$, is independent of r_1. This implies that the temporal fairness scheduling selects user 1 with probability f_1 disregarding the current channel state of user 1. So, the weak user does not get any wireless scheduling gain. On the other hand, the selection probability of user 2 for given $R_2 = \alpha r'_2$ becomes

$$\Pr(k^* = 2|R_2 = \alpha r'_2) = \Pr(R_1 + v_1 \leq \alpha r'_2 + v_2|R_2 = \alpha r'_2)$$

$$= F_{R_1}(\alpha r'_2 - v)$$

$$\to \begin{cases} 1, & r'_2 \geq F_{R'_2}^{-1}(f_1), \\ 0, & \text{otherwise}, \end{cases}$$

(5.35)

because $v/\alpha \to F_{R'_2}^{-1}(f_1)$. This implies that the scheduler selects user 2 when the rate of user 2 exceeds a threshold, disregarding the current channel state of user 1. In summary, the temporal fairness scheduling prefers strong users to weak users, so a weak user may get no multiuser diversity gain in the extreme case.

5.3.4 Utilitarian Fairness Scheduling

In order to maximize the total throughput under some fairness constraints, we may introduce the concept of *utility*. In the context of wireless packet scheduling, utility may represent the satisfaction of a user as a function of throughput T. The basic assumptions of utility function $U(x)$ are twofold. First, $U(x)$ is a differentiable strictly increasing function. Second, $U(x)$ is a concave function. The first assumption is obvious. The second assumption is derived from the human perception:for two users whose service throughputs are 1 Mbps and 10 Mbps, an additional 1 Mbps of throughput is no less valuable to the 1 Mbps user than to the 10 Mbps user. So, utility is defined as a concave function. Thus, it would be fairer to allocate this additional 1 Mbps to the 1 Mbps user who would value it higher. *Utilitarian fairness scheduling* tries to maximize the total utility of all users, which yields fair resource allocation.

Mathematically, we can express the total utility maximization by

$$\text{maximize} \sum_{k=1}^{K} U_k(T_k(n)) \tag{5.36}$$

where $U_k(x)$ is a utility function of user k and the throughput $T_k(n)$ is given by (5.25). Assume that time slot $n + 1$ is assigned to user k. Then the utility difference is expanded as the first-order Taylor series

$$U_k(T_k(n+1)) - U_k(T_k(n)) \approx U'_k(T_k(n))(T_k(n+1) - T_k(n))$$

$$= U'_k(T_k(n))\left(\frac{R_k(n) - T_k(n)}{t_c}\right), \tag{5.37}$$

where t_c is the window size used in (5.25). The objective is to allocate the time slot to a user who maximizes this utility difference. As $T_k(n)$ is given, the optimal allocation is to assign this time slot to the user whose $U'_k(T_k(n))R_k(n)$ is the largest; that is

$$k^*(n) = \arg\max_k U'_k(T_k(n))R_k(n). \tag{5.38}$$

Note that if the utility function is given by $U_k(x) = x$, then utilitarian fairness scheduling becomes the maximal-rate scheduling in (5.20). If $U_k(x) = \log x$, then it becomes the proportional fairness scheduling in (5.27). So, utilitarian fairness scheduling may be considered as a generalized packet scheduling.

When the user rates are stationary, each throughput $T_k(n)$ converges to a finite value T_k. We can define an achievable region $\mathfrak{R}$ to be a set of all possible throughput vectors that any scheduling algorithms can achieve. If we allow time-sharing of scheduling algorithms, the achievable region $\mathfrak{R}$ becomes a convex set. Now we assume that there is a differentiable function $g(\cdot)$ such that $g(\mathbf{T}) \leq 0$ if $\mathbf{T} = (T_1, T_2, \dots, T_K) \in \mathfrak{R}$, and $g(\mathbf{T}) > 0$ otherwise. Then we may rewrite the utility maximization problem in (5.36) with the constraint; that is

$$\text{maximize} \quad \sum_{k=1}^{K} U_k(T_k)$$
$$\text{subject to} \quad g(\mathbf{T}) \leq 0. \tag{5.39}$$

This is a convex optimization problem whose Lagrangian takes the form

$$L = \sum_{k=1}^{K} U_k(T_k) - \lambda g(\mathbf{T}), \tag{5.40}$$

where λ is a Lagrange multiplier. The KKT condition is given by

$$\frac{\partial L}{\partial T_k} = U_k'(T) - \lambda \frac{\partial}{\partial T_k} g(\mathbf{T}) = 0 \text{ for all } k. \tag{5.41}$$

The optimal solution $\mathbf{T}$ and the Lagrange multiplier λ satisfy the KKT condition in (5.41).

If user k is charged a price p_k per unit bit and is allowed to freely adjust throughput T_k for a given utility $U_k(x)$, then user k maximizes the utility as follows:

$$\underset{T_k}{\text{maximize}} \quad U_k(T_k) - p_k T_k. \tag{5.42}$$

Let the demand function $D_k(p_k)$ be the optimal solution to this problem. $D_k(p_k)$ is the throughput that user k would request if the price per unit bit were p_k. Since the optimality condition is

$$U_k'(T_k) - p_k = 0, \tag{5.43}$$

the demand function is given by $D_k(p_k) = U_k'^{-1}(p_k)$ if the inverse function $U_k'^{-1}(x)$ exists. If the base station receives a revenue p_k per unit bit from user k and is allowed to freely adjust the throughput, then the base station maximizes the sum of the weighted throughput as follows:

$$\text{maximize} \quad \sum_{k=1}^{K} p_k T_k$$
$$\text{subject to} \quad g(\mathbf{T}) \leq 0. \tag{5.44}$$

The KKT condition of the problem in (5.44) is given by

$$p_k - \mu \frac{\partial}{\partial T_k} g(\mathbf{T}) = 0. \tag{5.45}$$

It is not difficult to prove that there exists a price vector $(p_1, p_2, \ldots, p_K)$ such that the optimal throughput vector $\mathbf{T}$ in (5.39) is also the solution to the problems (5.42) and (5.44) (Kelly, 1997). In this case, the Lagrange multipliers λ and μ are the same. This implies that the utility

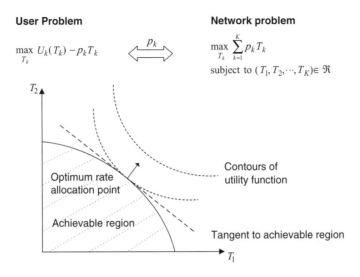

Figure 5.8 Optimal throughput vector on the boundary of the achievable region

maximization problem in (5.39) may be decomposed into two subproblems, the user problem in (5.42) and the base station problem in (5.44). Here, the parameter p_k – that is, $U'_k(T_k)$ – plays the role of a shadow price. In Section 4.3 it was called the *marginal utility*.

Let the boundary of the achievable region be a set of throughput vectors such that the throughput of a user cannot be increased while the throughputs of all other users remain fixed. Then, the throughput vector that maximizes the sum of the weighted throughput in (5.44) must be on the boundary of the achievable region. Conversely, for any throughput vector on the boundary of the achievable region, there exists a weight vector $(p_1, p_2, \ldots, p_K)$ such that the optimal solution to the problem in (5.44) yields that throughput vector. Figure 5.8 depicts an optimal throughput vector on the boundary of the achievable region for a two-user case. The contour of utility – that is, the curve connecting the points that have the same total utility value – is an *indifference curve* (see Section 4.3). At the optimal throughput vector, the indifference curve meets the achievable region. In addition, the associated weight vector $(p_1, p_2) = (U'_1(T_1), U'_2(T_2))$ is a normal vector to the contour line at the optimal throughput vector.

The boundary of the achievable region is fully characterized by the weight vectors because any optimal throughput on the boundary is the solution to (5.44) for some weight vector $(p_1, p_2, \ldots, p_K)$. So, if we want to allocate user throughput under some throughput constraint, it is sufficient to find a weight vector that brings forth the desired scheduling result. It is easy to show that, for a given weight vector $(p_1, p_2, \ldots, p_K)$, the time-slot allocation algorithm that yields the optimal solution to the problem in (5.44) is

$$k^*(n) = \arg\max_k p_k R_k(n). \tag{5.46}$$

When the reward to user k is p_k per unit bit, the scheduling algorithm in (5.46) is to allocate the current time slot to the user who would get the maximum reward. Note that (5.38) and (5.46) turn out to be the same if the relation $p_k = U'_k(T_k)$ is taken into account. The utilitarian fairness scheduling algorithm in (5.46) corresponds to $h_k(x) = p_k x$ in the memoryless scheduling in (5.19).

Some examples of packet scheduling under throughput constraints maximize the minimum relative throughput and guarantee the minimum throughput. The objective of maximizing the minimum relative throughput is to maximize $\min_k T_k / \alpha_k$, where α_k is the relative target throughput value (Borst and Whiting, 2003). To guarantee the minimum throughput, we should maximize throughput under the constraint $T_k \geq C_k$, where C_k is the minimum required throughput (Liu et al., 2003a; Liu et al., 2003b). The optimal scheduling algorithm to both of these resource allocation problems is given by the utilitarian fairness scheduling in (5.46). Various scheduling algorithms under the throughput constraints take the same form. The main difficulty of this utilitarian fairness scheduling is how to find the appropriate weight vector that yields the desired scheduling result. Also, for the given weight vectors, the scheduling results are expressed in multidimensional integration, so it is not feasible to calculate the throughput analytically. However, utilitarian fairness scheduling has a general form for throughput constraint.

5.3.5 Scheduling based on Cumulative Distribution Function

Consider the maximal-rate scheduling algorithm when all user rates are independent and identically distributed (i.i.d.). The user selection probability and the average throughput are given by (5.21) and (5.22), respectively. However, it is not feasible to evaluate the exact throughput of scheduling algorithms when the user rates are not i.i.d., as it would involve multidimensional integration. So, a scheduler that yields an exact analytic expression is very desirable, so that it is possible to predict scheduling performance before performing actual scheduling and possibly investigate the effect of schedulers on related protocols such as call admission control. *CDF-based scheduling* was introduced in answer to this demand.

We can assume that the cumulative density function of $R_k(n)$, $F_{R_k}(r)$, is known because we can estimate the channel distribution by using feedback data. It is possible to collect the feedback rate information to produce a histogram and regard that as the rate distribution. Before introducing CDF-based scheduling, we consider an ideal scheduler that schedules a time slot to user k only when its rate is near to its peak rate, while making the number of slots occupied by each user $1/K$, when the number of users is K. Then, as illustrated in Figure 5.9, user k would be

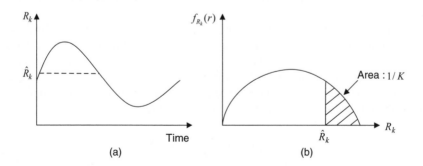

Figure 5.9 User selection by an ideal wireless scheduling algorithm: (a) instantaneous rate, and (b) probability density function.

served when its rate falls in the right tail of the probability distribution – more specifically, when its rate is above the threshold $\hat{R}_k$, where $\hat{R}_k$ is the root of $F_{R_k}(r) = 1 - 1/K$. However, it can happen that more than two users may be above their threshold at the same time or no users may be above the threshold. So, this ideal scheduling is unrealistic.

To imitate the ideal scheduler, we can select a user whose rate is high enough, but least probable to become higher. Mathematically, if $R_k(n)$ takes on a value r_k at slot n for $k = 1, 2, \ldots, K$, we can select a user for which $\Pr(R_k > r_k)$ is the smallest among all users; that is, the user whose rate is hardest to obtain. Noting that $\Pr(R_k > r_k) = 1 - F_{R_k}(r_k)$, CDF-based scheduling is given by (Park et al., 2005)

$$k^*(n) = \arg \max_k F_{R_k}(R_k(n))^{1/w_k}, \tag{5.47}$$

where w_k denotes a non-negative control parameter associated with user k by the relation $\sum_{k=1}^{K} w_k = 1$. The CDF-based scheduling in (5.47) corresponds to $h_k(x) = F_{R_k}(x)^{1/w_k}$ in the context of memoryless packet scheduling. If the rates are i.i.d. and $w_k = 1/K$, then CDF-based scheduling will be equivalent to MR scheduling.

For CDF-based scheduling, the probability that user k is selected for transmission, given that $R_k(n) = r_k$, is (Park et al., 2005)

$$\Pr(k^*(n) = k | R_k(n) = r_k) = F_{R_k}(r_k)^{1/w_k - 1}. \tag{5.48}$$

This implies that the probability of selecting a user for transmission depends on the user's own distribution and the control parameter w_k. Interestingly, it is independent of the probability distribution of other users. In general, this selection probability is expressed in a multidimensional integration in other scheduling algorithms.

When we compare the selection probabilities of MR scheduling for the i.i.d. case and CDF-based scheduling, we find that the number of users K in (5.21) is replaced with $1/w_k$ in (5.48). This means that the probability that user k is selected for transmission is the same as the case when the scheduler is MR scheduling, in which all user rates are i.i.d., and the number of users is $1/w_k$. In other words, CDF-based scheduling performs packet scheduling such that user k feels as if the distribution of other users were identical to its own distribution, and virtually $1/w_k$ users were competing for selection. This is a very desirable property, because it simplifies the performance analysis significantly, which was not possible in other scheduling algorithms.

For CDF-based scheduling, the time fraction that user k occupies for data transmission is given by w_k; that is

$$\Pr(k^*(n) = k) = w_k. \tag{5.49}$$

Note that the control parameter w_k can be set arbitrarily as long as $\sum_{k=1}^{K} w_k = 1$. For example, if $w_k = 1/K$, then all users get the same time fractions. Note that in the temporal fairness scheduling in (5.30), it was necessary to set the offset v_k appropriately to guarantee the given time-fraction assignment by using some stochastic approximation techniques. The average throughput of a user is given by

$$T_k = \int_0^1 u^{1/w_k - 1} F_{R_k}^{-1}(u) du, \tag{5.50}$$

where $F_{R_k}^{-1}(u)$ is the inverse function of the CDF. Note that the average throughput of a user is also independent of the probability distribution of other users' rates. So, it remains fixed as long as the rate distribution of the user is fixed, whether or not the rates of some other users are relatively better distributed to yield a higher throughput.

In practical communication systems, the transmission rates are discrete; that is, $R_k(n) \in \{r_1, r_2, \ldots, r_M\}$, with $r_1 < r_2 < \ldots < r_M$. Since the CDF-based scheduling in (5.47) is for continuous transmission rates, a simple modification is required for discrete rates. Assume that user k feeds back $m_k(n) \in \{1, 2, \ldots, M\}$ to the base station at time slot n if $r_{m_k(n)}$ is the maximum available transmission rate of user k. The base station can obtain empirical distribution of feedback rate information by collecting all feedback channel information. Denote by $p_{k,m} \equiv \Pr(R_k(n) = r_m)$ the probability density function, and

$$q_{k,m} \equiv \sum_{i=1}^{m} p_{k,i}$$

is the CDF of user k for $k = 1, 2, \ldots, K$, $m = 1, 2, \ldots, M$. For notational convenience we set $q_{k,0} = 0$. For the discrete transmission rate, we have CDF-based scheduling in the following procedure:

Step 1: Each user k reports the rate index $r_k(n)$.
Step 2: The scheduler generates a uniform random variable $U_k(n)$ in the interval $[q_{k,m_k(n)-1}, q_{k,m_k(n)})$.
Step 3: The scheduler selects the user for which the following metric is the largest

$$k^*(n) = \arg\max_{k} U_k(n)^{1/w_k}. \tag{5.51}$$

Step 4: The index of rate to be transmitted for time slot n is $m_{k^*(n)}(n)$. So, the base station serves user $k^*(n)$ with the transmission rate $r_{m_{k^*(n)}}(n)$.
Step 5: The scheduler updates the probabilities $p_{k,m}$ and $q_{k,m}$ as follows:

$$p_{k,m} \leftarrow \lambda p_{k,m} + (1-\lambda) 1_{m=m_k(n)}, \tag{5.52a}$$

$$q_{k,m} \leftarrow \sum_{i=1}^{m} p_{k,i}, \tag{5.52b}$$

where $0 < \lambda < 1$, and 1_A is an indicator that becomes 1 if condition A is met, and 0 otherwise.

For this scheduling algorithm, the probability that user k is selected for transmission, given that $m_k = m$, is

$$\Pr(k^* = k | m_k = m) = w_k \frac{q_{k,m}^{1/w_k} - q_{k,m-1}^{1/w_k}}{q_{k,m} - q_{k,m-1}}. \tag{5.53}$$

So, the probability of selecting a user for transmission is independent of the probability distribution of other users. In addition, the time fraction that user k occupies for data transmission is given by w_k; that is, $\Pr(k^*(n) = k) = w_k$. Note that this is the same for the continuous rate case as in (5.49). The average throughput of user k with CDF-based scheduling is

$$T_k = w_k \sum_{m=1}^{M} r_m \left(q_{k,m}^{1/w_k} - q_{k,m-1}^{1/w_k} \right). \tag{5.54}$$

Note that the average throughput is determined by (5.54) whether the channel is time-independent or time-correlated. This comes from the fact that the selection probability in (5.53) depends only on the marginal distribution, not on the joint distribution of the user channel.

CDF-based scheduling is useful for resource allocation in the sense that the scheduling performance can be predicted before performing actual scheduling. The parameter that the CDF-based scheduling controls is w_k, which directly controls the time fraction and the average throughput. Suppose that a newly arrived user k demands an average throughput T_k^{req}. With the other packet scheduling algorithms, there are no systematic ways to provide the requested throughput, because the average throughput is involved with a multidimensional integration. However, with CDF-based scheduling, we can calculate w_k that should be allocated to user k to satisfy $T_k^{req} \leq T_k$. As the sum of the time fractions cannot exceed 1, there is a limitation in admitting users. If the sum of time fractions does not exceed 1 even after adding this new w_k, user k can be accommodated without degrading other users' performance. Otherwise, this user cannot be accommodated. This demonstrates how we can allocate the scarce wireless resources so that the desired *quality of service* (QoS) can be met in CDF-based scheduling.

5.3.6 Comparison of Scheduling Algorithms

Scheduling algorithms have their own metrics intended to maximize throughput while achieving some fairness among users. They may come up with the same scheduling results in many time slots, as they are intended to pick up the user in the best channel condition to exploit multiuser diversity. On the other hand, they may also select different users in some time slots to meet fairness constraints.

An example can now illustrate the differences between the various scheduling algorithms. There are two students, David and Tom, each attending a different high school – school A for David and school B for Tom. They took math exams in their own schools, answering different problems, and got 85 and 75 points, respectively. The average points of the two schools were 50 points and 45 points, respectively (see Figure 5.10). Who did best in their math exam?

According to maximal-rate scheduling, David did better than Tom simply because his score is higher. Maximal-rate scheduling selects a user without considering the probability distribution of users. When the user rate distributions are not too different, maximal-rate scheduling allocates resources fairly while achieving maximum throughput. However, if some

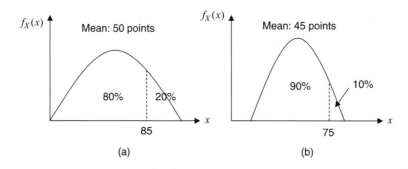

Figure 5.10 Example of math exam scores of two students: (a) histogram of school A, and (b) histogram of school B

users are near to the base station and the others are far away from the base station, the rate distributions are very different, which may cause a very unfair scheduling result.

Proportional fair scheduling compares scores using the average. So it is also in favor of David because his score normalized by the average is higher than Tom's normalized score. Proportional fair scheduling takes into account the average value of the score, not the whole distribution. Compared with maximal-rate scheduling, it can yield a fairer scheduling result. Note that the average score in this example does not exactly correspond to the average throughput in (5.27), but we take it for this meaning for illustration.

To determine whose score was hardest to get, we have to take account of the score distribution in each school. David's score belongs in the top 20% in school A and Tom's score in the top 10% in school B. In this sense, Tom's score was hardest to get. CDF-based scheduling selects a user whose rate is high enough and it is hard to become higher. So, in parallel with CDF-based scheduling, Tom did better than David, even though his score was lower in number than David's.

Table 5.1 lists various wireless packet scheduling algorithms in terms of memoryless packet scheduling function $h_k(x)$. Proportional fairness scheduling is also included in the table for comparison, even though it is not memoryless scheduling. (Note that the throughput T_k relies on the past history of user selection.) For all cases, the function $h_k(x)$ is monotonically increasing and associated with some control parameters. In essence, designing a new packet scheduling means determining some function $h_k(x)$ which yields some desirable fairness criterion as listed in Table 5.1.

Table 5.1 Comparison of various scheduling in terms of scheduling function $h_k(x)$

Scheduling	Function
Maximal rate scheduling	$h_k(x) = x$
Proportional fairness scheduling	$h_k(x) = x/T_k$
Temporal fairness scheduling	$h_k(x) = x + v_k$
Utilitarian fairness scheduling	$h_k(x) = p_k x$
CDF-based scheduling	$h_k(x) = F_{R_k}(x)^{1/w_k}$

Table 5.2 SNR versus transmission rate in the CDMA 1xEV-DO system (Reproduced with permission from P. Bender *et al.*, "CDMA/HDR: A bandwidth-efficient high-speed wireless data service for nomadic users," *IEEE Communications Magazine*, **38**, July, 70–77, 2000, IEEE. © 2000 IEEE)

SNR (dB)	Rate (kbps)
−12.5	38.4
−9.5	76.8
−8.5	102.6
−6.5	153.6
−5.7	204.8
−4.0	307.2
−1.0	614.4
1.3	921.6
3.0	1228.8
7.2	1843.2
9.5	2457.6

In wireless communication channels, multiple paths induce signal level fluctuations, and the time variation of the signal level is characterized by the *Doppler frequency*. Assume that the channel is a block-fading channel, which remains constant in each slot but varies slot to slot. We consider the SNR threshold levels $\gamma_0, \gamma_1, \ldots, \gamma_M$, such that $0 = \gamma_0 < \gamma_1 < \ldots < \gamma_M = \infty$. Then the fading channel is said to be in state s_m, $m = 1, 2, \ldots, M$, if the received SNR is in the interval $[\gamma_{m-1}, \gamma_m)$. If the channel of a user is in state s_m, then this user feeds back the number m to the BS. If this user is selected for transmission, then the BS transmits a packet of size r_m to that user, where r_m is determined by a lookup table. Table 5.2 lists an example of the SNR thresholds and associated rates in the CDMA2000 1xEv-DO system (Bender *et al.*, 2000).

Now compare the scheduling performances among four different types of packet scheduling: round-robin scheduling, proportional fairness scheduling, temporal fairness scheduling, and CDF-based scheduling. Assume that each user channel is an independent Rayleigh fading channel, and the instantaneous rate of each user is determined from the instantaneous SNR value, according to Table 5.2. Consider nine users in one cell and the SNR of these users are $(-10, -10, -10, 0, 0, 0, 10, 10, 10)$ dB. Figure 5.11 depicts the scheduling results in terms of time-fraction and average throughput. Since round-robin scheduling serves each user at a time in a deterministic order, the time fraction each user gets is exactly 1/9. As shown in Figure 5.11(a), temporal fairness scheduling and CDF-based scheduling achieve a fair time-slot allocation. However, the proportional fairness selects high-SNR users more frequently than low-SNR users. According to Figure 5.11(b), all three opportunistic scheduling algorithms yield higher user throughput than round-robin scheduling, which originates from the multiuser diversity gain. The temporal fairness and CDF-based scheduling algorithms achieve very similar levels of throughput. As proportional fairness scheduling serves 10 dB SNR users frequently, it yields higher throughput for those users than do other scheduling algorithms.

Figure 5.12 depicts scheduling performances when the SNRs of the nine users change to $(-4, -3, -2, 1, 0, 1, 2, 3, 4)$ dB. For the time fraction, proportional fairness scheduling prefers

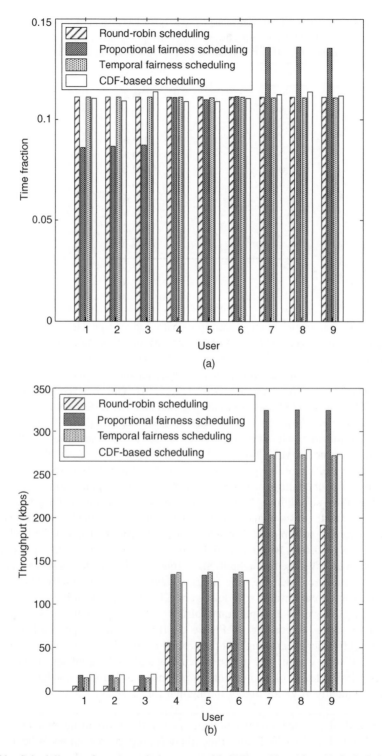

Figure 5.11 Scheduling performance of nine users with SNRs −10, −10, −10, 0, 0, 0, 10, 10, and 10 dB: (a) time fraction, and (b) throughput

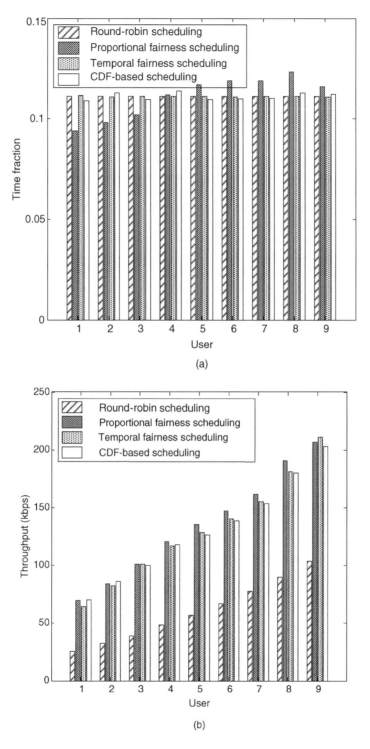

Figure 5.12 Scheduling performance of nine users with SNRs −4, −3, −2, 1, 0, 1, 2, 3, and 4 dB: (a) time fraction, and (b) throughput

high-SNR users, while the other three types allocate equal amounts of time slots to each user. According to Figure 5.12(b), three opportunistic scheduling algorithms obtain multiuser diversity gain over the round-robin scheduling and they achieve very similar user throughput performance.

5.4 Delay Performance of Wireless Schedulers

Multiuser diversity gain can be achieved by serving the user whose channel condition is the best among the users at each time slot. The basic assumption of multiuser diversity is that user queues are infinitely backlogged, meaning that there are always enough packets to send in each queue. Taking this into account, the wireless packet scheduling algorithms in Section 5.3 are designed for throughput maximization without considering any queueing dynamics. However, in real communication situations, packets arrive according to some probability distribution and are stored in buffers at the base station. So, if user selection is performed irrespective of user queues, there may be wastage of wireless resources. For example, even though a user is selected for transmission owing to its high channel gain, it is possible that there remain no packets to send in the queue of the selected user, or there remain fewer packets than can be transmitted through the allocated wireless channel. If the packet scheduler selects a user without considering the queue length, it may fail to keep queues bounded in some cases (Andrews, 2004). So, for optimal resource management, it is necessary to take into account queue status as well as user channel conditions.

As far as user queues are concerned, resource allocation problems become different from the opportunistic scheduling algorithm under the infinitely backlogged queue assumption. The main objective is to transmit all packets in the queues with bounded delay rather than to maximize the total throughput with a fair allocation among users. So, the optimal scheduler design focuses on keeping user queues stable, minimizing packet delays, and statistically guaranteeing packet delays of each user below a specified value, and so on.

This section considers a downlink system in which the base station (BS) selects one user for transmission at each time slot while maintaining the queues stable. Figure 5.6 depicts the block diagram of this wireless packet scheduling system. In addition to the system model in Section 5.3, packets come from the backbone networks and are stored in queues at the base station. The BS determines who will be served at the current time slot by considering both the channel condition and the queue length. We first describe the throughput optimality and then discuss optimal scheduling algorithms.

5.4.1 Throughput Optimality

Consider a queueing system with varying channel conditions. In the base station there are per-user queues for storing packets yet to be transmitted. Packets arrive according to some probability distribution and wait in queues until they are transmitted. Let $Q_k(n)$ denote the length of the queue of user k at time slot n. Then the queue length is updated as

$$Q_k(n+1) = \max(Q_k(n) + A_k(n) - S_k(n), 0), \qquad (5.55)$$

where $A_k(n)$ is the size of the arrival packets to user k at time slot n, and $S_k(n)$ denotes the amount of service; that is, the size of the departure packets of user k. Assume that the amount of arrival packets is not controllable at the BS and the average of the arrival process is given by $\mathbb{E}[A_k(n)] = A_k$. The BS can control user queues by determining the departure packet $S_k(n)$ of user k through a packet scheduling algorithm. The average of the departure process is given by $\mathbb{E}[S_k(n)] = S_k$.

Assume that there is a finite set $\{1, 2, \ldots, M\}$ of the channel states, which we denote by $\mathfrak{M}$. Let R_k^m denote the maximum available transmission rate that the base station can transmit to user k when the channel state is m. In other words, $R_k^m(n)$ is the maximum amount of data transmitted to user k when time slot n is fully assigned to user k. When the channel state is m, the amount of departure packets is given by $S_k(n) = R_k^m(n)1_{k=k^*(n)}$, where $k^*(n)$ is the index of the user selected for transmission at time slot n. The transmission rate vector $(R_1^m, R_2^m, \ldots, R_K^m)$ is associated with each channel state $m \in \mathfrak{M}$. The channel state process is assumed to be an irreducible discrete-time Markov chain with the finite state space $\mathfrak{M}$. The stationary distribution of this Markov chain is given by $(\pi_1, \pi_2, \ldots, \pi_M)$. The system state of the entire system is $X = (X(n), n = 0, 1, 2, \ldots)$, where

$$X(n) = \{m, (D_{k1}(n), D_{k2}(n), \ldots, D_{kQ_k(n)}(n)) \text{ for all } k\}. \tag{5.56}$$

$D_{kj}(n)$ denotes the current delay of the jth bit in the queue of user k. In each user queue, the bits are numbered in the order of their arrival. The scheduling rule for the system may be expressed as a mapping from a system state to a fixed probability distribution. For a scheduling rule H, user k is selected randomly according to the distribution $H(X(n))$. The vector of queue lengths, $Q(n) = (Q_1(n), Q_2(n), \ldots, Q_K(n))$, forms a K-dimensional Markov chain. The system is stable if

$$\lim_{q \to \infty} \lim_{n \to \infty} \Pr(Q_k(n) > q) = 0 \text{ for all } k. \tag{5.57}$$

The basic design principle of packet scheduling algorithms is to make the system stable by controlling the departure packets based on the information of the current queue status and the current channel condition.

Now consider a stochastic matrix $\phi = (\phi_{mk}, m \in M, k = 1, 2, \ldots, K)$ with $\phi_{mk} \geq 0$ for all m, k and $\sum_{k=1}^{K} \phi_{mk} = 1$ for all m. The *static service split* (SSS) scheduling algorithm is associated with the stochastic matrix ϕ (Andrews *et al.*, 2004). When the channel state is m, the SSS scheduling algorithm selects user k with probability ϕ_{mk}. This scheduling is considered as static because user selection depends only on the channel states, not on the queue states. The long-term average service rate of user k takes the form

$$T_k(\phi) = \sum_m \pi_m \phi_{mk} R_k^m, \tag{5.58}$$

and the average service rate vector is given by $T(\phi) = (T_1, T_2, \ldots, T_K)$.

Consider now a scheduling rule H under which the system is stable. Let $H_k(x)$ denote the probability of selecting user k when the system is in state x. Then, for any k and time slot n:

$$
\begin{aligned}
A_k &= \mathbb{E}[A_k(n)] \\
&\leq \mathbb{E}[S_k(n) + Q_k(n+1) - Q_k(n)] \\
&= \mathbb{E}[S_k(n)] = \sum_m \pi_m \mathbb{E}[S_k(n)|m(n) = m] \\
&= \sum_m \pi_m \sum_x \Pr(X(n) = x|m(n) = m) H_k(x) R_k^m \\
&= \sum_m \pi_m \phi_{mk} R_k^m = T_k(\phi),
\end{aligned}
\tag{5.59}
$$

where

$$
\phi_{mk} = \sum_x \Pr(X(n) = x|m(n) = m) H_k(x).
\tag{5.60}
$$

This implies that if the entire system is stable, then there exists a stochastic matrix ϕ such that

$$
A_k \leq T_k(\phi) \text{ for all } k.
\tag{5.61}
$$

In addition, if there exists a stochastic matrix ϕ such that

$$
A_k < T_k(\phi) \text{ for all } k,
\tag{5.62}
$$

then the system is stable (Andrews *et al.*, 2004). Define the stability region to be the set of all arrival rates $\mathbf{A} = (A_1, A_2, \ldots, A_K)$ for which there exists an SSS scheduling algorithm that satisfies (5.62).

The SSS scheduling algorithm associated with the stochastic matrix ϕ^* is said to be *maximal* if the service rate vector $T(\phi^*)$ is not dominated by the vector $T(\phi)$ for any other stochastic matrix ϕ. For a maximal SSS scheduling algorithm associated with a stochastic matrix ϕ^* and the service rate $T_k^* = T_k(\phi^*)$, there exists a set of positive constants $p_k, k = 1, 2, \ldots, K$, such that $\phi_{mj} > 0$ implies

$$
j \in \arg\max_k p_k R_k^m(n).
\tag{5.63}
$$

This can be shown by considering the following linear program (Andrews *et al.*, 2004):

$$
\begin{aligned}
\underset{\Lambda, \phi}{\text{maximize}} \quad & \Lambda \\
\text{subject to} \quad & \sum_{m=1}^{M} \pi_m R_k^m \phi_{mk} \geq \Lambda T_k^*, \quad k = 1, 2, \cdots, K, \\
& \sum_{k=1}^{K} \phi_{mk} = 1, \quad 0 \leq \phi_{mk} \leq 1, \quad \forall m, k.
\end{aligned}
\tag{5.64}
$$

From the definition of T_k^*, the solution to the problem in (5.64) is $\Lambda = 1$ and $\phi = \phi^*$. The Lagrangian of (5.64) with Lagrange multipliers p_k is given by

$$\begin{aligned}
\underset{\Lambda,\phi}{\text{maximize}} \quad & \Lambda + \sum_{k=1}^{K} p_k \left(\sum_{m=1}^{M} \pi_m R_k^m \phi_{mk} - \Lambda T_k^* \right) \\
\text{subject to} \quad & \sum_{k=1}^{K} \phi_{mk} = 1, 0 \leq \phi_{mk} \leq 1, \forall m, k,
\end{aligned}$$
(5.65)

and the objective can be rewritten as

$$\underset{\Lambda,\phi}{\text{maximize}} \quad \Lambda - \Lambda \sum_{k=1}^{K} p_k T_k^* + \sum_{m=1}^{M} \pi_m \sum_{k=1}^{K} p_k R_k^m \phi_{mk}.$$
(5.66)

To achieve the maximum at $\phi = \phi^*$, (5.63) should hold. So, a maximal SSS scheduling algorithm selects a user which maximizes $p_k R_k^m(n)$ at time slot n. The scheduling algorithm in (5.63) is nothing but the utilitarian fairness scheduling algorithm of Section 5.3.4. Note that the throughput vector obtained through the scheduling algorithm in (5.63) must be on the boundary of the achievable region. Consequently, the stability region is the achievable throughput region and the maximal SSS rule yields the service-rate vector on the boundary of the achievable throughput region. Conversely, the utilitarian fairness scheduling algorithm with parameters p_k can also be expressed in terms of the SSS scheduling algorithm by appropriately assigning the stochastic matrix ϕ. The parameters p_k may correspond to different maximal SSS scheduling algorithms due to various rules of tie breaking.

In summary, if the arrival rate vector $\mathbf{A} = (A_1, A_2, \ldots, A_K)$ is in the stability region, then there exists an SSS scheduling algorithm that stabilizes the user queues; or equivalently, there exists a utilitarian fairness scheduling algorithm that achieves the throughput vector $\mathbf{T} = (T_1, T_2, \ldots, T_K)$ such that $A_k \leq T_k$. The main problem of SSS scheduling and utilitarian fairness scheduling algorithms is to determine an appropriate stochastic matrix ϕ or a price vector p that stabilizes the user queues. As these parameters depend on *a priori* information such as the arrival rate vector and the user channel distribution, these scheduling algorithms are not universal. So, the desirable property of a scheduling algorithm is that it can keep user queues stable as long as the arrival rates are in the stability region and it does not depends on *a priori* information. The *throughput-optimal* scheduling algorithm is a scheduling algorithm that can make stable any system for which the condition in (5.62) holds. Any fixed maximal SSS scheduling cannot be throughput-optimal because it may fail to stabilize a system for some feasible arrival rate vector even though it can be stabilized under other SSS scheduling algorithms. The following sections deal with throughput-optimal scheduling algorithms.

5.4.2 Modified Largest-Weight-Delay-First (LWDF) Scheduling

As any fixed SSS scheduling algorithm is not throughput-optimal, it is necessary to adapt the stochastic matrix or the price vector according to the current states to make the system stable.

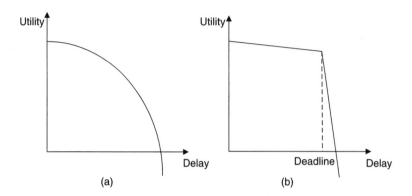

Figure 5.13 Examples of utility functions: (a) utility in smooth form, and (b) utility with delay deadline

For this reason, we consider the delays of HOL (head-of-line) packets or the current lengths of the user queues, $Q_k(n)$. The delay of a HOL packet is the current delay of the first packet present in the queue of user k; that is

$$W_k(n) = D_{k1}(n). \tag{5.67}$$

Consider first a utility-based scheduling algorithm that takes account of the channel conditions and the users' utility functions. Each user has a utility function $U_k(W_k)$ which is a decreasing function of delay W_k; that is, a longer delay lowers the utility (Liu *et al.*, 2003a; Liu *et al.*, 2003b). This utility represents the level of satisfaction of a user who experiences packet transmission delay W_k. Note that these utility functions are different from the throughput utility functions in Section 5.3.4. Figure 5.13 depicts two examples of utility functions (Liu *et al.*, 2003a; Liu *et al.*, 2003b). In general, the utility decreases as packets experience more delay, as shown in Figure 5.13(a). If there is a deadline for packets, the utility drops abruptly beyond the deadline, as shown in Figure. 5.13(b).

The utility-based scheduling algorithm aims to maximize total utility. For example, take a simple case where two users are waiting for services in queues. Let W_1 and W_2 respectively denote the current HOL packet delays of the two users, and let $U_1(W)$ and $U_2(W)$ be the associated utility functions. Then the utility-based scheduling algorithm selects a user whose rate multiplied by the utility derivative is the largest; that is (Liu *et al.*, 2003a; Liu *et al.*, 2003b)

$$k^*(n) = \arg\max_k |U_k'(W_k(n))| R_k(n). \tag{5.68}$$

This is similar to the scheduling algorithm based on the throughput utility in (5.38) except for the slight difference in the argument of the utility function. We can interpret the term $|U'_k(W_k(n))|$ as a reward per unit bit. So, utility-based scheduling selects a user who gets most reward at each time slot. When the utility function $U_k(W_k)$ is concave, the reward increases with the delay W_k, which reflects that a packet experiencing longer delay becomes more urgent. If $U_k(W) = -p_k W$ for $p_k > 0$, scheduling based on the delay utility in (5.68) becomes the same as scheduling based on the throughput utility in (5.38). In this case, the scheduler is not throughput-optimal because it does not rely on any queue-related information such as delay or queue length.

Consider next the utility function

$$U_k(W) = -\frac{\gamma_k}{\beta+1} W^{\beta+1} \quad \text{for } \gamma_k > 0 \text{ and } \beta > 0.$$

Then the scheduling algorithm becomes

$$k^*(n) = \arg\max_k \gamma_k (W_k(n))^\beta R_k(n). \tag{5.69}$$

This is *modified largest-weighted-delay-first* (M-LWDF) scheduling that selects a user whose rate multiplied by the HOL packet delay is the largest (Andrews *et al.*, 2004). The name M-LWDF comes from the LWDF scheduling in wired systems. Analogously, *modified largest-weighted-(unfinished)-work-first* (M-LWWF) scheduling is an algorithm that selects a user whose rate multiplied by the queue length is the largest; that is

$$k^*(n) = \arg\max_k \gamma_k (Q_k(n))^\beta R_k(n). \tag{5.70}$$

These two scheduling algorithms are known as *throughput-optimal scheduling rules* (Andrews *et al.*, 2004). Also, a more general form of the scheduling algorithm is throughput-optimal; that is

$$k^*(n) = \arg\max_k \gamma_k (V_k(n))^\beta R_k(n), \tag{5.71}$$

where $V_k(n) = \theta_k^{(W)} W_k(n) + \theta_k^{(Q)} Q_k(n)$. Here, $\theta_k^{(W)} \geq 0$ and $\theta_k^{(Q)} \geq 0$ are not equal to 0 at the same time. These parameters give more flexibility to control the delay and the queue length for QoS provision. For example, if user k is interested in tighter delay with low arrival rate, then the scheduler needs to pay more attention to the current delay by setting the weight $\theta_k^{(W)}$ larger than $\theta_k^{(Q)}$. Conversely, if the arrival rate of user k is very high and it is necessary to bound the queue length, the weight $\theta_k^{(Q)}$ should be set larger than $\theta_k^{(W)}$.

5.4.2.1 Throughput Optimality

To prove the stability of a system, it is sufficient to show a negative drift of a Lyapunov function for this system. Let $x = 0$ be an equilibrium point of the system described by

$$\dot{x} = f(x). \tag{5.72}$$

If there exists a continuously differentiable function $L(x)$ such that $L(x) \geq 0$ with equality if and only if $x = 0$ and $\dot{L}(x) \equiv \nabla L \cdot f(x) \leq 0$, then $L(x)$ is called a *Lyapunov function* and the system is stable in the sense of Lyapunov (Lyapunov, 1966). The existence of a Lyapunov function is sufficient to prove the stability of the system. For the M-LWDF and M-LWWF scheduling algorithms, a fluid limit technique has been used to prove throughput optimality (Andrews *et al.*, 2004). For analytical tractability, the packet flows are modeled as fluid and the queue length processes are scaled so that the limit processes are analyzed in the stability problem in the queueing networks. In this fluid limit, $W_k(n)$ and $Q_k(n)$ become proportional to each other, so the scheduling algorithms based on $W_k(n)$, $Q_k(n)$, or their linear combinations are indistinguishable.

The throughput optimality of the M-LWWF algorithm is shown in (5.70) with $\beta = 1$ without the support of a fluid limit technique. In a relaxed sufficient condition of stability, it suffices to show that there exists a compact region Σ and a constant $\alpha > 0$ such that (Kumar and Meyn, 1996)

$$
\begin{aligned}
&\mathbb{E}[L(X(n+1))|X(n)] < \infty, \quad \text{for all} \quad X(n), \\
&\mathbb{E}[L(X(n+1)) - L(X(n))|X(n)] < -\alpha \quad \text{for} \quad X(n) \notin \Sigma,
\end{aligned}
\tag{5.73}
$$

where $X(n)$ denotes the system state at time n. This implies that it suffices to show that there is a negative drift in the Lyapunov function only when the backlogs are sufficiently large. We assume that the system is in the stability region and consider the Lyapunov function

$$
L(\mathbf{Q}(n)) = \sum_{k=1}^{K} \gamma_k Q_k(n)^2
\tag{5.74}
$$

by setting $\mathbf{Q}(n)$ to the system state. The queue update equation in (5.55) satisfies the first condition in (5.73). For the second condition, we use the bound

$$
Q_k(n+1)^2 \leq Q_k(n)^2 + A_k(n)^2 + S_k(n)^2 - 2Q_k(n)(S_k(n) - A_k(n)).
\tag{5.75}
$$

From this bound, we have

$$
\begin{aligned}
\mathbb{E}[L(\mathbf{Q}(n+1)) - L(\mathbf{Q}(n))|\mathbf{Q}(n)] &\leq \sum_{k=1}^{K} \gamma_k \mathbb{E}[A_k(n)^2|\mathbf{Q}(n)] + \sum_{k=1}^{K} \gamma_k \mathbb{E}[S_k(n)^2|\mathbf{Q}(n)] \\
&\quad - 2\sum_{k=1}^{K} \gamma_k Q_k(n)\mathbb{E}[S_k(n) - A_k(n)|\mathbf{Q}(n)] \\
&\leq B - 2\sum_{k=1}^{K} \gamma_k Q_k(n)(\mathbb{E}[S_k(n)|\mathbf{Q}(n)] - A_k),
\end{aligned}
\tag{5.76}
$$

where

$$
B = \sum_{k=1}^{K} \gamma_k \mathbb{E}[A_k(n)^2|\mathbf{Q}(n)] + \sum_{k=1}^{K} \gamma_k R_{\max}^2
$$

and R_{max} is the largest transmission rate that is assumed to be finite. Since the system is in the stability region, there exists a stochastic matrix ϕ such that $A_k \leq T_k(\phi)$. If we assume, for a simple proof, a more strict condition $A_k < T_k(\phi)$ for some stochastic matrix ϕ, then we have (Ren *et al.*, 2004)

$$
\begin{aligned}
\sum_{k=1}^{K} \gamma_k Q_k(n) A_k &< \sum_{k=1}^{K} \gamma_k Q_k(n) T_k(\phi) \\
&= \sum_m \pi_m \sum_{k=1}^{K} \phi_{mk} \gamma_k Q_k(n) R_k^m(n) \\
&\leq \sum_m \pi_m \max_{k=1,\ldots,K} \{\gamma_k Q_k(n) R_k^m(n)\} \\
&= \sum_{k=1}^{K} \gamma_k Q_k(n) \mathbb{E}[S_k(n)|\mathbf{Q}(n)],
\end{aligned}
\tag{5.77}
$$

where the last equality holds due to (5.70) with $\beta = 1$. Since $A_k < T_k(\phi)$, $A_k + \varepsilon$ is also less than $T_k(\phi)$ for some small ε. Therefore, the inequality in (5.77) holds even if we replace A_k with $A_k + \varepsilon$, and we can obtain the following bound:

$$
\sum_{k=1}^{K} \gamma_k Q_k(n)(\mathbb{E}[S_k(n)|\mathbf{Q}(n)] - A_k) \geq \varepsilon \sum_{k=1}^{K} \gamma_k Q_k(n);
\tag{5.78}
$$

and we finally get

$$
\mathbb{E}[L(\mathbf{Q}(n+1)) - L(\mathbf{Q}(n))|\mathbf{Q}(n)] \leq B - 2\varepsilon \sum_{k=1}^{K} \gamma_k Q_k(n).
\tag{5.79}
$$

For any positive α, we can define a compact region

$$
\Sigma = \left\{ \mathbf{Q}(n) \,\middle|\, \sum_{k=1}^{K} Q_k(n) \leq \frac{B + \alpha}{2\varepsilon} \right\}.
\tag{5.80}
$$

If $\mathbf{Q}(n) \notin \Sigma$, then $\mathbb{E}[L(\mathbf{Q}(n+1)) - L(\mathbf{Q}(n))|\mathbf{Q}(n)] \leq -\alpha$, which satisfies the second condition in (5.73) (Ren *et al.*, 2004). This shows that the M-LWWF scheduling algorithm is throughput-optimal.

5.4.2.2 Quality of Service

It is an important issue of wireless networks to provide QoS. Since the wireless channel is time-varying in nature, it is hard to guarantee strict QoS for each user packet. For a loose QoS requirement, we may consider statistical guarantee for delay requirement, that is, most of packets experience a specified delay. More specifically, suppose that we have the statistical delay requirement

$$\Pr(W_k \geq d_k) < \varepsilon_k, \tag{5.81}$$

where d_k is the delay threshold and ε_k is the maximum probability of exceeding the delay threshold. Then we consider how we can meet this QoS requirement by using the M-LWDF scheduling, which takes the expression

$$k^*(n) = \arg \max_k \gamma_k W_k(n) R_k(n). \tag{5.82}$$

There is a set of parameters $\{\gamma_k\}$ which affect the packet delay distributions. If we take a higher value of γ_k, while keeping other γ_i fixed, then user k is selected more frequently and consequently the packet delay of this user is reduced at the expense of the increased delay of other users. In this way, we can control the delay distribution.

Even though the M-LWDF scheduling in (5.82) is throughput optimal, it does not guarantee that the QoS requirement is automatically satisfied as given in (5.81). So, an appropriate parameter setting is needed to satisfy the QoS. A rule of thumb for the QoS provision is

$$\gamma_k = \frac{-\log \varepsilon_k}{T_k d_k}, \tag{5.83}$$

where T_k is the average throughput of user k when all wireless channels are assigned to that user. Note that a small ε_k or a small d_k increases the parameter γ_k. For this parameter setting, the M-LWDF scheduling prefers the user whose delay requirement is tighter. In this way, the M-LWDF scheduling algorithm can balance the delay bound violation probability to meet the QoS requirement in (5.81).

5.4.3 Exponential Rule Scheduling

In order to keep user queues stable, it is necessary to take account of the user queue status as well as the current channel condition. The fixed SSS scheduling is not throughput optimal because it may not make system stable for some arrival traffic even though this traffic can be stabilized for some other SSS scheduling algorithms. In the previous section, we discussed that the M-LWWF and its variants are throughput optimal. These scheduling algorithms take the channel rate multiplied by the HOL packet delay (or queue size) as the scheduling metrics. One may wonder whether these are the only scheduling metrics to achieve throughput optimality. In fact, there are exponential rule based ones that are also throughput optimal, as introduced below.

The *exponential queue-length rule* (EXP-Q) scheduling chooses a user based on

$$k^*(n) = \arg\max_k \gamma_k R_k(n) \exp\left(\frac{a_k Q_k(n) - \overline{Q}(n)}{\beta + [\overline{Q}(n)]^\theta}\right),\tag{5.84}$$

where $\overline{Q}(n) = \frac{1}{K}\sum_{k=1}^{K} a_k Q_k(n)$. Similarly, the *exponential waiting-time rule* (EXP-W) scheduling chooses a user based on

$$k^*(n) = \arg\max_k \gamma_k R_k(n) \exp\left(\frac{a_k W_k(n) - \overline{W}(n)}{\beta + [\overline{W}(n)]^\theta}\right),\tag{5.85}$$

where $\overline{W}(n) = \frac{1}{K}\sum_{k=1}^{K} a_k W_k(n)$. Such EXP rule scheduling strives to equalize the weighted queue length $a_k Q_k(n)$ or the weighted delay $a_k W_k(n)$ within the difference in the order of $[\overline{Q}(n)]^\theta$ or $[\overline{W}(n)]^\theta$. In the case of the EXP-W scheduling, if some weighted delay is larger for a user than for the others by more than the order of $[\overline{W}(n)]^\theta$, then the exponent term becomes large and that user gets scheduling priority. In this way, the EXP-Q and EXP-W equalize the weighted queue length or the weighted delay. We can drop the terms $\overline{Q}(n)$ and $\overline{W}(n)$ in the exponents of (5.84) and (5.85) without changing the scheduling decision because they are common for all users. When there is a small weighted delay difference, the exponential term is not significant and the scheduling algorithm turns into a throughput-maximization scheduling. It is proved by using a fluid limit argument that the EXP (EXP-Q and EXP-W) scheduling, with any fixed positive parameter β, $\theta \in (0, 1)$, and γ_k and a_k, is throughput optimal (Shakkottai and Stolyar, 2000a, 2000b).

We compare the delay performance of various scheduling algorithms through an example. We consider an opportunistic transmission scheme with 5 users. We assume that the channels are independent Rayleigh fading and the SNR of users are (5, 6, 7, 8, 9) dB, respectively. For the packet arrival, we take a Bernoulli process with a mean rate of 350 kbps. Figure 5.14 depicts the delay distribution tails for the best user (i.e., the 9 dB user) and the worst user (i.e., the 5 dB user). We observe that the EXP rule and the M-LWDF scheduling provide a good delay performance, keeping the delays of the best and the worst users close. However, the PF scheduling does not provide a good delay distribution for the worst user because it schedules users independently of the queue status. So, if the mean arrival rate increases slightly, the PF may fail to stabilize the user queues while the EXP rule and the M-MWDF scheduling still stabilize the queues.

5.5 QoS in Wireless Scheduling and Admission Control

A best effort service does not provide any guarantees on packet delivery. It is analogous to the ordinary postal service which does not provide any guarantee on the delivery of a letter. However, if the sender pays extra for a certified mail, the carrier provides a prioritized treatment on the letter and sends back a proof of delivery. QoS is basically related to the prioritization of

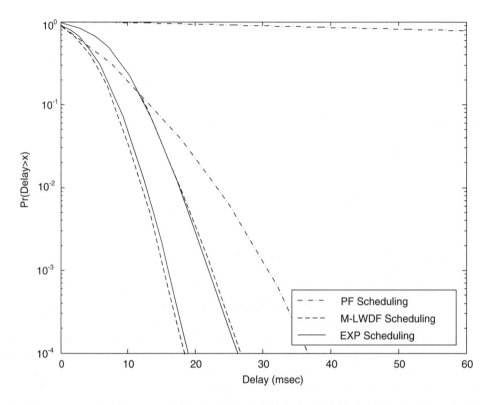

Figure 5.14   Delay performance of various scheduling algorithms

network traffic. QoS is especially important for real-time multimedia applications such as voice over IP, streaming video, and other multimedia services. The goal of QoS is to guarantee the ability of network to provide a predictable service including dedicated bandwidth, controlled latency and jitter, and improved error rate, and so on. QoS involves resource reservation control mechanisms to ensure the provision of the target service in a certain quality level.

Providing QoS in wireless networks is not an easy task, because the time-varying nature of wireless channel may hinder predicting performance and cause severe QoS violation. So, it may not be feasible to guarantee a deterministic delay bound on the maximum delay that all packet experience. It is more practical to require that 99% of the packets should not experience more than a specified delay and, in this case, a stochastic delay bound is more appropriate.

We consider a multiuser system which has per-user queues and serves a user in a time-slotted fashion as shown in Figure 5.6. We assume that the traffic of user k has an average arrival rate A_k and the queue size is infinite, so that no packet loss occurs due to buffer overflow. If a specified delay bound is d_k, the delay bound violation probability of user k is given by

$$\Pr(D_k(\infty) \geq d_k) \leq \varepsilon_k, \tag{5.86}$$

where $D_k(\infty)$ is the delay experienced by a packet of user k at the steady state. So, the user QoS may be specified by the triplet $\{A_k, d_k, \varepsilon_k\}$. We discuss a QoS provision with the aid of the CDF-based scheduling algorithm introduced in Section 5.3.5 (Park and Lee, 2006).

5.5.1 Effective Bandwidth and Effective Capacity

We consider the tail probability of a random variable in the form of (5.86). The large deviation theory is concerned with the rare events that happen with probability on the tail of a probability distribution (Lewis and Russell, 1997). We consider bounded the i.i.d. random variables $X_1, X_2, \ldots, X_n$ and their empirical mean M_n, interrelated by

$$M_n = \frac{1}{n}(X_1 + X_2 + \cdots + X_n). \tag{5.87}$$

If we take the Chernoff bound on the tail probability $\Pr(M_n \geq x)$, we get

$$
\begin{aligned}
\Pr(M_n \geq x) &= \Pr(nM_n \geq nx) \\
&= \mathbb{E}[1_{nM_n \geq nx}] \\
&\leq e^{-\theta nx} \mathbb{E}e^{\theta nM_n} \\
&= e^{-\theta nx} \mathbb{E}e^{\theta(X_1 + X_2 + \cdots + X_n)},
\end{aligned} \tag{5.88}
$$

where we use the inequality $1_{X \geq x} \leq e^{X-x}$. Since the inequality holds for all positive θ, there exists a θ that minimizes the upper bound, that is,

$$\Pr(M_n \geq x) \leq \min_{\theta > 0} e^{-n(\theta x - \Lambda(\theta))} = \exp(-n \max_{\theta > 0}\{\theta x - \Lambda(\theta)\}), \tag{5.89}$$

where

$$\Lambda(\theta) = \frac{1}{n} \ln\mathbb{E}\left[e^{\theta(X_1 + X_2 + \cdots + X_n)}\right]. \tag{5.90}$$

Consequently, the tail probability is bounded by

$$\Pr(M_n \geq x) \leq e^{-n\Lambda^*(x)}, \tag{5.91}$$

where $\Lambda^*(x)$ is the Legendre transform of $\Lambda(\theta)$,

$$\Lambda^*(x) = \max_{\theta > 0}\{\theta x - \Lambda(\theta)\}. \tag{5.92}$$

So, the tail probability decays exponentially with the decay rate $\Lambda^*(x)$. Since the Legendre transform is a point-wise maximum of the affine functions of x, $\Lambda^*(x)$ is a convex function without regard to the convexity of $\Lambda(\theta)$.

We consider the tail probability of the queue length. Let $A(s, t)$ and $S(s, t)$ be the amount of packet arrival and the amount of packet departure, in bits, over time interval $[s, t)$, respectively. If we set $Z(n) = A(n-1, n) - S(n-1, n)$, the queue length at time n is given by

$$
\begin{aligned}
Q(n) &= \max\{0, Z(n) + Q(n-1)\} = \max\{0, Z(n) + \max\{0, Z(n-1) + Q(n-2)\}\} \\
&= \max\{0, Z(n), Z(n-1) + Z(n) + Q(n-2)\}.
\end{aligned} \tag{5.93}
$$

By letting $V(m) = Z(n) + Z(n-1) + \cdots + Z(n-m+1)$ with $V(0) = 0$ and repeating the above recursion, we get

$$Q(n) = \max\{V(0), V(1), \ldots, V(n-1), V(n)\} \qquad (5.94)$$

by assuming an empty initial queue, that is, $Q(0) = 0$. $V(m)$ is called the *workload process* which refers to the accumulation of the difference between the amount of packet arrival and departure made during the last m time slots. If the steady state queue length $Q(\infty)$ exists, then we get from (5.94)

$$\Pr(Q(\infty) \geq x) = \lim_{n \to \infty} \max_n \Pr(V(n) \geq x). \qquad (5.95)$$

By assuming an i.i.d. workload process, we can apply the large deviation theory to the workload process as it is expressed as the accumulation of a random variable characterizing the difference between packet arrival and packet departure at each time. So, we have

$$\Pr(V(n) \geq x) \leq e^{-n\Lambda^*(x/n)} = e^{-x(\Lambda^*(c)/c)}, \qquad (5.96)$$

where $c = x/n$. $\Lambda^*(x)$ takes the expression in (5.92) and $\Lambda(\theta)$ is given by

$$\Lambda(\theta) = \lim_{n \to \infty} \frac{1}{n} \ln \mathbb{E}\left[e^{\theta \sum_{m=0}^{n} Z(m)}\right] = \lim_{n \to \infty} \frac{1}{n} \ln \mathbb{E}\left[e^{\theta(A(0,n) - S(0,n))}\right]. \qquad (5.97)$$

As the right-hand side of (5.96) is the optimized upper bound, we can use it as the approximation of the tail probability of the workload process. Then, the steady state queue length can be expressed as

$$\Pr(Q(\infty) \geq x) = \lim_{n \to \infty} \max_n \Pr(V(n) \geq x) \approx e^{-\theta^* x}, \qquad (5.98)$$

where

$$\theta^* = \min_c \frac{\Lambda^*(c)}{c}. \qquad (5.99)$$

The definition in (5.99) implies that (Chang and Thomas, 1995; Lewis and Russell, 1997)

$$\theta^* \leq \Lambda^*(c)/c \quad \text{for all } c \Leftrightarrow \max_c\{\theta^* c - \Lambda^*(c)\} = \Lambda(\theta^*) \leq 0, \qquad (5.100)$$

where the last equality holds for the involutional property of the Legendre transform.[4] Moreover, there exists a $c^* > 0$ such that $\theta^* = \Lambda^*(c^*)/c^*$ by (5.99) and for the c^* we get

$$\Lambda^*(\theta^*) = \max_c\{\theta^* c - \Lambda^*(c)\} \geq \theta^* c^* - \Lambda^*(c^*) = 0. \qquad (5.101)$$

---

[4] If a function $f(x)$ is convex, its Legendre transform is inverted by repeating it again; that is, $L(L(f(x))) = f(x)$ for the Legendre transform operator $L(\cdot)$. This is called the *involutional property*.

By connecting the two inequalities in (5.100) and (5.101), we get $\Lambda(\theta^*) = 0$. This implies that the tail probability is exponentially decaying with the rate θ^* which is given by the zero value of $\Lambda(\theta) = 0$.

We consider the case that the capacity of wireless link is constant, that is, the packet departure rate is constant to make $S(0, n) = \mu n$. Then, we can rewrite (5.97) as

$$
\begin{aligned}
\Lambda(\theta) &= \lim_{n \to \infty} \frac{1}{n} \ln \mathbb{E}\left[e^{\theta(A(0,n) - \mu n)}\right] \\
&= \lim_{n \to \infty} \frac{1}{n} \ln \mathbb{E}\left[e^{\theta A(0,n)}\right] - \theta \mu \\
&= \Lambda_A(\theta) - \theta \mu,
\end{aligned}
\tag{5.102}
$$

where

$$
\Lambda_A(\theta) = \lim_{n \to \infty} \frac{1}{n} \ln \mathbb{E}\left[e^{\theta A(0,n)}\right].
\tag{5.103}
$$

By (5.100)–(5.102), if θ^* is a solution of $\Lambda_A(\theta)/\theta = \mu$, then the tail probability follows the exponential decay with the decaying rate θ^* as in (5.98). Since μ is a capacity of wireless link, $\Lambda_A(\theta)/\theta$ may be interpreted as the effective arrival rate that the wireless link can support while keeping the queue length probability in (5.98) with the exponent $\theta^* = \theta$. So, we can define the *effective bandwidth* of the arrival process by

$$
a(\theta) = \frac{\Lambda_A(\theta)}{\theta}.
\tag{5.104}
$$

By using the concept of effective bandwidth, we can calculate the required minimum capacity of the wireless link to guarantee that the probability of the buffer overflow is less than ε. The required θ^* is given by $\theta^* = -\ln \varepsilon/B$ due to the approximation $\Pr(Q(\infty) \geq B) \approx e^{-\theta^* B} = \varepsilon$, where B is a queue length threshold. Then, the required minimum capacity of the wireless link is nothing but the effective bandwidth $a(\theta^*)$. The effective bandwidth is usually larger than the mean bandwidth of arrival. As a dual case, we also consider the case when the arrival process is constant bit rate traffic, that is, the packet arrival rate is constant to make $A(0, n) = \lambda n$. Then, we can similarly define the *effective capacity* of the wireless link by

$$
b(\theta) = -\frac{\Lambda_S(-\theta)}{\theta},
\tag{5.105}
$$

where

$$
\Lambda_S(-\theta) = \lim_{n \to \infty} \frac{1}{n} \ln \mathbb{E}\left[e^{-\theta S(0,n)}\right].
\tag{5.106}
$$

We consider how to guarantee the buffer overflow probability to be less than ε when neither the arrival process nor the wireless link capacity is constant. The required θ^* is given by $\theta^* = -\ln \varepsilon/B$. If the effective bandwidth $a(\theta^*)$ is less than $b(\theta^*)$, then the buffer overflow

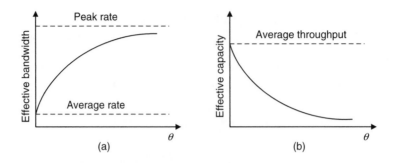

Figure 5.15 Typical curves for (a) effective bandwidth, and (b) effective capacity

probability becomes less than ε. So, the effective bandwidth and the effective capacity play key roles in providing QoS.

Figure 5.15 shows typical curves for the effective bandwidth and the effective capacity with respect to the QoS exponent θ (Wu and Negi, 2003). In the figures $a(0)$ is equal to the average data rate of the arrival traffic while $a(\infty)$ corresponds to the peak data rate. The effective bandwidth is a monotonically increasing function. In the case of effective capacity, $b(0)$ is equal to the average throughput of wireless link, while $b(\infty)$ is 0. The effective capacity is monotonically decreasing.

5.5.2 QoS Provision

We consider a multiuser system which has per-user queues and serves a user in a time-slotted fashion as in Figure 5.6. We assume that the traffic of user k has an average arrival rate A_k with a delay bound d_k, and the delay bound violation probability ε_k, as in (5.86). By applying the large deviation theory, the probability that the queue length of user k at the steady state, $Q_k(\infty)$, exceeds a queue length threshold B_k follows the relation

$$\Pr(Q_k(\infty) \geq B_k) \approx e^{-\theta_k B_k} \tag{5.107}$$

as $B_k \to \infty$. According to the Little's theorem (Little, 1961), the average number of packets in a stable system (i.e., average queue length) is equal to their average arrival rate multiplied by their average time in the system (i.e., average delay). If we apply the Little's lemma and the relation $B_k = A_k d_k$, we get

$$\Pr(D_k(\infty) \geq d_k) \approx e^{-\theta_k^D d_k}, \tag{5.108}$$

where $D_k(n)$ is the delay experienced by a packet of user k arriving at the steady state and $\theta_k^D = A_k \theta_k$ is the asymptotic decay rate of the waiting time tail probability. For a given QoS triplet $\{A_k, d_k, \varepsilon_k\}$, we want to check the feasibility of QoS by using the effective bandwidth and the effective capacity.

In the Bernoulli process with an arrival rate A_k, a packet with size A_k/p_a arrives with an arrival probability p_a, and no packet arrives with a probability $1 - p_a$. The *constant bit rate*

(CBR) traffic can be regarded as a special case of Bernoulli traffic with $p_a = 1$. The effective bandwidth of the Bernoulli traffic with the average arrival rate A_k and the arrival probability p_a is

$$a(\theta) = \frac{1}{\theta} \ln(p_a \exp(\theta A_k / p_a) + 1 - p_a). \tag{5.109}$$

As the traffic becomes burstier (i.e., the arrival probability p_a is lowered), the effective bandwidth increases. The effective bandwidth of the CBR traffic is just the arrival rate A_k.

The effective capacity is determined by the packet scheduling and the channel conditions. As long as the effective bandwidth $a(\theta_k)$ does not exceed the effective capacity $b(\theta_k)$, the queue length tail probability asymptotically follows the relation in (5.107). As $a(\theta)$ and $b(\theta)$ are monotonically increasing and decreasing functions, respectively, there is a unique solution θ_k to the equality $a(\theta_k) = b(\theta_k)$. This implies that the queue length is bounded with the decay rate θ_k if θ_k is the solution to $a(\theta_k) = b(\theta_k)$. We can determine the feasibility of the QoS triplet $\{A_k, d_k, \varepsilon_k\}$ by checking if the relation $\theta_k^D \geq -\ln\varepsilon_k / d_k$ is met.

Since the system operates in a time-division fashion, the base station transmits packets to only one user at each time slot. So, the packets of user k departed until time N are given by

$$S_k(0, N) = \sum_{n=1}^{N} R_k(n) 1_{k^*(n)=k}, \tag{5.110}$$

where $k^*(n)$ denotes the index of the user to whom the base station transmits packets at time slot n. If we employ the CDF-based scheduling, it is possible to obtain the closed form expression of the effective capacity. By using (5.53), we have

$$\begin{aligned}
&\Pr(R_k(n) 1_{k^*(n)=k} = r_m) \\
&= \Pr(k^*(n) = k | m_k(n) = m) \Pr(m_k(n) = m) \\
&= w_k \left(q_{k,m}^{1/w_k} - q_{k,m-1}^{1/w_k} \right),
\end{aligned} \tag{5.111}$$

for $m = 1, 2, \ldots, M$, and

$$\Pr(R_k(n) 1_{k^*(n)=k} = 0) = 1 - w_k. \tag{5.112}$$

So, the effective capacity of user k is

$$b(\theta) = -\frac{1}{\theta} \ln \left(1 - w_k + w_k \sum_{m=1}^{M} e^{-\theta r_m} \left(q_{k,m}^{1/w_k} - q_{k,m-1}^{1/w_k} \right) \right). \tag{5.113}$$

It is easy to verify that $b(0)$ is equal to the average throughput T_k in (5.54). This implies that if the arrival rate A_k is equal to the average throughput T_k, then θ becomes 0 and the delay bound violation probability in (5.108) becomes one for any delay bound d_k. So, the queue overflow occurs unless the arrival rate A_k is strictly less than the average throughput T_k. We also have $b(\infty) = 0$. When an infinite value of θ is required (i.e., zero delay violation probability is required), the only allowable rate is zero. This implies that a deterministic delay bound is meaningless in time-varying wireless channels. From these two extreme cases, we realize that if the arrival rate (more exactly, the effective bandwidth) lies between zero and the throughput T_k, then θ is positive and the queue becomes bounded.

Figure 5.16 shows the delay bound violation probability of the Bernoulli traffic with 300 kbps average rate and various arrival probabilities. The predicted delay bound violation probability is obtained through (5.108) by using the effective capacity and the effective bandwidth. From the figure, we can make the following observations: First, the simulated delay bound violation probability decays exponentially with the delay bound d_k. This confirms that the waiting time tail probability is exponentially decaying in (5.108). Second, as the arrival probability p_a decreases, the traffic gets burstier and has a larger effective bandwidth, which makes the delay bound violation probability larger. Third, the predicted delay bound violation probability is very close to the simulation value, which verifies the usefulness of the effective capacity analysis. Therefore, the delay bound violation probability is easily obtained analytically by using the effective capacity and we may use this effective capacity model to provided pre-specified user QoS in a simple manner.

The effective capacity in (5.113) is monotonically increasing with the control parameter w_k. So, we can adjust the control parameter w_k until the effective capacity exceeds the effective

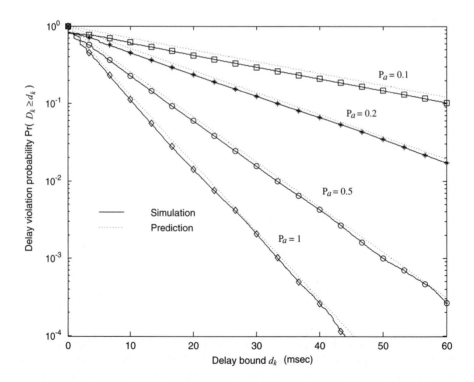

Figure 5.16 Example of delay bound violation probability (Reproduced with permission from D. Park and B.G. Lee, "QoS support by using CDF-based wireless packet scheduling in fading channels," *IEEE Transactions on Communications*, **54**, no. 11, 2051–2061, Nov. 2006. © 2006 IEEE)

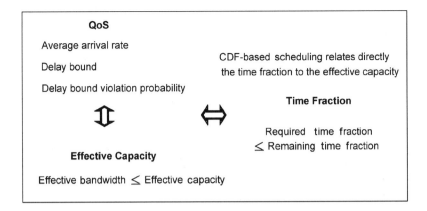

Figure 5.17 Relationships between QoS provisioning, effective capacity, and time fraction in CDF-based scheduling (Reproduced with permission from D. Park and B.G. Lee, "QoS support by using CDF-based wireless packet scheduling in fading channels," *IEEE Transactions on Communications*, **54**, no. 11, 2051–2061, Nov. 2006. © 2006 IEEE)

bandwidth in order to meet the QoS requirement. Figure 5.17 illustrates the relationship among the QoS provisioning, effective capacity, and time fraction in the CDF-based scheduling. As long as the effective capacity exceeds the effective bandwidth, the delay bound violation probability can be made smaller than the target probability. This is equivalent to the condition that the required time fraction is less than the remaining time fraction in the CDF-based

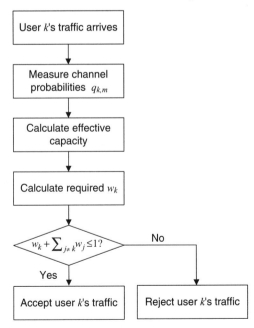

Figure 5.18 Flow chart for the resource allocation procedure with admission control (Reproduced with permission from D. Park and B.G. Lee, "QoS support by using CDF-based wireless packet scheduling in fading channels," *IEEE Transactions on Communications*, **54**, no. 11, 2051–2061, Nov. 2006. © 2006 IEEE)

scheduling algorithm. Consequently, the QoS can be guaranteed by adjusting the time fraction in the CDF-based scheduling.

In order to provide QoS guarantees in wireless networks, it is necessary to take account of admission control and packet scheduling jointly. When a new user requests some amount of wireless resources, the admission controller accepts the new request if the system can meet the request without degrading the QoS of the existing services, and rejects otherwise. So, it is very crucial to determine whether or not the packet scheduler could meet the demand while not degrading the QoS of the currently served users.

Figure 5.18 shows the flow chart of the resource allocation procedure, with the admission control conducted based on the CDF-based scheduling algorithm. Suppose that user k demands some QoS triplet $\{A_k, d_k, \varepsilon_k\}$. First, the scheduler estimates that user's channel probabilities $q_{k,m}$ as in (5.52). Then, the scheduler calculates the effective capacity based on those probabilities. Next, it calculates the smallest time fraction of user k, w_k, that makes the effective capacity exceed the effective bandwidth. Finally it checks if the sum of the time fractions of all users including user k is less than one and then admits user k if so and rejects otherwise.

References

Andrews, M. (2004) Instability of the proportional fair scheduling algorithm for HDR. *IEEE Transactions on Wireless Communications*, **3**, 1422–1426.

Andrews, M. *et al.* (2004) Scheduling in a queueing system with asynchronously varying service rates. *Probability in the Engineering and Informational Sciences*, **18**, 191–217.

Bender, P., Black, P., Grob, M., Padovani, R., Sindhushayana, N. and Viterbi, A. (2000) CDMA/HDR: A bandwidth-efficient high-speed wireless data service for nomadic users. *IEEE Communications Magazine*, **38**, 70–77.

Bhagwat, P., Bhattacharya, P., Krishna, A. and Tripathi, S. (1996) Enhancing throughput over wireless LAN's using channel state dependent packet scheduling. In: Proceedings IEEE Infocomm, San Francisco, CA, 1133–1140.

Borst, S. and Whiting, P. (2003) Dynamic channel-sensitive scheduling algorithms for wireless data throughput optimization. *IEEE Transactions on Vehicular Technology*, **52**, 569–586.

Caire, G., Müller, R.R. and Knopp, R. (2007) Hard fairness versus proportional fairness in wireless communications: the single-cell case. *IEEE Transactions on Information Theory*, **53**, 1366–1385.

Cao, Y. and Li, V.O.K. (2001) Scheduling algorithms in broad-band wireless networks. *Proceedings of the IEEE*, **89**, 76–87.

Chang, C.-S. and Thomas, J.A. (1995) Effective bandwidth in high-speed digital networks. *IEEE Journal on Selected Areas in Communications*, **13**, 1091–1100.

Goyal, P., Vin, H.M. and Chen, H. (1996) Start-time fair queueing: A scheduling algorithm for integrated services. In: Proceedings ACM Sigcomm, 157–168.

Jalali, A., Padovani, R. and Pankaj, R. (2000) Data throughput of CDMA-HDR a high efficiency-high data rate personal communication wireless system. In: Proceedings IEEE Vehicular Technology Conference, Tokyo, 1854–1858.

Kelly, F.P. (1997) Charging and rate control for elastic traffic. *European Transactions on Telecommunications*, **8**, 33–37.

Knopp, R. and Humblet, P.A. (1995) Information capacity and power control in single-cell multiuser communications. In: Proceedings IEEE International Conference on Communications, Seattle, WA, 331–335.

Kumar, P.R. and Meyn, S.P. (1996) Duality and linear programs for stability and performance analysis of queueing networks and scheduling policies. *IEEE Transactions on Automatic Control*, **41**, 4–17.

Lewis, J. and Russell, R. (1997) *An Introduction to Large Deviations for Teletraffic Engineers*, Dublin Institute for Advanced Studies.

Little, J.D.C. (1961) A proof of the queueing formula $L = \lambda W$. *Operations Research*, **9**, 383–387.

Liu, X., Chong, E.K.P. and Shroff, N.B. (2001) Opportunistic transmission scheduling with resource-sharing constraints in wireless networks. *IEEE Journal on Selected Areas in Communications*, **19**, 2053–2064.

Liu, X., Chong, E.K.P. and Shroff, N.B. (2003a) A framework for opportunistic scheduling in wireless networks. *Elsevier Computer Networks*, **41**, 451–474.

Liu, P., Berry, R. and Honig, M.L. (2003b) Delay-sensitive packet scheduling in wireless networks. In: Proceedings IEEE WCNC, New Orleans, LA, 1627–1632.

Lu, S., Bharghavan, V. and Srikant, R. (1999) Fair scheduling in wireless packet networks. *IEEE/ACM Transactions on Networking*, **7**, 473–489.

Lyapunov, A.M. (1966) *Stability of Motion*, Academic Press.

Ng, T.S.E., Stoica, I. and Zhang, H. (1998) Delay-sensitive packet scheduling in wireless networks. In: Proceedings IEEE Infocomm, San Francisco, CA, 1103–1111.

Ng, T.S.E., Stoica, I. and Zhang, H. (1998) Packet fair queueing algorithms for wireless networks with location-dependent errors, Carnegie Mellon University. Available at ftp.cs.cmu.edu/user/hzhang/INFOCOM98at.ps.Z.

Parekh, A.K. and Gallager, R.G. (1993) A generalized processor sharing approach to flow control in integrated services networks: the single-node case. *IEEE/ACM Transactions on Networking*, **1**, 344–357.

Park, D. and Lee, B.G. (2006) QoS support by using CDF-based wireless packet scheduling in fading channels. *IEEE Transactions on Communications*, **54**, 2051–2061.

Park, D., Seo, H., Kwon, H. and Lee, B.G. (2005) Wireless packet scheduling based on cumulative distribution function of user transmission rates. *IEEE Transactions on Communications*, **53**, 1919–1929.

Ren, T., La, R.J. and Tassiulas, L. (2004) Optimal transmission scheduling with base station antenna array in cellular networks. In: Proceedings IEEE Infocomm, Hong Kong, 1684–1693.

Shakkottai, S. and Stolyar, A.L. (2000a) Scheduling for multiple flows sharing a time-varying channel: exponential rule. Bell Laboratories technical report.

Shakkottai, S. and Stolyar, A.L. (2000b) Scheduling algorithms for a mixture of real-time and non-real-time data in HDR. Bell Laboratories technical memorandum.

Wu, D. and Negi, R. (2003) Effective capacity: A wireless link model for support of quality of service. *IEEE Transactions on Wireless Communications*, **2**, 630–643.

Further Reading

Andrews, M. *et al.* (2001) Providing quality of service over a shared wireless link. *IEEE Communications Magazine*, **39**, 150–154.

6

Transmission Power Management

Transmission power management refers to techniques for determining the transmission power level adequate to achieve system objectives in a given communication environment. As was discussed in Chapter 3, transmission power has several limiting factors. The power level is constrained by limited battery life, by the implementation cost of high-output power amplifiers, and by interference to neighboring transmitter–receiver pairs. So the power should be managed such that the achieved power efficiency can be maximized. Consequently, *power efficiency* (or energy efficiency) is the most important issue in this context. It is addressed in problem formulations in the form of maximizing the performance (e.g., total throughput) for a given transmission power (or energy), or minimizing power consumption while achieving the required performance goal.

One distinctive characteristic of transmission power, as a wireless resource, is that it is related to capacity in a nonlinear manner, as expressed by the capacity equation (3.1). This indicates that the contribution of a unit increment of transmission power to an increase of capacity may be different in different environments. The nonlinear relation is governed by the channel gain: the same transmission power yields a higher received power with a higher channel gain. However, due to the concavity of the relation, the increment of capacity is smaller than the increment of the received power level. As the channel condition may be different for different users and for different channels, at different times, the transmission power management problem should be formulated carefully to reflect this nonlinear characteristic. In addition, the problem becomes more complicated in a multiuser environment as an increase of transmission power for one user means an increase of interference to other users.

Such complications make it difficult to derive a unified solution that can maximize system efficiency while meeting all the performance requirements. As a result, each transmission power management scheme focuses on its own objectives according to the underlying system model. In the following, transmission power management issues are considered in three different aspects. First, the mutually interfering effect of transmission power in the multiuser environment is dealt with. Several management schemes are discussed that consider the harmonious coexistence of multiple transmitter–receiver pairs in non-orthogonal channels. Second, the chapter investigates the relation between transmission power and the available transmission rate, and discusses how to distribute a given amount of transmission power over

Wireless Communications Resource Management   Byeong Gi Lee, Daeyoung Park, and Hanbyul Seo
© 2009 John Wiley & Sons (Asia) Pte Ltd

multiple orthogonal subchannels. Finally, the chapter examines the time-varying nature of the communication channel environment caused by the randomness of traffic arrivals or the fading process, and discusses some power-efficient methods that adaptively determine the transmission power according to the time-varying environment.

6.1 Transmission Power Management for Interference Regulation

The first reason why transmission power needs to be managed properly is that the transmission of one user is likely to interfere with the transmissions of other users. If, for example, we consider a common channel shared by multiple users in a non-orthogonal way (e.g., CDMA with non-orthogonal spreading codes), an increase of transmission power of one user may contribute to strengthening the link quality of that particular user but will cause interference to all the other links, thereby deteriorating the overall performance. To guarantee a harmonious coexistence of multiple transmissions from multiple users in non-orthogonal channels, it is essential to incorporate a transmission power management mechanism that can regulate each transmission power such that each constituent user can get an acceptable level of link quality. The power management performed for such a purpose is called *power control*, to differentiate it from power management schemes for different purposes.

In CDMA systems, especially in uplink channels, power control is a key feature which dictates the system performance by regulating the interference level generated by each constituent user. Traditionally, the power-related problem in CDMA systems has been called the *near–far problem*, which expresses the fact that a user located near a base station (BS) causes serious interference to a user located far from the BS unless the transmission power of each user is properly controlled. This near–far problem can be solved by the BS controlling the transmission power of each individual user to a level that can maintain the link quality while limiting interference to a tolerable level. In this way, power control may be viewed as a mechanism for performance enhancement, as well as a means for QoS support or fairness provision. On the other hand, power control may also be considered in the context of orthogonal multiple access, such as FDMA and TDMA, or an issue closely related to *channel reuse*. If the transmission power of a link can be controlled to a proper level, the channel may be available for use by another link located far from that transmitter, provided the interference level is tolerable to each. So, a well-managed power control may offer a better chance to reuse a given bandwidth geographically, in addition to providing a solution to the near–far problem. This matter is covered in detail in Section 8.1 in relation to spectrum allocation and other issues. Here we concentrate on determining an adequate level of transmission power for each user. Various types of power-control schemes are discussed, after introducing the performance metric for power control.

Performance Metric for Power Control

To support later discussion of power-control schemes, we first consider the metric that represents link quality in the presence of interference. A typical metric is the *carrier-to-interference-and-noise ratio* (CINR), which is defined as the ratio of the received power of the desired signal to the power of noise-plus-interference from other transmitters. Consider a wireless system containing K users, which complies with the system model depicted in Figure 6.1. Denote by P_k and σ_k^2 the transmission power and noise variance, respectively, of

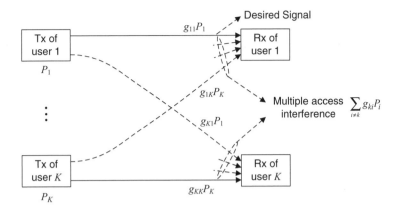

Figure 6.1 System model for power control

user k in the system.[1] The CINR of user k is given by

$$\gamma_k = \frac{g_{kk}P_k}{\sigma_k^2 + \sum\limits_{i=1,i\neq k}^{K} g_{ki}P_i},\tag{6.1}$$

where g_{ki} denotes the channel gain between the transmitter of user i and the receiver of user k.[2] The CINR in (6.1) may well describe the link quality in FDMA or TDMA systems, but it cannot be used in CDMA systems where each signal is spread by its own sequence. In this case, the *signal-to-interference-and-noise ratio* (SINR) is applicable, which is defined as the ratio of signal power to noise-plus-interference power after despreading. If a spreading code $\mathbf{c}_k$ with a spreading factor N is assigned to user k, the SINR is given by

$$\Gamma_k = \frac{g_{kk}P_k|\mathbf{c}_k|^4}{\sigma_k^2|\mathbf{c}_k|^2 + \sum\limits_{i=1,i\neq k}^{K} g_{ki}P_i|\mathbf{c}_i\cdot\mathbf{c}_k|^2}\tag{6.2}$$

As shown in (6.2), the SINR depends on the cross-correlation among the spreading codes. Thus, in order to determine the SINR of each user, it is necessary to specify the cross-correlation property of the employed spreading code. This complicates the analysis of user performance,

---

[1] In the literature of CDMA, the transmission power may be interpreted as the energy delivered on a *chip*. Then, the noise variance may be defined in the *chip level* and, in the case of white noise, it is equal to N_0W for noise power spectral density N_0 and bandwidth of the spread spectrum W.

[2] Note that the definition of the channel gain, g_{ki}, can encompass the multi-cell environment where different users may be connected to different base stations. The BS to which user k is connected becomes the receiver of user k in the uplink case and the transmitter of user k in the downlink case.

since the cross-correlation between two spreading codes is associated with the class of the spreading code, the indices of the considered codes, and the time shift between the codes. To get around this difficulty, many researchers on power control in CDMA systems adopt the following approximation.

First, let the spreading code $\mathbf{c}_k = [c_{1k}, c_{2k}, \ldots, c_{Nk}]$ be an independent equiprobable binary random sequence; that is, c_{nk} takes on 1 or -1 with probability 1/2, respectively. By dividing the denominator and the numerator by $|\mathbf{c}_k|^4 = N^2$, the SINR in (6.2) becomes

$$\Gamma_k = \frac{g_{kk}P_k}{\dfrac{\sigma_k^2}{N} + \displaystyle\sum_{i=1,i\neq k}^{K} g_{ki}P_i \frac{1}{N^2}|\mathbf{c}_i \cdot \mathbf{c}_k|^2}. \tag{6.3}$$

For those binary random spreading codes, it holds that

$$\frac{1}{N^2}|\mathbf{c}_i \cdot \mathbf{c}_k|^2 - \frac{1}{N} \to 0 \tag{6.4}$$

with probability 1 (i.e., almost surely) for $i \neq k$ as the spreading factor N increases to infinity (Zhang and Chong, 2000). This implies that, for almost every realization of the spreading codes, the cross-correlation of any two user codes goes to zero at the rate of $1/N$. Second, a lower bound of the peak cross-correlation, which is called the *Welch bound*, is developed by (Proakis, 1995)

$$\max|\mathbf{c}_i \cdot \mathbf{c}_k| \geq N\sqrt{\frac{M-1}{MN-1}} \approx \sqrt{N} \tag{6.5}$$

for M different binary spreading codes with the spreading factor N. The approximation holds for large values of N and M. In fact, optimal codes have been developed that achieve this lower bound and, for them, the cross-correlation term can be approximated by[3]

$$\frac{1}{N^2}|\mathbf{c}_i \cdot \mathbf{c}_k|^2 \approx \frac{1}{N} \tag{6.6}$$

for a large spreading factor. Applying this approximation to (6.3), we get the following SINR approximation, assuming a large spreading factor:

$$\Gamma_k \approx \frac{g_{kk}P_k}{\dfrac{\sigma_k^2}{N} + \displaystyle\sum_{i=1,i\neq k}^{K} \frac{1}{N}g_{ki}P_i} = N\gamma_k, \tag{6.7}$$

which is equal to the CINR scaled up by the spreading factor. This approximation is widely used in research on power control of CDMA systems, as it relates the SINR directly to the CINR without requiring any knowledge about the cross-correlation property of the applied spreading codes. This approximation is used in the following subsections.

---

[3] In reality, it is a slight overestimation because the Welch bound is established for the worst-case cross-correlation.

6.1.1 Power Control with Strict SINR Requirement

The most straightforward formulation of the power-control problem requires the SINR of each user to be above a certain level. Then a link is perceived to be satisfactory by the corresponding user if its SINR is higher than the requirement, but is regarded useless otherwise. This is called power control with a *strict SINR requirement*.

6.1.1.1 Distributed Power Control Algorithm

Since the SINR is proportional to the CINR, as described by (6.7), we may formulate the requirement on the CINR of each user by

$$\gamma_k \geq \gamma_k^{\text{req}} \tag{6.8}$$

for the CINR requirement γ_k^{req} of user k. We rearrange (6.1) and apply to (6.8) to get the matrix expression

$$(\mathbf{I}_K - \mathbf{A})\mathbf{P} \geq \mathbf{u}, \tag{6.9}$$

where $\mathbf{P} = [P_1, P_2, \ldots, P_K]^{\text{T}}$ is the column vector of transmission powers;

$$\mathbf{u} = \left[\frac{\gamma_1^{\text{req}} \sigma_1^2}{g_{11}}, \frac{\gamma_2^{\text{req}} \sigma_2^2}{g_{22}}, \ldots, \frac{\gamma_K^{\text{req}} \sigma_K^2}{g_{KK}} \right]^{\text{T}} \tag{6.10}$$

is the column vector of noise variances normalized by the ratio of the channel gain to the CINR requirement; matrix $\mathbf{A}$ is a K-by-K matrix with

$$A_{ij} = \begin{cases} 0, & \text{if } i = j, \\ \dfrac{\gamma_i^{\text{req}} g_{ij}}{g_{ii}}, & \text{if } i \neq j, \end{cases} \tag{6.11}$$

and $\mathbf{I}_K$ is the K-by-K identity matrix. In this formulation, the objective of power-control schemes is to determine a power vector $\mathbf{P}$ that meets the inequality in (6.9).

It is obvious that a power vector satisfying (6.9) exists only when the CINR requirement is satisfied with equality for every user. Then, the solution to the power-control problem is determined to be $\mathbf{P}^* = (\mathbf{I}_K - \mathbf{A})^{-1}\mathbf{u}$ as long as the matrix $(\mathbf{I}_K - \mathbf{A})^{-1}$ exists. Recalling that

$$(\mathbf{I}_K - \mathbf{A})(\mathbf{I}_K + \mathbf{A} + \mathbf{A}^2 + \cdots) = \mathbf{I}_K - \mathbf{A} + \mathbf{A} - \mathbf{A}^2 + \mathbf{A}^2 - \mathbf{A}^3 + \cdots = \mathbf{I}_K,$$

we rewrite the matrix inversion by

$$(\mathbf{I}_K - \mathbf{A})^{-1} = \sum_{i=0}^{\infty} \mathbf{A}^i. \tag{6.12}$$

Thus, the existence of $(\mathbf{I}_K - \mathbf{A})^{-1}$ is equivalent to the existence of $\sum_{i=0}^{\infty} \mathbf{A}^i$. It is well known that this infinite sum of matrices has a finite value if and only if $\mathbf{A}^i \rightarrow 0$ as $i \rightarrow \infty$, which implies

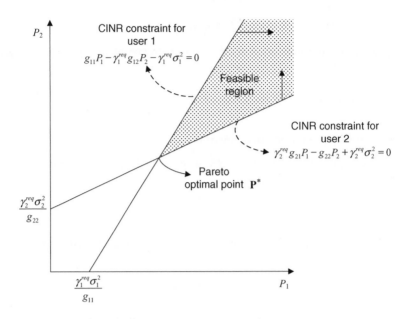

Figure 6.2 Feasible region and the Pareto-optimal point for the two-user case

that the matrix $(\mathbf{I}_K - \mathbf{A})^{-1}$ exists if and only if $|\rho_A| < 1$ for the maximum modulus eigenvalue of $\mathbf{A}$, ρ_A.[4] In summary, the following three statements are equivalent:

1. There exists a power vector satisfying $(\mathbf{I}_K - \mathbf{A})\mathbf{P} \geq \mathbf{u}$.
2. $|\rho_A| < 1$.
3. There exists the inverse matrix $(\mathbf{I}_K - \mathbf{A})^{-1} = \sum_{i=0}^{\infty} \mathbf{A}^i$.

Figure 6.2 illustrates the feasible region of the power-control problem in (6.9) for a two-user case. The two lines in the figure are the sets of the power vectors $[P_1, P_2]^T$ that satisfy the CINR requirement in (6.8) with equality. The line which starts from the horizontal axis is for user 1 and the other line starting from the vertical axis is for user 2. The shaded region indicates the feasible region, which is the set of the power vectors satisfying (6.9) with inequality. The most effective power vector in the feasible region is the *Pareto-optimal point* $\mathbf{P}^* = (\mathbf{I}_K - \mathbf{A})^{-1}\mathbf{u}$ where all the CINR requirements are met with equality. Any other feasible solution point consumes more transmission power than the Pareto-optimal point. The condition for the feasible region not to be empty is that the slope of the constraint line for user 1 is larger than that for user 2; that is

$$\frac{g_{11}}{\gamma_1^{req} g_{12}} > \frac{\gamma_2^{req} g_{21}}{g_{22}} \Leftrightarrow \frac{\gamma_1^{req} \gamma_2^{req} g_{12} g_{21}}{g_{11} g_{22}} < 1. \tag{6.13}$$

---

[4] The maximum modulus eigenvalue refers to the eigenvalue that is the largest in absolute value among all the eigenvalues of a matrix.

One can easily show that this is equivalent to the condition $|\rho_A| < 1$ as the eigenvalues are given by $\pm\sqrt{\gamma_1^{req}\gamma_2^{req}g_{12}g_{21}/g_{11}g_{22}}$ in the two-user case.

The above analysis leads to the Pareto-optimal power vector that satisfies each user's SINR requirement with minimal power consumption. However, it is too complicated to derive the optimal vector directly as it contains a matrix inversion process whose computation is hardly tractable as the number of users increases. Furthermore, it is necessary to provide all the channel information for the matrix inversion, which causes excessive signaling overheads.

As an alternative, consider an iterative control method to solve the problem in (6.9) (Foschini and Miljanic, 1993). We can update the transmission power of each user iteratively by

$$\mathbf{P}(n+1) = \mathbf{A}\mathbf{P}(n) + \mathbf{u}, \tag{6.14}$$

where $\mathbf{P}(n)$ denotes the power vector at the nth iteration. For a given initial power vector $\mathbf{P}(0)$, we can rewrite the recursive form as

$$\mathbf{P}(n) = \mathbf{A}^n\mathbf{P}(0) + \left[\sum_{i=0}^{n-1} \mathbf{A}^i\right]\mathbf{u}. \tag{6.15}$$

If the power-control problem (6.9) has a solution, the power-control vector converges to

$$\lim_{n\to\infty} \mathbf{P}(n) = \lim_{n\to\infty} \mathbf{A}^n\mathbf{P}(0) + \lim_{n\to\infty}\left[\sum_{i=0}^{n-1} \mathbf{A}^i\right]\mathbf{u} = 0 + \left[\sum_{i=0}^{\infty} \mathbf{A}^i\right]\mathbf{u} = (\mathbf{I}-\mathbf{A})^{-1}\mathbf{u} \equiv \mathbf{P}^*, \tag{6.16}$$

since $\lim_{n\to\infty} \mathbf{A}^n = 0$ for $|\rho_A| < 1$. This implies that, regardless of the initial power vector, the iterative power-control scheme in (6.14) converges to the Pareto-optimal solution as long as the problem is feasible.

One important observation is that the power control in (6.14) may be implemented in a fully distributed manner even though it yields the Pareto-optimal solution. In the iterative power-control scheme, user k updates its transmission power according to

$$P_k(n+1) = \frac{\gamma_k^{req}}{g_{kk}}\left(\sum_{i=1,i\neq k}^{K} g_{ki}P_i(n) + \sigma_k^2\right) = \gamma_k^{req}\left(\frac{\sum_{i=1,i\neq k}^{K} g_{ki}P_i(n) + \sigma_k^2}{g_{kk}P_k(n)}\right)P_k(n). \tag{6.17}$$

Note that the term in brackets on the right-hand side is the inverse of the CINR of user k at the nth iteration. Thus, by expressing the CINR by $\gamma_k(n)$, we get

$$P_k(n+1) = \frac{\gamma_k^{req}}{\gamma_k(n)}P_k(n), \tag{6.18}$$

for user k. It is noteworthy that the power control in (6.18) requires only local information. The current transmission power, $P_k(n)$, and the CINR, $\gamma_k(n)$, can be measured locally and made available to the transmitter of user k without any cooperation with the other users. Each link may increase its power independently when its current CINR is below its requirement and may decrease it otherwise. This distributed nature renders an easy and simple deployment, and thus the scheme is called the *distributed power-control* (DPC) algorithm. There have been reported a number of modifications to the DPC algorithm, including the algorithm integrated with base station assignment (Yates and Huang, 1995; Hanly, 1995), the macro-diversity based approach (Hanly, 1996), and a scheme considering the constrained power case (Grandhi and Zander, 1994).

6.1.1.2 Generalized Framework for Power Control

We next consider how to arrange a generalized framework for power control based on Yates (). For a given transmission power vector $\mathbf{P}$, the SINR requirement of each user in (6.8) may be described by a vector inequality

$$\mathbf{P} \geq \mathbf{F}(\mathbf{P}), \tag{6.19}$$

where $\mathbf{F}(\mathbf{P}) = [F_1(\mathbf{P}), F_2(\mathbf{P}), \ldots, F_K(\mathbf{P})]^T$ is the *interference function*, with each element, $F_k(\mathbf{P})$, denoting the *effective* interference from the other users which user k has to overcome using its own transmission power. In view of (6.1) and (6.8), one can easily derive the relation

$$F_k(\mathbf{P}) = \frac{\gamma_k^{\text{req}}}{g_{kk}} \left(\sigma_k^2 + \sum_{i=1, i \neq k}^{K} g_{ki} P_i \right), \tag{6.20}$$

which is equivalent to the noise-plus-interference power rescaled by the CINR requirement and the channel gain of the desired signal.

In this framework, an interference function $\mathbf{F}(\mathbf{P})$ is said to be *standard* if it satisfies the following three properties for all power vectors $\mathbf{P} \geq 0$:

- *Positivity:* $\mathbf{F}(\mathbf{P}) > 0$
- *Monotonicity:* If $\mathbf{P} \geq \mathbf{P}'$, then $\mathbf{F}(\mathbf{P}) \geq \mathbf{F}(\mathbf{P}')$
- *Scalability:* For an $\alpha > 1$, $\alpha \mathbf{F}(\mathbf{P}) > \mathbf{F}(\alpha \mathbf{P})$.

The vector inequality $\mathbf{x} > \mathbf{x}'$ is a strict inequality for all components; that is, $x_k > x'_k$ for all k. The positivity property implies non-zero background noise. The monotonicity property means that any increment in transmission power vector increases the interference to overcome. In other words, when the other users increase their transmission power, the transmission power required to overcome the resulting increased interference increases accordingly. The scalability property implies that if $P_k \geq F_k(\mathbf{P})$ then $\alpha P_k \geq \alpha F_k(\mathbf{P}) > F_k(\alpha \mathbf{P})$ for $\alpha > 1$. That is, if a user has an acceptable connection under a power vector $\mathbf{P}$, then the user will still have an acceptable connection even when all powers are scaled up by the same ratio. It is trivial to show that the interference function in (6.20) is standard.

Now consider a power-control scheme which updates each user's transmission power iteratively by

$$\mathbf{P}(n+1) = \mathbf{F}(\mathbf{P}(n)). \tag{6.21}$$

This power update rule implies that each user takes the level that can overcome the currently experiencing interference as its transmission power at the next iteration. The power control in (6.21) is called a *standard power control* if the interference function is standard.

Standard power-control algorithms have the following two desirable properties:

1. *Uniqueness of fixed point:* A standard power-control algorithm has a unique fixed point, that is, there exists only one power vector $\mathbf{P}$ at which $\mathbf{P} = \mathbf{F}(\mathbf{P})$.

   We can prove this by contradiction. Suppose $\mathbf{P}$ and $\mathbf{P}'$ are two distinct fixed points. From the positivity, we have $P_k > 0$ and $P'_k > 0$ for all k. Without loss of generality, we assume $\max_k P'_k / P_k = \alpha > 1$. Then we have $\alpha \mathbf{P} \geq \mathbf{P}'$ and $\alpha P_j = P'_j$ for $j = \arg \max_k P'_k / P_k$. By monotonicity and scalability, we have

   $$P'_j = F_j(\mathbf{P}') \leq F_j(\alpha \mathbf{P}) < \alpha F_j(\mathbf{P}) = \alpha P_j, \tag{6.22}$$

   which contradicts $\alpha P_j = P'_j$.

2. *Convergence to unique fixed point:* If the SINR requirement in (6.19) is feasible, then for any initial power vector $\mathbf{P}(0)$, the standard power control in (6.21) converges to a unique fixed point $\mathbf{P}^*$.

Since $P_k^* > 0$ for all k, we can find an $\alpha \geq 1$ such that $\alpha \mathbf{P}^* \geq \mathbf{P}(0)$. Then, by monotonicity, we have

$$\mathbf{F}^n(\mathbf{0}) \leq \mathbf{F}^n(\mathbf{P}(0)) \leq \mathbf{F}^n(\alpha \mathbf{P}^*), \tag{6.23}$$

where $\mathbf{F}^n(\mathbf{P})$ denotes the power vector obtained after n iterations of (6.21) starting from the power vector $\mathbf{P}(0)$. We prove the convergence by mathematical induction. When $n = 1$, we get $\mathbf{0} < \mathbf{F}(\mathbf{0}) \leq \mathbf{F}(\mathbf{P}^*) = \mathbf{P}^*$ by the positivity and monotonicity. Suppose it holds for $2, 3, \ldots, n-1$. Then we get

$$\mathbf{F}^{n-1}(\mathbf{0}) = \mathbf{F}(\mathbf{F}^{n-2}(\mathbf{0})) \leq \mathbf{F}(\mathbf{F}^{n-1}(\mathbf{0})) = \mathbf{F}^n(\mathbf{0}) \leq \mathbf{F}^n(\mathbf{P}^*) = \mathbf{P}^*. \tag{6.24}$$

This implies that the sequence $\mathbf{F}^n(\mathbf{0})$ is non-decreasing and bounded above by $\mathbf{P}^*$. Then, due to the uniqueness property, $\mathbf{F}^n(\mathbf{0})$ converges to $\mathbf{P}^*$. We can show the convergence of $\mathbf{F}^n(\alpha \mathbf{P})$ similarly. Since $\alpha \geq 1$, we have $\mathbf{F}(\alpha \mathbf{P}^*) \geq \mathbf{F}(\mathbf{P}^*) = \mathbf{P}^*$ and, from scalability, we get $\mathbf{F}(\alpha \mathbf{P}^*) \leq \alpha \mathbf{F}(\mathbf{P}^*) = \alpha \mathbf{P}^*$. Then, by monotonicity, we get

$$\mathbf{F}^{n-1}(\alpha \mathbf{P}^*) = \mathbf{F}(\mathbf{F}^{n-2}(\alpha \mathbf{P}^*)) \geq \mathbf{F}(\mathbf{F}^{n-1}(\alpha \mathbf{P}^*)) = \mathbf{F}^n(\alpha \mathbf{P}^*) \geq \mathbf{F}^n(\mathbf{P}^*) = \mathbf{P}^*. \tag{6.25}$$

This implies that the sequence $\mathbf{F}^n(\alpha\mathbf{P}^*)$ is non-increasing and bounded below by $\mathbf{P}^*$. Thus, $\mathbf{F}^n(\alpha\mathbf{P}^*)$ also converges to $\mathbf{P}^*$ by the uniqueness of the fixed point. Therefore, it is proved that the convergence of $\mathbf{F}^n(\mathbf{P}(0))$ is guaranteed regardless of the initial power vector.

The above two properties establish that a standard power-control algorithm converges to the unique fixed point from any initial power vector if the CINR requirement described in (6.19) is feasible. Obviously, the unique fixed point is the Pareto-optimal power vector in the sense that it minimizes each user's transmission power among all the power vectors satisfying the feasibility condition.

The standard property is maintained in the presence of maximum and minimum constraints. If an interference function $\mathbf{F}(\mathbf{P})$ is standard, then the interference functions

$$\hat{F}_k(\mathbf{P}) = \min\{P_k^{\max}, \quad F_k(\mathbf{P})\}, \tag{6.26}$$

$$\tilde{F}_k(\mathbf{P}) = \max\{P_k^{\min}, \quad F_k(\mathbf{P})\} \tag{6.27}$$

are standard too. In addition, if two interference functions $\mathbf{F}(\mathbf{P})$ and $\mathbf{F}'(\mathbf{P})$ are standard, the component-wise minimum and the maximum of them, $\mathbf{F}^{\min}(\mathbf{P})$ and $\mathbf{F}^{\max}(\mathbf{P})$, defined by

$$F_k^{\min}(\mathbf{P}) = \min\{F_k(\mathbf{P}), \quad F_k'(\mathbf{P})\} \quad \text{and} \quad F_k^{\max}(\mathbf{P}) = \max\{F_k(\mathbf{P}), \quad F_k'(\mathbf{P})\} \tag{6.28}$$

are standard too. This result implies that the DPC equipped with the capability of base station selection is standard as well. Consider the DPC algorithm in which each user is allowed to select the base station to which the user is connected at each iteration step. Then, it is optimal to select the base station at which the user experiences the least interference since the CINR requirement can be met with the smallest transmission power. In this case, the interference to overcome is given by the minimum among the interferences at the base stations to select. As this minimum interference function is also standard by the above property, the DPC algorithm in the presence of base station selection is also a standard power control whose convergence is guaranteed.[5]

It is noteworthy that being a standard interference function is a sufficient condition to the convergent power control. In other words, it guarantees the convergence of the power-control algorithm. However, this does not necessarily mean that a non-standard power control does not converge. For example, the generalized constrained power control in (Berggren *et al.*, 2001), which is proposed for energy efficient operation in the presence of a maximum power limit, the interference function is not standard but the convergence is proven in the paper.

6.1.1.3 Active Link Protection

The DPC algorithm in (Foschini and Miljanic, 1993) provides a very effective way of regulating *multiple-access interference* (MAI). It operates in a fully distributed manner and eventually converges to the Pareto-optimal power vector at which each user's SINR requirement is satisfied. However, this algorithm does not guarantee the SINR requirement during the

---

[5] Refer to (Yates, 1995), (Hanly, 1995), (Hanly, 1996), (Grandhi and Zander, 1994) for more standard extensions of DPC.

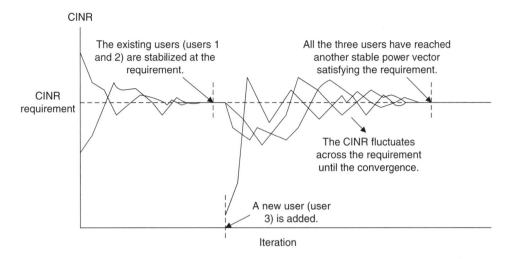

CINR

The existing users (users 1 and 2) are stabilized at the requirement.

All the three users have reached another stable power vector satisfying the requirement.

CINR requirement

The CINR fluctuates across the requirement until the convergence.

A new user (user 3) is added.

Iteration

Figure 6.3 CINR evolution without active link protection

convergence process; that is, the SINR of a user may fluctuate across the requirement boundary before every transmission power is stabilized. Thus, a newly incoming user may cause an instantaneous SINR drop to existing users, and the SINR may not be kept above the requirements until all the constituent users adjust power towards a new Pareto-optimal power vector.

Figure 6.3 illustrates the above process. The existing users 1 and 2 maintain their SINR requirements after the DPC procedure. An incoming user (3) initially sets its transmission power at a relatively low level so as not to induce sudden interference to existing users. Unfortunately, however, the SINR of the existing users are damaged by the power update of this new user. An important observation is that the new user may set its transmission power to a very high level at the first iteration of the power update in (6.18) because its SINR is very low when compared with its requirement at the initial state. This suddenly increases the interference level of existing users, and the power vector dynamically changes in proportion to the gap between the current SINR and the requirement. The CINR of each user fluctuates dynamically but eventually meets the CINR requirement in the steady state.

As a solution to the above problem, we consider the concept of *active link protection* (ALP) which preserves the SINR requirement during the power update procedure (Bambos *et al.*, 2000). The main idea is that the current users shoot for an enhanced target while newly incoming ones gradually increase their transmission power. The power-control algorithm with ALP is given by

$$
P_k(n+1) = \begin{cases} \dfrac{\delta \gamma_k^{\text{req}}}{\gamma_k(n)} P_k(n), & \text{if} \quad \gamma_k(n) \geq \gamma_k^{\text{req}}, \\[2ex] \delta P_k(n), & \text{if} \quad \gamma_k(n) < \gamma_k^{\text{req}}, \end{cases} \tag{6.29}
$$

for a control parameter δ which is slightly larger than 1. This algorithm employs a separate power update according to the current SINR level. The *active users* whose SINR requirement is currently met update their power by the original DPC algorithm in (6.18) with an enhanced target $\delta\gamma_k^{req}$. This provides a protection margin to active users and allows absorbing the degrading effect of the new users powering up in the channel. The *inactive users* having an SINR below the requirement power up gradually, inducing a limited degradation on the active users per step and allowing them enough time to react. From a different viewpoint, this algorithm may be rewritten as

$$P_k(n+1) = \min\left\{\frac{\delta\gamma_k^{req}}{\gamma_k(n)}P_k(n), \quad \delta P_k(n)\right\}. \tag{6.30}$$

This implies that the overshoot of the power update algorithm is bounded by the parameter δ.

One important property of (6.29) is the SINR protection of active users. By (6.30), the MAI after an iteration step is not larger than δ times the current level. Since each active user shoots for an SINR target with the enhancement of δ, its SINR requirement is still met even with an increased MAI during the power update procedure. Suppose that a user k is active in the iteration n; that is, $g_{kk}P_k(n)/I_k(n) \geq \gamma_k^{req}$. Then, since the increment of the overall MAI is limited by $I_k(n + 1) \leq \delta I_k(n)$, we get

$$\frac{g_{kk}P_k(n+1)}{I_k(n+1)} \geq \frac{g_{kk}P_k(n+1)}{\delta I_k(n)} = \frac{g_{kk}}{\delta I_k(n)}\frac{\delta\gamma_k^{req}}{\gamma_k(n)}P_k(n) = \gamma_k^{req}. \tag{6.31}$$

This certifies that a user who was active at iteration n remains active at iteration $n + 1$; that is, $\gamma_k(n) \geq \gamma_k^{req} \Rightarrow \gamma_k(n+1) \geq \gamma_k^{req}$. This means that a link remains active throughput the power update procedure if it is initially active or once it becomes active in the middle. Therefore, in the DPC with ALP, the set of active users is non-decreasing during the whole process. In addition, the SINR of inactive users is non-decreasing as the power update proceeds; that is, $\gamma_k(n) \leq \gamma_k$ $(n + 1)$ for all users satisfying $\gamma_k(n) < \gamma_k^{req}$. This is because any inactive user increases its transmission power by δ, which is the maximum possible MAI increment. This implies that the SINR of a currently inactive user eventually rises above the required level, in which case the user becomes active and remains so throughout the service time.

By virtue of the above properties, the DPC with ALP algorithm allows smooth accommodation of newly incoming users as illustrated in Figure 6.4. As all the existing users achieve an enhanced target, their SINRs decrease but may not drop below the requirement by the interference induced by the initial power of the newly incoming user. Obviously, this result can be obtained by a proper control of the initial power and the system parameter δ, as discussed in (Bambos et al., 2000). In addition, due to the active-link protection property, the CINR of an active user may fluctuate but does not fall below the requirement even during the converging iterations.

The power update rule in (6.29) and (6.30) is not standard in the framework of (Yates, 1995) as the positivity condition is not met for the zero vector. However, with a slight modification

$$F_k^{ALP}(\mathbf{P}) = \min\{\delta F_k(\mathbf{P}+\varepsilon), \quad \delta P_k + (\delta-1)\varepsilon_k\} \tag{6.32}$$

for a positive constant vector $\boldsymbol{\varepsilon} = [\varepsilon_1, \ldots, \varepsilon_K]^T$ (> 0), the DPC with ALP becomes a standard power-control algorithm as long as the interference function $\mathbf{F}(\mathbf{P})$ is standard. In (6.32), the constant vector ε can be assumed to be very small as it has no practical significance but to avoid

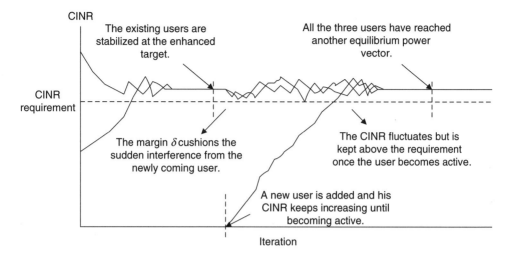

Figure 6.4   CINR evolution with active link protection

the zero-vector equilibrium.[6] Then, the standard power-control algorithm $\mathbf{P}(n + 1) = \mathbf{F}^{ALP}$ $(\mathbf{P}(n))$ also has the active link protection property (Yates, 1995), which can be applied to various standard power-control algorithms specified in (Yates and Huang, 1995) and (Grandhi and Zander, 1994).

6.1.2 Utility-based Power Control

The strict SINR requirement is a straightforward but effective formulation in supporting delay-sensitive services like voice traffic. Once a frame of delay-sensitive traffic is transmitted, there is no way to recover any erroneous transmission owing to the hard delay requirement. Since the utility function for a hard real-time application is a step function (see Section 3.1.2), one cannot utilize another opportunity in time to enhance the reliability of the service. In this case, the reliability of the service is determined solely by the SINR as long as all the other resources like bandwidth or antenna are readily allocated to each user. As a consequence, it is quite natural to formulate the strict SNIR requirement for delay-sensitive traffics.

In contrast, the SINR requirement may not be strictly satisfied for delay-tolerable services like best-effort data traffic, even though such traffic requires a much higher level of reliability. This is because the delay-tolerable nature allows transmission error recovery at the cost of bandwidth reduction. One typical method is to employ an ARQ procedure. The reliability of services can be improved by retransmitting an erroneous data frame; that is, by sacrificing the bandwidth allocated to newly incoming data frames. As the utility function for data traffic gradually increases along with the transmission rate, bandwidth reduction does not degrade user satisfaction as much as it does in the case of the delay-sensitive traffic.

---

[6] In the original DPC/ALP in (6.16) or (6.17), the constant vector ε is set to the zero vector. In this case, the zero vector is an equilibrium because the DPC/ALP sets the transmission power at the next iteration step to zero if the current one is zero. The convergence of the original DPC/ALP is proved in (Leung *et al.*, 2004).

From the above observation, we can claim that it is not desirable to apply power-control algorithms developed for supporting strict SINR requirements to delay-tolerable services. When the SINR requirement is not feasible, the algorithms with strict requirements will have no choice but to exclude several users from the channel. Such an operation may provide good connections to the users in service but may be inefficient in terms of network-wise performance, such as total throughput or aggregated user satisfaction. In this case, it may yield better performance if the requirement level is lowered and all users access the channel with a lowered SINR level. This arrangement may degrade the performance of some users but will improve the network-wise performance since more users will be admitted for transmission. In this sense, the adaptation of the SINR target may be interpreted as a *softened SINR requirement*. Such a softened SINR requirement can be derived from the utility function of each user which defines the effect of SINR enhancement and the degradation of user satisfaction. Therefore the power-control algorithm with softened SINR requirement may be called *utility-based power control*.

6.1.2.1 Softened SINR Requirement via Utility Maximization

Utility-based power control may be formulated in terms of microeconomics and game theory. Each user has its own utility function which quantifies the degree of satisfaction from service quality and the communication environment. As the utility is determined not only by the user's transmission power but also by interference from other users, the utility function of a user k is generally denoted by $U_k(P_k, \mathbf{P}_{-k})$, where $\mathbf{P}_{-k} = [P_1, \ldots, P_{k-1}, P_{k+1}, \ldots, P_K]^T$ is the power vector of all users except for user k.

In order to ensure a distributed operation of utility-based power control, we assume that each user tries to maximize its utility regardless of what happens to the other users. This type of formulation is easy to implement as no information is required about the other users in determining each user's transmission power. However, this operation may cause the undesirable result that every user sets transmission power as high as possible. This happens because, at any given interference level, each user would expect more satisfaction with a higher transmission power. As a consequence, all the users keep increasing their transmission power to ever higher levels, thereby inducing excessive interference.

The above problem can be resolved in a distributed manner by introducing the concept of *pricing* for usage of transmission power. A pricing mechanism is introduced in such a way that a cost is charged to each user against inducing interference to the other users. Each user will then increase its transmission power only when the improvement of satisfaction is larger than the cost that would be charged. The pricing function, $C_k(P_k)$, may be a function of environmental factors, such as user location and the number of admitted users. With the pricing mechanism, each user tries to maximize the net utility, which is the difference between the utility and the cost, by

$$\underset{P_k}{\text{maximize}} \quad \{U_k(P_k, \mathbf{P}_{-k}) - C_k(P_k)\}. \tag{6.33}$$

Now each user can increase its transmission power to a proper level as long as the utility and the cost function are designed such that the increment of the utility is relatively marginal when compared with that of the cost function.

Utility-based power-control algorithms in the form of (6.33) update each user's transmission power at iteration $(n + 1)$ by maximizing the net utility for the current interference $\mathbf{P}_{-k}(n)$. This rule may be expressed by

$$P_k(n+1, \mathbf{P}_{-k}(n)) = \arg \max_{P_k}\{U_k(P_k, \mathbf{P}_{-k}(n)) - C_k(P_k)\}, \tag{6.34}$$

where the notation $P_k(n + 1, \mathbf{P}_{-k}(n))$ emphasizes that the transmission power at the next iteration depends on the current interference. In the case of the DPC algorithm in (6.18), the target CINR is expressed by

$$\gamma_k^{\text{req}} = \frac{g_{kk}}{I_k(n)} P_k(n+1, \mathbf{P}_{-k}(n)), \tag{6.35}$$

where $I_k(n)$ denotes the interference and noise power at iteration n. The relation in (6.35) implies that the target CINR becomes a time-varying value depending on $\mathbf{P}_{-k}(n)$ unless the transmission power $P_k(n + 1, \mathbf{P}_{-k}(n))$, determined by (6.34), is proportional to the current interference level.[7] As the marginal increase of the utility and the cost function are different in general, the determined transmission power is not proportional to the current interference level for many utility functions considered in this book. Therefore, a well-designed utility-based power-control algorithm has the *target CINR adaptation capability* which softens the SINR requirement as discussed above.

We can consider the target CINR adaptation capability by examining the following case. As a typical formulation of the net utility maximization, consider a utility determined by the CINR and a cost proportional to the transmission power. In this case, the net utility maximization in (6.33) is formulated by

$$\underset{P_k}{\text{maximize}}\{U_k(\gamma_k) - c_k P_k\}, \tag{6.36}$$

where c_k is the price coefficient (the cost for a unit of transmission power). The term $U_k(\gamma_k)$, which is an alternative form of $U_k(P_k, \mathbf{P}_{-k})$, implies that the utility is determined solely by the CINR γ_k. By differentiating (6.36) with respect to P_k, we get

$$\frac{d(U_k(\gamma_k) - c_k P_k)}{dP_k} = U_k'(\gamma_k)\frac{d\gamma_k}{dP_k} - c_k = U_k'\left(\frac{g_{kk}P_k}{I_k}\right)\frac{g_{kk}}{I_k} - c_k, \tag{6.37}$$

where $U_k'(\gamma_k)$ is the derivative of $U_k(\gamma_k)$, and $I_k = \sum_{i=1,i\neq k}^{K} g_{ki}P_i + \sigma_k^2$ denotes the interference-plus-noise power. The last equality in (6.37) comes from the linear relation between CINR and transmission power in (6.1). The solution of the utility-maximization problem (6.36), P_k^*, makes the derivative in (6.37) zero. So we get

$$U_k'\left(\frac{g_{kk}P_k^*}{I_k}\right) = U_k'(\gamma_k^*) = \frac{c_k I_k}{g_{kk}}. \tag{6.38}$$

---

[7] For example, this can happen with $U_k = -(g_{kk}P_k/I_k - \alpha)^2$ and $C_k(P_k) = 0$ for a positive constant α.

This implies that the solution of (6.36) achieves the CINR value for which the derivative of the utility equals the price coefficient rescaled by the interference power and the channel gain. Then, by (6.38), the optimal power holds the relation

$$P_k^* = \gamma_k^* \frac{I_k}{g_{kk}},\tag{6.39}$$

where γ_k^* is the CINR value which meets the relation (6.38).

Now, based on (6.38), we can write the power update rule of the utility maximization problem in the form

$$P_k(n+1) = \gamma_k^*(n) \frac{I_k(n)}{g_{kk}} = \gamma_k^*(n) \frac{I_k(n)}{g_{kk} P_k(n)} P_k(n) = \frac{\gamma_k^*(n)}{\gamma_k(n)} P_k(n).\tag{6.40}$$

As the target CINR $\gamma_k^*(n)$ is determined by (6.38), it is a function of the interference power normalized by the channel gain which varies in time. Comparing this with the original DPC algorithm in (6.18), we can observe that utility-based power control is equivalent to the original DPC algorithm for the time-varying CINR requirement $\gamma_k^*(n)$ (Xiao et al., 2003). This observation exactly coincides with the discussion about the softened SINR requirement in the sense that utility-based power control adjusts the target CINR according to the communication environment, such as the channel gain g_{kk} and the interference level $I_k(n)$, by (6.38).

The specific property of the power-control algorithm is determined by the shape of the utility function and the pricing mechanism. In the following, we discuss those properties with respect to several specific utility-based power-control algorithms.

6.1.2.2 Sigmoid Utility Function

Utility-based power control can also be formulated assuming a sigmoid utility function, as illustrated in Figure 6.5 (Xiao et al., 2003). The sigmoid utility function is a monotonically increasing function of SINR, which is convex in the low-SINR region and concave in the high-SINR region. If we consider a linear cost mechanism, the net utility maximization problem takes the form in (6.36).

In this case, the net utility-maximization problem may be graphically interpreted as in Figure 6.6. The figure overlays the sigmoid utility function curve and two cost lines. As the cost is written by $c_k P_k = c_k (I_k/g_{kk}) \gamma_k$ by (6.39), the slope of the cost line is $c_k(I_k/g_{kk})$. Then, the net utility of a specific transmission power is equal to the difference between the utility function and the cost line at the CINR value that the transmission power achieves. In the figure, the optimality condition in (6.38) implies that net utility is maximized at the CINR at which the slope of the line tangential to the utility function, $U_k'(\gamma_k^*)$, is equal to the slope of the cost line, $c_k I_k/g_{kk}$. There may be two points of CINR that meet the condition in (6.38), but the figure shows the one located in the concave region, since the point in the convex region always renders a negative net utility. The optimal CINR at which the net utility is maximized decreases as the slope of the cost line increases. Figure 6.7 illustrates this. This implies that the power update rule in (6.40) reduces the target CINR level $\gamma_k^*(n)$ when the communication environment becomes hostile; that is, when the price c_k and the interference level I_k are high or when the channel gain g_{kk} is low. When the slope of the cost line becomes steeper than that of the dotted line in Figure 6.6, it is impossible to

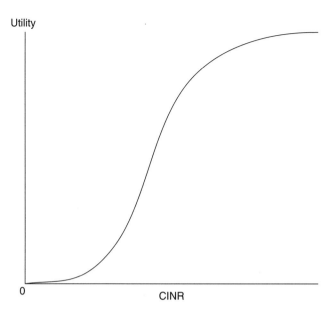

Figure 6.5 Sigmoid utility function

attain a positive net utility. Then, the optimal solution to (6.36) is to use zero transmission power by turning off the transmission, which yields zero net utility. In this sense, we may claim that utility-based power control using a sigmoid utility function has the capability of target CINR adaptation according to the communication environment.

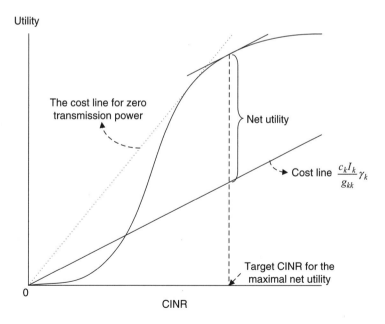

Figure 6.6 Graphical interpretation of net utility maximization for the sigmoid utility function

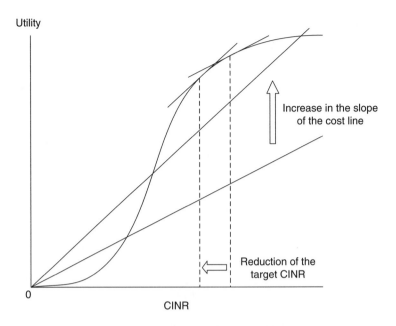

Figure 6.7   Relation between the change in the cost line and the target CINR

It is easy to show that sigmoid utility-based power control is a standard power control if the function $B_k(x) = (x/c_k)b_k^{-1}(x)$ for $b_k(x) = U_k'(x)$ is an increasing function in the concave part of the utility function (Xiao *et al.*, 2003). By (6.38), we may rewrite the optimal power in (6.39) by

$$P_k^* = \gamma_k^* \frac{I_k}{g_{kk}} = b_k^{-1}\left(\frac{c_k I_k}{g_{kk}}\right)\frac{I_k}{g_{kk}} = B_k\left(\frac{c_k I_k}{g_{kk}}\right) \equiv F_k(\mathbf{P}). \tag{6.41}$$

Note that the last equality is obtained by letting the optimal transmission power be an interference function of a standard power control. Thus the condition that $B_k(x)$ is an increasing function may be interpreted as follows. Suppose that the interference level I_k increases while c_k and g_{kk} are fixed. Then, the target CINR decreases by the softening effect as discussed above, and as a result the increment of the transmission power should be less than that of the interference. An increasing $B_k(x)$ means that the optimal transmission power P_k^* also increases to maintain the monotonicity in spite of the decrement of the target CINR. To help understand this, consider the following example. Suppose that the optimal transmission power is 1 for an interference level of 1 to yield a target CINR of 1. If the interference level is doubled, the optimal power also increases (e.g., to 1.6) by the increasing property of $B_k(x)$, but the target CINR should decrease (e.g., to 0.8). However, if $B_k(x)$ were not an increasing function, the target CINR might decrease too much (e.g., to 0.4) and the transmission power would decrease to 0.8, which is less than that at the previous iteration step – thereby violating the monotonicity property of the standard power control.

Obviously, the condition of an increasing $B_k(x)$ is to satisfy the monotonicity condition of the standard interference function. By this condition, the power update in (6.41) implies that an increase in the other users' power leads to an increase of the transmission power. The scalability condition holds by, for an $\alpha > 1$

$$F_k(\alpha\mathbf{P}) = b_k^{-1}\left(\frac{c_k\left(\frac{\sum_{i\neq k}g_{ki}\alpha P_i + \sigma_k^2}{g_{kk}}\right)\frac{\sum_{i\neq k}g_{ki}\alpha P_i + \sigma_k^2}{g_{kk}}}{}\right)$$

$$< b_k^{-1}\left(\frac{c_k\left(\frac{\sum_{i\neq k}g_{ki}P_i + \sigma_k^2}{g_{kk}}\right)\frac{\sum_{i\neq k}g_{ki}\alpha P_i + \sigma_k^2}{g_{kk}}}{}\right) \qquad (6.42)$$

$$< \alpha b_k^{-1}\left(\frac{c_k\left(\frac{\sum_{i\neq k}g_{ki}P_i + \sigma_k^2}{g_{kk}}\right)\frac{\sum_{i\neq k}g_{ki}P_i + \sigma_k^2}{g_{kk}}}{}\right) = \alpha F_k(\mathbf{P}),$$

where the first inequality holds because $b_k^{-1}(x)$ is decreasing in the region where the utility function is concave. The positivity condition holds by constraining the operation region such that the target CINR is always larger than the value that yields zero transmission power.

The above assumption on $B_k(x)$ is acceptable since a power control scheme is required to overcome an increment in interference by using more transmission power (but with a lowered target CINR to meet feasibility) until the interference level exceeds a given upper limit. This upper limit is represented by the slope of the dotted line in Figure 6.5 in the case of the sigmoid utility function. To meet the sufficient condition to be a standard power control, the utility function should be designed such that it does not increase too slowly in the operating CINR region. If the increasing rate is too low, a small increase in interference may cause significant reduction of the target CINR, which results in a decrease of the transmission power; that is, it makes $B_k(x)$ a decreasing function. In Figure 6.7 which shows the relation between the change of the cost line and the target CINR reduction, observe that the same amount of increase in the slope of the cost line would yield more reduction in the target CINR if the utility increases more slowly in the concave region. With the utility function designed under this guideline, the sigmoid utility-based power-control algorithm can be analyzed in the framework of standard power control and its convergence is guaranteed.

6.1.2.3 Pricing Mechanism based on KKT Condition

We next consider modifying the pricing mechanism of a utility-based power control (Sung and Wong, 2003). In order to devise a more intuitive pricing mechanism, we need to impose several restrictions on the system model. First, we consider a *concave* utility function in the uplink channel of a single-cell system. We assume that a user k has a spreading factor N_k and has a limit on the maximum transmission power. As there is only one receiver in the uplink channel in a single-cell system, we may set all the channel gains for a particular transmitter to be the same;

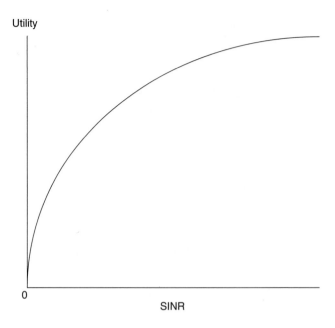

Figure 6.8 Concave utility function

that is, $g_{1k} = g_{2k} = \cdots = g_{kk}$. If we represent all those channel gains by g_k, we get the SINR expression

$$\Gamma_k = \frac{N_k Q_k}{\sum_{i=1, i \neq k}^{K} Q_i + \sigma^2},\tag{6.43}$$

where $Q_k = g_k P_k$ indicates the power of user k received at the common receiver (i.e., the base station). Then, the utility function of user k may be expressed by

$$U_k(\Gamma_k) = R_k f(\Gamma_k)\tag{6.44}$$

for a common mapping from the SINR to the utility, $f(\Gamma_k)$, and the weighting factor, R_k.[8] We assume that $f(\Gamma_k)$ is an increasing concave function as illustrated in Figure 6.8. Then the utility function in (6.44) becomes a special case of the sigmoid utility function in Figure 6.5.

One important implication of the approach is that the pricing mechanism is derived from the utility-maximization problem (Sung and Wong, 2003). Consider the following problem:

$$\begin{aligned} \text{maximize} \quad & \sum_{k=1}^{K} U_k(\Gamma_k) \\ \text{subject to} \quad & \sum_{k=1}^{K} Q_k = Q. \end{aligned}\tag{6.45}$$

---

[8] For example, $f(\Gamma_k)$ may be the function that maps the SINR to the frame success probability. If all the users have the same modulation and coding, the frame success probabilities of two users will be the same provided they have the same SINR.

This aims to maximize the sum of each user's utility while keeping the sum of the received power at the common receiver at level Q. From (6.43) and the constraint in (6.45), we get

$$\frac{\Gamma_k}{\Gamma_k + N_k} = \frac{Q_k}{Q + \sigma^2} \tag{6.46}$$

for the noise power at the base station, σ^2. By summing up this for all K users and applying the result to the constraint in (6.45), we get

$$\sum_{k=1}^{K} \frac{\Gamma_k}{\Gamma_k + N_k} = \frac{Q}{Q + \sigma^2}. \tag{6.47}$$

With this modified constraint, the maximization problem (6.45) becomes an optimization over SINR vectors. Then, we may apply the Lagrange multiplier method by defining a Lagrangian

$$L = \sum_{k=1}^{K} U_k(\Gamma_k) - \lambda \left(\sum_{k=1}^{K} \frac{\Gamma_k}{\Gamma_k + N_k} - \frac{Q}{Q + \sigma^2} \right). \tag{6.48}$$

On taking the partial differentiation of L for each k and setting it to zero, we obtain the KKT condition

$$\frac{\partial L}{\partial \Gamma_k} = R_k f'(\Gamma_k) - \frac{\lambda N_k}{(\Gamma_k + N_k)^2} = 0. \tag{6.49}$$

By (6.43), we get $\partial \Gamma_k / \partial Q_k = N_k / I_k$ for $I_k = \sum_{i=1, i \neq k}^{K} Q_i + \sigma^2$. Applying this relation, we can rewrite the KKT condition as

$$0 = R_k f'(\Gamma_k) \frac{N_k}{I_k} - \frac{\lambda N_k}{(\Gamma_k + N_k)^2} \frac{N_k}{I_k} = R_k f'(\Gamma_k) \frac{\partial \Gamma_k}{\partial Q_k} - \frac{\lambda I_k}{(Q_k + I_k)^2}. \tag{6.50}$$

Note that the optimization problem in (6.45) is not a convex problem as the equality constraint is not affine. So, the KKT condition cannot guarantee optimality; it provides only the upper bound (actually, an optimized upper bound) of the solution based on weak duality. The optimality issue is discussed further below.

Observe from (6.50) that the last expression is the derivative of the function

$$U_k(\Gamma_k) - \frac{\lambda Q_k}{Q_k + I_k} \equiv U_k(P_k, \mathbf{P}_{-k}) - C_k(P_k) \tag{6.51}$$

with respect to the received power Q_k. If we compare it with (6.33), the equation in (6.51) turns out to be the net utility defined for a nonlinear pricing function $C_k(P_k) = \lambda g_k P_k / (g_k P_k + I_k)$. This means that the KKT condition in (6.49) is equivalent to the condition of the optimal SINR that maximizes (6.51). By considering it and comparing (6.36) and (6.51), we may interpret the term $\lambda Q_k/(Q_k + I_k)$ in (6.51) as the cost paid for using transmission power. This modified pricing implies that each user's cost is determined by the proportion of the power received from that user to the total received signal plus the noise power at the common receiver. In other words, each user's cost increases in proportion to the received power level normalized by the total power at the receiver. The rationale of this normalization is that the degree of harm caused by a user depends on the total power: a small increment of transmission power will induce serious interference to the others if the total power is low, but its effect will be marginal if the total power is high. In this insight, the Lagrange multiplier λ may be interpreted as the *pricing parameter* which operates as a global constant determining the cost from the normalized received power. As the pricing term in (6.51) is equivalent to $\lambda \Gamma_k/(\Gamma_k + N_k)$, the net utility-maximization problem may be written as

$$\text{maximize}\left\{ U_k(\Gamma_k) - \frac{\lambda \Gamma_k}{\Gamma_k + N_k} \right\}. \tag{6.52}$$

Figure 6.9 illustrates this operation graphically. The target SINR is determined such that the marginal increments of the utility and the cost functions are equalized.

Based on the above discussion of the pricing mechanism, we can analyze the power control algorithm in which each user tries to maximize its net utility in (6.51). In the following it is shown that, under the assumption of $2f'(\Gamma_k) + (\Gamma_k + N_k)f''(\Gamma_k) < 0$, the power-control algorithm with a properly given pricing factor λ has a unique Nash equilibrium point. We also show that the above assumption guarantees the uniqueness of the received power Q_k that meets the condition in (6.49).

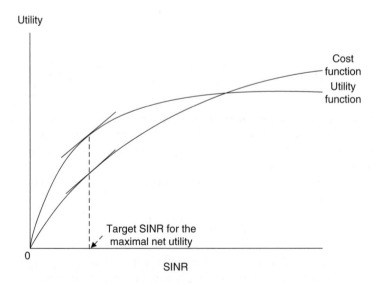

Figure 6.9 Net utility maximization for concave utility functions

First rewrite (6.49) in the form

$$d_k(\Gamma_k) \equiv (\Gamma_k + N_k)^2 f'(\Gamma_k) = \frac{\lambda N_k}{R_k}, \tag{6.53}$$

and derive

$$d_k'(\Gamma_k) \equiv (\Gamma_k + N_k)\{2f'(\Gamma_k) + (\Gamma_k + N_k)f''(\Gamma_k)\} \tag{6.54}$$

with $d_k(0) = N_k^2 f'(0) > 0$. Then, the condition $2f'(\Gamma_k) + (\Gamma_k + N_k)f''(\Gamma_k) < 0$ yields a monotonically decreasing $d_k(\Gamma_k)$ and, for a properly given λ, the received power which meets the condition in (6.49) becomes unique. Moreover, this unique equilibrium point is a local optimum. From (6.49), the second-order derivatives are

$$\frac{\partial^2 L}{\partial \Gamma_k^2} = R_k f''(\Gamma_k) + \frac{2\lambda N_k}{(\Gamma_k + N_k)^3} \tag{6.55}$$

and

$$\frac{\partial^2 L}{\partial \Gamma_k \partial \Gamma_l} = 0 \quad \text{for} \quad k \neq l. \tag{6.56}$$

By inserting (6.49), which is the condition of the equilibrium point Γ^*, into (6.55), we get

$$\frac{\partial^2 L}{\partial \Gamma_k^2}\Big|_{\Gamma^*} = R_k f''(\Gamma_k^*) + \frac{2\lambda N_k}{(\Gamma_k^* + N_k)^3} = R_k f''(\Gamma_k^*) + \frac{2R_k}{\Gamma_k^* + N_k} f'(\Gamma_k^*) < 0, \tag{6.57}$$

where the last inequality holds by the assumption of $2f'(\Gamma_k) + (\Gamma_k + N_k)f''(\Gamma_k) < 0$. From (6.56) and (6.57), we observe that a slight change in the SINR vector at the equilibrium decreases the overall utility. Thus, the unique Nash equilibrium is guaranteed to be a local optimum.[9]

It is noteworthy that the power control schemes in (6.36) and (6.52) can be regarded as examples of the hybridization of the centralized and distributed resource management discussed in Section 3.1.2. In both cases, the transmission power – that is, the amount of resource used in each transmitter–receiver pair – is determined locally in a distributed manner. Each pair adjusts its own transmission power according to the SINR measurement but there is no centralized controller which determines each pair's transmission power. However, the pricing factor, such as c_k in (6.36) and λ in (6.52), should be determined in a centralized manner. As discussed above, these parameters affect the stability and the convergence of the power update procedures, so they determine the overall system performance. Thus, in these

---

[9] The next subsection provides more detailed discussion of the global optimality of the sum utility maximization problem, by introducing an information-theoretic capacity as the concave utility function.

utility-based power-control schemes, a centralized controller is required to set the pricing factors intelligently according to the communication environments. Based on the discussions in Section 3.1.2, we can regard those pricing factors as the reference information that is distributed from the central controller to each transmitter–receiver pair.

6.1.3 Power Control along with Rate Control

So far, power-control methods have been discussed assuming that the transmission rate of each user is fixed regardless of the communication environments. The objective was to maintain the link quality of each user at an acceptable level to support the given transmission rate. Now we extend the discussion to cases where it is possible to adjust each user's transmission rate according to the communication environment. This type of rate control is effective especially for the best-effort data services which do not require any specific QoS on the transmission rate provided the data traffic is delivered reliably.

Suppose that a user with best-effort data service faces a very hostile communication environment. If the power-control schemes discussed in the previous subsections are adopted, they will try to increase the transmission power to overcome the environment, inducing severe interference to other users. We can try to adjust the target SINR by adopting the utility-based power-control schemes, but this cannot be a fundamental solution as a certain level of SINR is still required even for minimal reliability. However, if it is assumed instead that a best-effort service does not require maintenance of a given transmission rate all the time, the above arrangement is found to be unnecessary. Instead of increasing the transmission power, alternative approaches can be adopted that change the transmission rate according to the interference level in the hostile environment.

As an alternative approach, consider the *rate-control* mechanism which adjusts the transmission rate of each user according to its communication environment. The delivery of information can be made more reliable by reducing the transmission rate of a user. This is because the power budget for each information bit increases as the transmission rate decreases for a fixed transmission power. As a best-effort service has no requirement on the transmission rate, this rate control is an effective solution to overcoming the interference. Rate control does not involve a power increase, so it can be performed without inducing interference to other users. Therefore, rate control contributes to enhancing total throughput by lowering the transmission rate and thereby preventing transmission failures. Thus, throughput maximization becomes the typical objective of rate control.

Rate control inherently necessitates a mechanism to determine each user's transmission rate. This mechanism may be regarded as a scheduling process since it may exclude a user from transmission (i.e., set its transmission rate to zero) when its transmission does not contribute to total throughput improvement. Such scheduling is a centralized operation and typically deals with users within a single cell. So we may consider only the uplink of a single-cell case when dealing with power-control schemes along with rate control.[10] In this case there is only one receiver, the base station, in the uplink channel, so all the channel gains of each transmitter are

---

[10] The downlink case is the same if the constituent users are not orthogonal. Otherwise, the problem becomes one of power allocation over parallel orthogonal channels, to be discussed in Section 6.2.

the same; that is, $g_{1k} = g_{2k} = \cdots = g_{kk}$. When the channel gain is g_k, the CINR of user k is written as

$$\gamma_k = \frac{g_k P_k}{\sum_{i \neq k} g_i P_i + \sigma^2}, \tag{6.58}$$

where σ^2 is the noise variance at the base station.

Basically, there are two ways of controlling the transmission rate in a single cell. One is to adjust the spreading factor of a CDMA system, and the other is to employ the adaptive modulation and coding method. Yet there is a third way of opportunistically allocating more transmission power to a user with lower interference. These three rate-control techniques are discussed next.

6.1.3.1 Rate Control via Spreading-factor Control

With CDMA systems it is possible to adjust each user's transmission rate by changing the spreading factor. Assume that the spreading bandwidth (i.e., the chip rate) is fixed at W. If a user k sets its transmission rate to R_k in the set of the supportable transmission rates $R = \{R^{(1)}, \ldots, R^{(M)}\}$, the spreading factor of user k becomes $N_k = W/R_k$. Then, the SINR of user k is given by

$$\Gamma_k = \frac{W}{R_k} \gamma_k = \frac{W}{R_k} \frac{g_k P_k}{\sum_{i \neq k} g_i P_i + \sigma^2}. \tag{6.59}$$

In order to support reliable communications with rate R_k, we need to constrain the SINR by $\Gamma_k \geq \Gamma_k^{\text{req}}$ with the transmission power constrained by $0 \leq P_k \leq P_k^{\text{max}}$.

Since there could be different feasible power vectors for a given target transmission rate of each user, an optimal power vector needs to be chosen. A straightforward guideline in choosing one among transmission power vectors with the same transmission rate is to select the one that minimizes the sum of each user's transmission power. This approach is reasonable as it is able to reduce inter-cell interference and extend the limited battery lifetime.

It is obvious that the SINR constraint is met with equality at the optimal power vector. Thus, we may write the SINR as

$$\Gamma_k = \frac{W}{R_k} \gamma_k = \Gamma_k^{\text{req}}. \tag{6.60}$$

We may replace the SINR constraint by the CINR constraint $\gamma_k \geq \gamma_k^{\text{req}} = \Gamma_k^{\text{req}} R_k / W$ and rearrange the problem in a matrix form as

$$(\mathbf{I}_K - \mathbf{A})\mathbf{P} = \mathbf{u}, \tag{6.61}$$

where

$$\mathbf{u} = \left[\frac{\gamma_1^{\text{req}} \sigma^2}{g_1}, \frac{\gamma_2^{\text{req}} \sigma^2}{g_2}, \ldots, \frac{\gamma_K^{\text{req}} \sigma^2}{g_K} \right]^T, \tag{6.62}$$

$$A_{ij} = \begin{cases} 0, & \text{if } i = j, \\ \dfrac{\gamma_i^{\text{req}} g_j}{g_i}, & \text{if } i \neq j. \end{cases} \cdot \tag{6.63}$$

Actually, this is a special case of the formulation in (6.9) with g_{ij} replaced by g_j, so it can be solved by the same method as in Section 6.1.1. However, in a single-cell uplink case, the problem can be solved more directly. For user k, we can write the equation in (6.61) as

$$\frac{1}{\gamma_k^{\text{req}}} g_k P_k - \sum_{i \neq k} g_i P_i = \left(\frac{1}{\gamma_k^{\text{req}}} + 1 \right) g_k P_k - \sum_i g_i P_i = \sigma^2. \tag{6.64}$$

Since

$$\frac{\gamma_k^{\text{req}}}{\gamma_k^{\text{req}} + 1} = \frac{g_k P_k}{\sum_i g_i P_i + \sigma^2}, \tag{6.65}$$

we get

$$\sum_j \frac{\gamma_j^{\text{req}}}{\gamma_j^{\text{req}} + 1} = \sum_j \left(\frac{g_j P_j}{\sum_i g_i P_i + \sigma^2} \right) = \frac{\sum_j g_j P_j}{\sum_i g_i P_i + \sigma^2}. \tag{6.66}$$

On the other hand, from (6.64) we have

$$\left(\frac{1}{\gamma_k^{\text{req}}} + 1 \right) g_k P_k = \sum_i g_i P_i + \sigma^2. \tag{6.67}$$

On inserting this into (6.66) we get

$$\sum_i g_i P_i = \left(\frac{1}{\gamma_k^{\text{req}}} + 1 \right) g_k P_k \sum_i \frac{\gamma_i^{\text{req}}}{\gamma_i^{\text{req}} + 1}. \tag{6.68}$$

Finally, by adding σ^2 to (6.68) and equating it with (6.67), we get the closed form of the solution:

$$P_k = \frac{\sigma^2}{1 - \sum_i \dfrac{\gamma_i^{\text{req}}}{\gamma_i^{\text{req}} + 1}} \frac{\gamma_k^{\text{req}}}{g_k (\gamma_k^{\text{req}} + 1)} \tag{6.69}$$

For the constraints to be feasible, the optimal power in (6.69) should be in the range $[0, P_k^{\max}]$ for all k. By arranging all those conditions, we get the following feasibility condition:

$$\sum_{k=1}^K \frac{\gamma_k^{\text{req}}}{\gamma_k^{\text{req}} + 1} \leq 1 - \max_k \left\{ \frac{\sigma^2 \gamma_k^{\text{req}}}{P_k^{\max} g_k (\gamma_k^{\text{req}} + 1)} \right\}. \tag{6.70}$$

By the above process, it is possible to check the feasibility of a given transmission-rate vector and calculate the optimal transmission power that achieves the given rate vector with a minimal total transmission power (Sampath *et al.*, 1995).

Now that each user's transmission power is determined for a given rate vector, the remaining task is to select the transmission-rate vector that maximizes total throughput among the feasible rate vectors. In general, it is too difficult to provide a global solution to the rate-maximization problem as it cannot be formulated as a tractable optimization problem. However, if all the users require the same link quality (i.e., the SINR requirements are identical), some interesting results can be obtained as discussed below. Note that this situation can be regarded as the case where the same modulation and coding scheme is applied to all the constituent users.

Consider the case with $\Gamma_1^{\text{req}} = \ldots = \Gamma_K^{\text{req}} = \Gamma^{\text{req}}$. Then, from the SINR constraint in (6.60), we get the relation of the SINR and the transmission rate as

$$R_k = \frac{W}{\Gamma^{\text{req}}} \gamma_k^{\text{req}}. \tag{6.71}$$

This implies that there is a linear relation between the CINR requirement and the transmission rate. By this relation, maximizing the total throughput is equivalent to maximizing the sum of CINR requirements. Then, the optimization problem is formulated as

$$\begin{aligned} \underset{\gamma_k^{\text{req}}}{\text{maximize}} \quad & \frac{W}{\Gamma^{\text{req}}} \sum_{k=1}^{K} \gamma_k^{\text{req}} \\ \text{subject to} \quad & \sum_{k=1}^{K} \frac{\gamma_k^{\text{req}}}{\gamma_k^{\text{req}}+1} \leq 1 - \max_k \left\{ \frac{\sigma^2 \gamma_k^{\text{req}}}{P_k^{\max} g_k (\gamma_k^{\text{req}}+1)} \right\} \\ & \gamma_k^{\text{req}} \in \Gamma, \end{aligned} \tag{6.72}$$

where $\Gamma = \{\gamma^{(1)}, \ldots, \gamma^{(M)}\}$ is the set of CINR levels with $\gamma^{(m)} = R^{(m)} \Gamma^{\text{req}} / W$.

In investigating the problem in (6.72), we assume that the maximum transmission power limit is identical for all the users; that is, $P_1^{\max} = P_2^{\max} = \cdots = P_K^{\max}$. This assumption does not violate the generality of the derived solution as each user's channel gain can be normalized without loss of generality. Suppose that user k has the normalized channel gain $\tilde{g}_k = g_k P_k^{\max} / P_1^{\max}$ and its transmission power is rewritten by $\tilde{P}_k = P_k P_1^{\max} / P_k^{\max}$ with the maximum power limit $P_1^{\max}$. Then, a feasible CINR requirement vector $\gamma = [\gamma_1^{\text{req}}, \gamma_2^{\text{req}}, \ldots, \gamma_K^{\text{req}}]$ is feasible if and only if it is feasible for this normalized case; that is

$$P_k = \frac{\sigma^2}{1 - \sum_i \dfrac{\gamma_i^{\text{req}}}{\gamma_i^{\text{req}}+1}} \frac{\gamma_k^{\text{req}}}{g_k(\gamma_k^{\text{req}}+1)} \in \left[0, P_k^{\max}\right]$$

$$\Leftrightarrow \tilde{P}_k = \frac{P_1^{\max}}{P_k^{\max}} P_k = \frac{\sigma^2}{1 - \sum_i \dfrac{\gamma_i^{\text{req}}}{\gamma_i^{\text{req}}+1}} \frac{\gamma_k^{\text{req}}}{(\gamma_k^{\text{req}}+1) g_k P_k^{\max} / P_1^{\max}} \tag{6.73}$$

$$= \frac{\sigma^2}{1 - \sum_i \dfrac{\gamma_i^{\text{req}}}{\gamma_i^{\text{req}}+1}} \frac{\gamma_k^{\text{req}}}{\tilde{g}_k(\gamma_k^{\text{req}}+1)} \in \left[0, P_1^{\max}\right].$$

Based on this result, we restrict our consideration to the case where the maximum transmission power is limited by the same level $P^{\max}$ for all the users.

First, observe from (6.72) that permuting a given CINR vector does not change the objective (i.e., the total throughput) but affects only the feasibility of the CINR requirement. In addition, observe from (6.73) that, if we permute a given CINR vector such that a higher target CINR is assigned to a user with a higher channel gain, it becomes more probable to make the CINR vector feasible. This is because the required transmission power is inversely proportional to the channel gain while the term $\sigma^2 / \left(1 - \sum_i (\gamma_i^{\text{req}} / \gamma_i^{\text{req}} + 1) \right)$ is not affected by the CINR permutation. This observation may be formalized by the following proposition. If we assume that the user index is arranged by $g_1 \geq g_2 \geq \ldots \geq g_K$, then, for any CINR vector, it is not feasible if its permutated version $\gamma_1^{\text{req}} \geq \gamma_2^{\text{req}} \geq \ldots \geq \gamma_K^{\text{req}}$ is not feasible. In other words, the permutation of a feasible CINR vector in a descending order is always feasible as follows. If a feasible CINR vector $\gamma = [\gamma_1^{\text{req}}, \gamma_2^{\text{req}}, \ldots, \gamma_K^{\text{req}}]$ with $\gamma_a^{\text{req}} < \gamma_b^{\text{req}}$ for $a < b$, then a new CINR vector $\tilde{\gamma}$, which is obtained by exchanging the target CINR for the users a and b, is also feasible, since

$$
\begin{aligned}
\tilde{P}_a &= C \frac{\gamma_b^{\text{req}}}{g_a(\gamma_b^{\text{req}} + 1)} \leq C \frac{\gamma_b^{\text{req}}}{g_b(\gamma_b^{\text{req}} + 1)} = P_b \leq P^{\max}, \\
\tilde{P}_b &= C \frac{\gamma_a^{\text{req}}}{g_b(\gamma_a^{\text{req}} + 1)} \leq C \frac{\gamma_b^{\text{req}}}{g_b(\gamma_b^{\text{req}} + 1)} = P_b \leq P^{\max},
\end{aligned}
\tag{6.74}
$$

where

$$
C \equiv \sigma^2 / \left(1 - \sum_i (\gamma_i^{\text{req}} / \gamma_i^{\text{req}} + 1) \right)
$$

and the inequality comes from the feasibility of γ. Thus, it suffices to restrict our consideration to the target CINR vectors with $\gamma_1^{\text{req}} \geq \gamma_2^{\text{req}} \geq \ldots \geq \gamma_K^{\text{req}}$ for $g_1 \geq g_2 \geq \ldots \geq g_K$. This observation can be interpreted such that an increased throughput can be achieved by allowing a higher CINR for a stronger user. As a result, the optimal solution of the problem in (6.72) always assigns a higher transmission power to a stronger user having a higher channel gain.

Second, observe that it cannot be optimal if a feasible rate allocation for a user is split and a part of it is allocated to another user. For example, consider the following two different rate allocations. One sets the CINR of the stronger user to $\gamma_1 + \gamma_2$, while setting that of the weaker user to zero; and the other allocates γ_1 and γ_2 to the stronger and weaker users, respectively. On calculating the difference between the terms $\sum_i (\gamma_i^{\text{req}} / \gamma_i^{\text{req}} + 1)$ for the two rate allocations, we get

$$
\frac{\gamma_1 + \gamma_2}{1 + \gamma_1 + \gamma_2} - \left(\frac{\gamma_1}{1 + \gamma_1} + \frac{\gamma_2}{1 + \gamma_2} \right) \leq 0,
\tag{6.75}
$$

where the inequality comes from the concavity of $\gamma / (1 + \gamma)$. Since the feasibility condition in (6.73) requires $\sum_i (\gamma_i^{\text{req}} / \gamma_i^{\text{req}} + 1) < 1$, the inequality in (6.75) implies that rate splitting makes the formulation more infeasible while rendering the same throughput as long as the CINR requirement $\gamma_1 + \gamma_2$ can be supported by the maximum transmission power limit. Thus, in order to achieve optimality, the target CINR of each user should be set as high as possible under the feasibility condition.

The above relations between the channel gain and the optimal CINR target can be interpreted in the context of the transmission power as follows. For a clearer insight, assume that the transmission rate is continuous with upper limit $R^{(M)}$. There are three factors that limit the increase of each user's rate, which are the upper limit on the transmission rate, the upper limit on the transmission power, and the feasibility condition in (6.70). For several users, the limiting factor is the upper bound on the transmission rate $R^{(M)}$. In order words, those users can achieve the maximum rate with a power less than the upper limit. Obviously, they have channel gains higher than the other users because $\gamma_1^{req} \geq \gamma_2^{req} \geq \ldots \geq \gamma_K^{req}$. We may say that those users belong to group 1. For some other users who have channel gains lower than those in group 1, packing the transmission rate is limited by the upper bound on the transmission power.[11] In other words, they use their maximum transmission power but achieve transmission rates less than the maximum transmission rate $R^{(M)}$. We may say those users belong to group 2. There may exist some users whose transmission rate is limited by the feasibility condition; we may say those users belong to group 3. One interesting observation is that there is at most one user who has a positive rate which is limited neither by the bound on the transmission rate nor by that on the transmission power. If more than two such users exist, it can reduce the total transmission power by packing the weaker user's transmission rate to the stronger user. Therefore, the number of users in group 3 is zero or one. All the users who are not in groups 1, 2, or 3 do not transmit at all; they may be said to be group 4. In summary, for optimal rate allocation with continuous transmission rate, the best k_1 users have the maximum transmission rate (group 1), the next best k_2 users use the maximum transmission power (group 2), and there exists at most one user who does not use the maximum rate or maximum transmission power (group 3). The remaining users do not transmit at all (group 4) (Jafar and Goldsmith, 2003).

Figure 6.10 illustrates the above interpretation. There are 14 users and their user indices are sorted by the channel gain. All the users are assumed to have the same maximum power limit. In the figure, there are three users in group 1 (users 1–3) who achieve the upper limit on the transmission rate. As these users have the same target CINR but the channel gain decreases as the user index increases, the transmission powers are set in an increasing order. There are five users in group 2 (users 4–8). They use the maximum transmission power and their transmission rate is in a decreasing order. User 9 is the user in group 3: both the transmission rate and power are below the upper limits. The remaining users (users 10–14) who do not transmit anything are in group 4.

Based on this interpretation, a *greedy rate-packing* (GRP) algorithm is proposed (Berggren and Kim, 2004). This algorithm determines the rate of each user as follows. First, it sorts the user indices in decreasing order of channel gain, and it sets the rate of user 1 such that its target CINR is maximized under the feasibility condition in (6.70). Then it determines the rate of user 2 such that its target CINR is also maximized while guaranteeing the feasibility with the target CINR of user 1. This algorithm recursively sets each user's transmission rate to its maximum while satisfying the feasibility condition with the currently determined CINR levels. It is easy to see that the resulting CINR vector satisfies the relation $\gamma_1^{req} \geq \gamma_2^{req} \geq \ldots \geq \gamma_K^{req}$. This GRP

---

[11] This is the reason why a continuous rate is assumed here. These users do not necessarily set their power to the upper limit if the transmission rate set is discrete. There may be no transmission rate that is achieved by the maximum transmission power in the discrete-rate case.

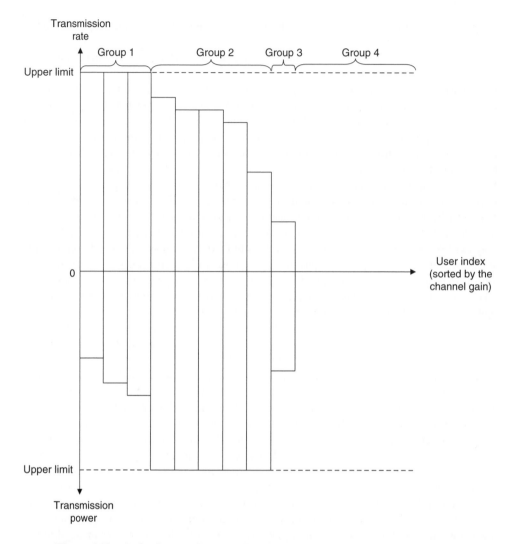

Figure 6.10   Optimal rate and power allocation under the continuous-rate assumption

algorithm does not necessarily find the solution to the optimization problem in (6.72) because the rate set is discrete and the users in group 2 may not use the maximum transmission power. However, the GRP achieves the optimality for a well-designed rate set.[12]

So far, we have discussed the property that arranging the target CINR along with the channel gain reduces the total transmission power. This property is preserved under minimum-rate constraints (or minimum-CINR constraints, equivalently). In this case, users with low channel gain target the minimum CINR requirements while stronger users are allocated higher rates according to their channel gains. Thus, a simple modification of GRP can be applied to the case with minimum-rate constraints.

---

[12] Refer to (Berggren and Kim, 2004) for a detailed discussion.

6.1.3.2 Rate Control via AMC

Another method to control the transmission rate is to adopt *adaptive modulation and coding* (AMC); that is, to change the employed modulation and coding scheme according to the CINR level. A useful way to relate the CINR to the transmission rate is to employ the information-theoretic capacity

$$R_k = W \log_2(1 + \gamma_k), \tag{6.76}$$

as discussed in Section 2.3.3.[13] With this relation, AMC provides a more effective rate-control method than spreading-factor control because it can utilize the spreading bandwidth in a more efficient manner. Suppose that the maximum transmission rate R with full spectrum utilization is attained at the CINR γ by $R = W \log(1 + \gamma)$. If the CINR reduces to $1/N$, spreading factor control uses the spreading factor N to maintain the SINR at γ, thus rendering the transmission rate

$$R_{\text{SFC}} = \frac{W}{N} \log_2(1 + \gamma), \tag{6.77}$$

which can be interpreted as a reduction of the bandwidth by a factor of N. On the other hand, AMC still utilizes the whole bandwidth but reduces the spectral efficiency for reliable transmission by

$$R_{\text{AMC}} = W \log_2(1 + \gamma/N). \tag{6.78}$$

By the concavity of the logarithmic function depicted in Figure 6.11, we have $R_{\text{SFC}} < R_{\text{AMC}}$ for $N > 1$.[14] Obviously, this performance gain is achieved at the cost of employing multiple modulators and channel encoder–decoder pairs.

By the relation in (6.76), the throughput maximization problem is formulated as

$$
\begin{aligned}
\underset{\gamma_k^{\text{req}}}{\text{maximize}} \quad & \sum_{k=1}^{K} W \log_2(1 + \gamma_k^{\text{req}}) \\
\text{subject to} \quad & \sum_{k=1}^{K} \frac{\gamma_k^{\text{req}}}{\gamma_k^{\text{req}} + 1} \leq 1 - \max_k \left\{ \frac{\sigma^2 \gamma_k^{\text{req}}}{P^{\max} g_k (\gamma_k^{\text{req}} + 1)} \right\}, \\
& \gamma_k^{\text{req}} \in \Gamma.
\end{aligned}
\tag{6.79}
$$

For simplicity we consider a continuous data rate set Γ. The optimal power control in (6.69) and the feasibility condition in (6.70) are also applicable to this AMC case, provided the target CINR vector is determined by solving (6.79). In fact, spreading-factor control and AMC rate control share the same relation of the transmission power in (6.69) and the feasibility condition in (6.70) for a given target CINR vector; the only difference is the transmission rate obtained from the determined CINR level. Therefore, the following proposition derived for spreading

---

[13] This formulation can include the introduction of the SNR gap discussed in Section 2.3.3 by regarding the channel gain of each user as the one normalized by the SNR gap.

[14] As shown in the figure, the relation $R_{SFC} < R_{AMC}$ holds as long as the mapping from CINR to the spectral efficiency is concave.

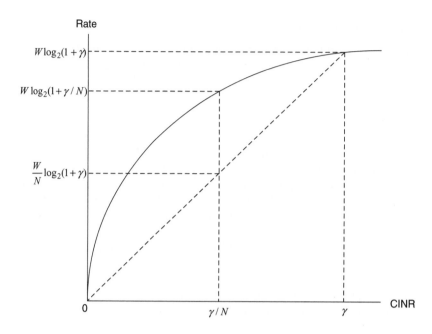

Figure 6.11 Comparison of spreading factor control and AMC-based rate control

factor control is also valid for the AMC case. If the user index is arranged as $g_1 \geq g_2 \geq \ldots \geq g_K$, then, for any feasible target CINR vector, the total power is minimized by arranging it such that $\gamma_1^{\text{req}} \geq \gamma_2^{\text{req}} \geq \ldots \geq \gamma_K^{\text{req}}$. This proposition is valid since it is derived by determining the optimal transmission power that minimizes the total power while achieving a given target CINR vector. This implies that a stronger user is allocated a higher rate in the AMC case, too.

In addition, it happens that there exists at most one user whose transmission rate is positive but less than the upper limit if there is no upper limit in the rate set (Oh and Soong, 2006). Suppose that, for the optimal allocation, two users m and n $(m < n)$ use transmission power less than the limit by $P_m < P^{\text{max}}$ and $P_n < P^{\text{max}}$. Then, by setting

$$P'_m = \frac{P_m + \Delta}{g_m}, \quad P'_n = \frac{P_n - \Delta}{g_n} \tag{6.80}$$

we can increase the power of user m while keeping the total received power unchanged. The CINR for the other users does not change as the total received power is the same. So the throughput difference obtained from the above modification is given by

$$
\begin{aligned}
D = {} & \log_2\left(1 + \frac{g_m P_m + \Delta}{I + g_n P_n - \Delta}\right) + \log_2\left(1 + \frac{g_n P_n - \Delta}{I + g_m P_m + \Delta}\right) \\
& - \log_2\left(1 + \frac{g_m P_m}{I + g_n P_n}\right) - \log_2\left(1 + \frac{g_n P_n}{I + g_m P_m}\right),
\end{aligned}
\tag{6.81}
$$

where I is the noise-plus-interference power from users other than m and n. This can be rearranged to get

$$D = \log_2((I + g_m P_m)(I + g_n P_n)) - \log_2((I + g_m P_m + \Delta)(I + g_n P_n - \Delta)). \qquad (6.82)$$

Since the optimal solution allocates a higher CINR to a stronger user, we get $g_m P_m \geq g_n P_n$. Then, the throughput difference in (6.82) is larger than zero because the function $x(\alpha - x)$ for a positive α is a decreasing function for $x \geq \alpha/2$. This contradicts the optimality of the initial setting.[15]

The above discussion implies that the optimal solution to (6.79) allocates the maximum transmission power to several strong users and there exists at most one user who uses a power less than the upper limit. This result exactly coincides with the result for spreading-factor control. We can regard the user whose power is less than the upper limit as the user in group 3. As no upper limit is assumed in the transmission rate in the formulation, group 1 is nonexistent in this framework (Oh and Soong, 2006).[16] Since all the users (there may exist one exception in group 3) act at one of the two extremes (i.e., the maximum transmission power or zero power), the power and rate allocation scheme is called the *semi-bang-bang strategy*.

6.1.3.3 Opportunistic Power Control

In the previous subsections it has been established that, for a system equipped with rate-control capability, it improves the overall throughput to give more chance to the users with stronger channel gain. The system throughput can be enhanced by applying the concept of *opportunistic transmission* discussed in Chapter 5 to power control. As this approach discriminates the constituent users by their channel gains, a centralized controller is necessary to achieve the optimal rate and power allocation.

We next consider a distributed way of implementing the opportunistic transmission in power control (Sung and Leung, 2005). The resulting strategy, called *opportunistic power control*, allocates more transmission power to the users with lower interference. Consider a power update rule similar to that of the framework of standard power control in Section 6.1.1; that is

$$\mathbf{P}(n+1) = \mathbf{F}(\mathbf{P}(n)). \qquad (6.83)$$

A power control for user k is described as being *opportunistic* if the iterative function $F_k(\mathbf{P})$ is a decreasing function of the effective interference $\left(\sum_{i \neq k} g_{ki} P_i(n) + \sigma_k^2\right)/g_{kk}$. This means that opportunistic power control increases the transmission power to allocate more resources to the user when the communication environment becomes favorable; that is, when the interference from the other users is relatively low when compared with its channel gain. An

---

[15] Refer to (Oh and Soong, 2003) and (Oh and Soong, 2006) for a detailed proof.

[16] If the transmission rate is limited, it is optimal for the users who can achieve the maximum rate with a power level less than the maximum power limit to use the maximum rate with a reduced transmission power. This is equivalent to group 1 in the case of the spreading factor control.

example of opportunistic power control is to set the transmission power inversely proportional to the effective interference level, so that

$$F_k(\mathbf{P}) = \frac{\beta_k g_{kk}}{\sum_{i \neq k} g_{ki} P_i(n) + \sigma_k^2} \tag{6.84}$$

for a positive constant β_k. Observe that the power update in (6.83) with (6.84) allocates more transmission power to a user with a higher channel gain and a lower interference. Note that opportunistic power-control algorithms including that in (6.84) are not limited to the single-cell case as the formulation assumes multiple channel gains from each user.

The standard power control in the previously discussed framework cannot be opportunistic since an opportunistic power control violates the monotonicity condition. So we define a new framework for the opportunistic standard power control as follows. An interference function $\mathbf{F}(\mathbf{P})$ is said to be *type-II standard* if it satisfies the following properties for all power vectors $\mathbf{P} \geq 0$:

- *Positivity:* $\mathbf{F}(\mathbf{P}) > 0$
- *Type-II monotonicity:* If $\mathbf{P} \geq \mathbf{P}'$, then $\mathbf{F}(\mathbf{P}) \leq \mathbf{F}(\mathbf{P}')$
- *Type-II scalability:* For all $\alpha > 1$, $(1/\alpha)\mathbf{F}(\mathbf{P}) < \mathbf{F}(\alpha\mathbf{P})$.

The type-II monotonicity property means that any decrement in the transmission power vector increases the transmission power of each user to exploit the opportunistic transmission. This property contrasts with the monotonicity of the framework in Section 6.1.1 and coincides with the definition of opportunistic power control. The type-II scalability property implies a lower limit on transmission power reduction in response to the increase of interference. If all the transmission powers are uniformly scaled up by a factor $\alpha > 1$, each user's power decreases by the type-II monotonicity; that is, the resultant power is $\mathbf{F}(\alpha\mathbf{P}) \leq \mathbf{F}(\mathbf{P})$. However, this power vector is larger than the power vector scaled down by factor α, $(1/\alpha)\mathbf{F}(\mathbf{P})$. It is trivial to show that the opportunistic power control in (6.84) is type-II standard.

We can easily show that the composite function $\mathbf{F}(\mathbf{F}(\mathbf{P})) = \mathbf{F}^{(2)}(\mathbf{P})$ of a type-II standard is standard in the sense of the framework in Section 6.1.1. The positivity and monotonicity directly come from the definition of the type-II standard. The scalability holds as follows. By the definition of type-II scalability, we get $\mathbf{F}(\mathbf{P}) < \alpha\mathbf{F}(\alpha\mathbf{P})$ for $\alpha > 1$. Then, by the type-II monotonicity, we get $\mathbf{F}^{(2)}(\mathbf{P}) \geq \mathbf{F}(\alpha\mathbf{F}(\alpha\mathbf{P}))$ and, by applying the type-II scalability again, we get $\mathbf{F}(\alpha\mathbf{F}(\alpha\mathbf{P})) > (1/\alpha)\mathbf{F}^{(2)}(\alpha\mathbf{P})$, which implies the scalability of $\mathbf{F}^{(2)}(\mathbf{P})$. This result renders a convergence property of the opportunistic power control: if an interference function $\mathbf{F}(\mathbf{P})$ is type-II standard and has a fixed point, then that fixed point is unique. Moreover, the power vector converges to the unique fixed point for any initial power vector (Sung and Leung, 2005).

This desirable property can be extended to a mixture of standard and type-II standard power control. An interference function $\mathbf{F}(\mathbf{P})$ is said to be *wide-sense standard* if each of its components is either standard or type-II standard. Then, the uniqueness of the fixed point and the convergence to the fixed point can be proved by applying the convergent properties of standard and type-II standard power control schemes.[17]

---

[17] See (Sung and Leung, 2005) for a detailed proof.

By introducing an opportunistic power control scheme, we can apply the opportunistic resource allocation which gives more chance to stronger users in a distributed manner. The desirable property of opportunistic power control guarantees stable operation and an improved total throughput. However, the distributed operation may not able to achieve the optimal solution provided by the power control schemes discussed above: a centralized rate and power allocation is still necessary to maximize the total throughput.

6.1.4 Power Control for Hybrid ARQ

As discussed in Section 2.3.2, hybrid ARQ is a very effective technology that brings throughput improvement. The main idea of HARQ is to "reuse" an erred packet of the initial transmission in recovering the failure after receiving the retransmissions. By virtue of this information reuse, HARQ improves the probability of error recovery and thereby reduces the interference caused by retransmissions in mutually interfering multiuser scenarios. The effectiveness of HARQ in the presence of MAI has been analyzed by (Bigloo *et al.*, 1997), (Zhang *et al.*, 1999), (Tuninetti and Caire, 2002), and (Cai *et al.*, 2003).

An interesting observation of HARQ is that the probability of a successful transmission for an initially transmitted packet can be made much smaller than that for a retransmitted packet. Clearly, this is because a retransmitted packet can be decoded with the help of the information contained in the previously received packets. This implies that the receiver already has much useful information about the packets to be retransmitted, so only a smaller amount of valid information may suffice to recover the packet error. For that reason, retransmission is likely to induce unnecessarily high interference to other users if it uses the same transmission power as was used in the initial transmission.

Based on the above reasoning, the retransmission power can be adjusted in such a way that only a small amount of information adequate to recover the corrupted packet is retransmitted (Seo *et al.*, 2005). For example, the retransmission power can be reduced to about 20% of the initial transmission power to deliver only a small amount of additional information. This *retransmission power adjustment* (RPA) helps to prevent unnecessary interferences and thus improves whole system performance.

Figure 6.12 illustrates the advantage of RPA. Suppose that a packet is erroneously received with the initial transmission. The amount of additional power required to recover this packet is determined based on the level of interference experienced. The HARQ protocol accumulates additional power through the subsequent retransmissions and the packet is recovered from error when the accumulated power becomes larger than a certain required level. Since this required power level is random and is not known to the transmitter, a higher retransmission power may possibly deliver more power than is required (see the case without RPA). However, the excessive amount of additional power is wasteful and induces unnecessary interferences to other transmissions. On the other hand, if the retransmission power is lowered by RPA, a finer tuning is possible in delivering the additional information, so preventing the use of an excessive amount of power which induces unnecessary interferences (see the case with RPA). A similar retransmission power adjustment is considered in (Kwon *et al.*, 2005), with the conclusion that reducing the retransmission power yields a better performance in terms of packet error rate and average delay.

Figure 6.13 shows how the effective throughput changes with RPA for different offered loads (Seo *et al.*, 2005). P_2 denotes the ratio of the retransmission power to the initial transmission

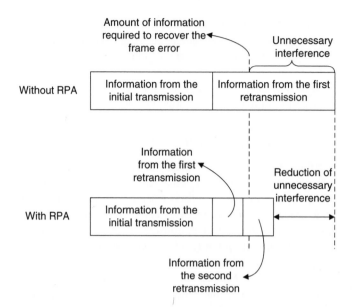

Figure 6.12   Reduction of unnecessary interference by RPA

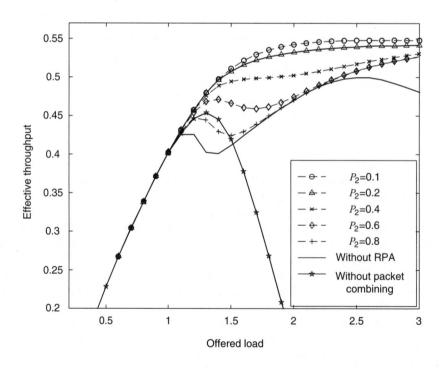

Figure 6.13   Effective throughput for different retransmission power adjustments in CDMA ALOHA systems with Chase combining (Reproduced with permission from H. Seo, S. Park, and B.G. Lee, "A Retransmission power adjustment scheme for performance enhancement in DS/SSMA ALOHA with packet combining," *Journal of Communications and Networks*, **7**, March, 36–44, 2005. © 2005 IEEE)

power. It is assumed that each user accesses the channel with the spreading factor 31 according to a Poisson process. A Chase combining method is used for HARQ. The figure demonstrates that the total throughput can be enhanced and the system can be made more stable by properly adjusting the retransmission power. It is noteworthy that a smaller retransmission power renders a better performance: this is because a lower power retransmission enables a finer tuning of additional information delivery.

The performance gain obtained from RPA is the improvement in success probability of the initial transmission. RPA reduces the interference induced to the initially transmitted packets, thereby improving the total throughput and the average delay. However, it has a disadvantage in that it decelerates the recovery of the erred packets. This is because only a small amount of information is delivered in retransmissions with RPA, so several retransmissions may be needed to recover an erred packet. For example, Figure 6.12 shows that only one retransmission can recover the initial transmission error in the case without RPA but one more retransmission is required in the case with RPA.

Figure 6.14 plots the ratio of long-delayed (by more than four transmissions) packets out of all the successful transmissions (Seo *et al.*, 2005). Note that, as the retransmission power decreases, an erred packet is likely to wait for more retransmissions even though the overall throughput is improved. Therefore there is a tradeoff between the total throughput and the speed of error recovery in controlling the retransmission power.

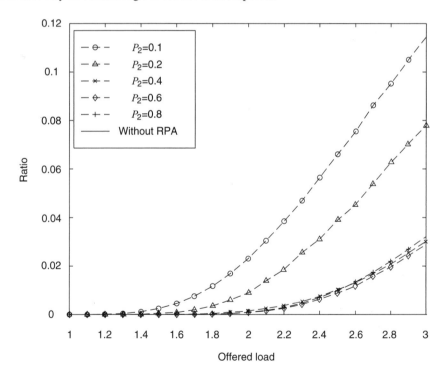

Figure 6.14 Ratio of the long-delayed packets for different retransmission power adjustments (Reproduced with permission from H. Seo, S. Park, and B.G. Lee, "A Retransmission power adjustment scheme for performance enhancement in DS/SSMA ALOHA with packet combining," *Journal of Communications and Networks*, **7**, March, 36–44, 2005. © 2005 IEEE)

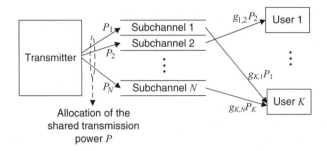

Figure 6.15   System model for power allocation for parallel subchannels

6.2 Transmission Power Management for Multiple Parallel Subchannels

Another purpose of transmission power management is to allocate the transmission power to multiple parallel channels. For systems that have multiple parallel subchannels (which are orthogonal to one another), such as an OFDMA system or a MIMO system equipped with the spatial multiplexing technique, the channel gain and noise power are different among the subchannels.[18] Such multichannel systems are advantageous in that they allow more flexible resource management: the shared bandwidth and transmission power can be properly allocated to each user according to the channel condition of each subchannel.

In the multichannel case, the contribution of unit transmission power to the overall performance is also different for different subchannels. Thus, it is necessary to devise a transmission power management algorithm that can effectively determine how much power each subchannel may use for transmission. Such an algorithm would distribute the total transmission power shared by the constituent subchannels in such a way that the overall performance is maximized. In this sense, the transmission power management for multiple parallel subchannels is distinguished from the power control discussed in the previous section. In contrast to the term "power control" used previously, we use the term *power allocation* to designate the algorithm that tries to maximize the efficiency of the power utilization in multiple parallel subchannels.

Assume that there are N orthogonal subchannels of equal bandwidth W/N and K users in the subchannel system, as depicted in Figure 6.15. Denote by $g_{k,n}$ the channel gain of user k at subchannel n. Since the subchannels are orthogonal and MAI does not exist, signal-to-noise ratio is the measure of the quality of each subchannel. To determine the SNR of each subchannel, it is necessary to determine which user is assigned to each subchannel. The procedure for determining the allocation of subchannels may be regarded as an extension of the scheduling considered in Chapter 5, so it is called *subchannel allocation*. Denote by k_n^* the user index to whom the subchannel n is allocated as the result of subchannel allocation. If the

---

[18] In OFDMA systems, a subchannel may be a subcarrier or a group of subcarriers. See Section 7.1 for detailed explanations of subchannel parallelization of MIMO systems.

transmission power is set to P_n for subchannel n, its SNR is given by

$$\gamma_n = \frac{g_{k,n}P_n}{\sigma_{k,n}^2} \quad \text{for } k = k_n^*. \tag{6.85}$$

This relation implies that, for a multichannel system, the system performance is determined after both the subchannel and the transmission power allocations are completed. In addition, the power allocation is strongly related to the subchannel allocation because the transmission power is distributed according to the channel information of each subchannel. However, this information becomes available only after the subchannel allocation. Thus, the two allocations are inseparable, so will be discussed together in this section.

In order to fully exploit the potential of power allocation, assume that the AMC capability is incorporated in the system. The transmission rate of each subchannel is determined by selecting a proper AMC mode which allows the maximum rate of information bits with an acceptable reliability at the given SNR of the subchannel. The relation between the SNR and the transmission rate depends on the system specification and is very complicated in general. Thus, for a simplified analysis, we use the information-theoretic model discussed in Section 2.3.3, assuming that the transmission rate of each channel is given by

$$R_n = \frac{W}{N}\log_2(1+\gamma_n). \tag{6.86}$$

This formulation reflects the SNR gap discussed in Section 2.3.3 by regarding the noise power of each user as rescaled by the SNR gap; that is, by letting $\sigma_{k,n}^2 = \Gamma\tilde{\sigma}_{k,n}^2$, for the noise power measurement $\tilde{\sigma}_{k,n}^2$ and the SNR gap Γ. We also assume a continuous transmission rate set, for a simplified analysis.

As discussed above, the objective of the power and subchannel allocation is to maximize the overall performance such as the total throughput. In the case with multiple users, however, QoS provision capability should be additionally incorporated to ensure a fair share of wireless resources among different users.

6.2.1 Single-user Case

First consider the case where a single user occupies all the subchannels and transmission power. In this case, no QoS issue arises and no subchannel allocation is required. Thus, it is sufficient to maximize the total transmission rate by properly determining the transmission power of each subchannel under the limit of the total transmission power P. In this case, the total transmission rate is given by

$$\sum_{n=1}^{N} R_n = \sum_{n=1}^{N} \frac{W}{N}\log_2\left(1 + \frac{P_n g_n}{\sigma_n^2}\right). \tag{6.87}$$

Note that no user index appears in the equation because only one user exists in the system. The optimal power allocation is formulated by the following optimization problem:

$$\text{maximize} \quad \sum_{n=1}^{N} \frac{W}{N} \log_2 \left(1 + \frac{P_n g_n}{\sigma_n^2} \right)$$

$$\text{subject to} \quad \sum_{n=1}^{N} P_n \leq P, \quad P_n \geq 0. \tag{6.88}$$

6.2.1.1 Water-filling Power Allocation

It is possible to explicitly determine the optimal power allocation for the problem in (6.88). Since the objective is a concave function of the powers, we can use the Lagrange multiplier method to find the optimal solution. The Lagrangian of (6.88) is

$$L = \sum_{n=1}^{N} \frac{W}{N} \log_2 \left(1 + \frac{P_n g_n}{\sigma_n^2} \right) - \lambda \left(\sum_{n=1}^{N} P_n - P \right), \tag{6.89}$$

for the Lagrange multiplier λ. By differentiating it, we get the KKT condition as discussed in Section 4.1. The resulting optimal power allocation becomes

$$P_n^* = \left[\frac{1}{\lambda'} - \frac{\sigma_n^2}{g_n} \right]^+, \tag{6.90}$$

for $\lambda' = \lambda N \ln 2/W$ and $[x]^+ = \max\{x, 0\}$. The Lagrange multiplier λ is determined by the KKT condition

$$\frac{\partial L}{\partial \lambda} = \sum_{n=1}^{N} P_n - P = 0, \tag{6.91}$$

which implies that the power constraint is met with equality. If we apply the allocation to (6.90), λ is chosen such that

$$\sum_{n=1}^{N} \left[\frac{1}{\lambda'} - \frac{\sigma_n^2}{g_n} \right]^+ = P. \tag{6.92}$$

Figure 6.16 illustrates this optimal power-allocation strategy. The horizontal axis is for the subchannel index n and the vertical axis is for the inverse of the channel gain normalized by the noise variance σ_n^2/g_n. The curve is obtained by plotting those values for each subchannel. Now, think of a vessel whose bottom is formed by this curve and assume that water of P units is poured into the vessel. Then, equations (6.90) and (6.92) state that the depth of the water at each subchannel is equal to the optimal power allocated to the channel, and $1/\lambda'$ is the height of the water surface. For this reason, this optimal power allocation is called *water-filling* or *water-pouring*.

The water-filling strategy tends to allocate more power to the stronger subchannels as the depth of the bottom is inversely proportional to the channel gain g_n. In other words, this optimal power allocation takes advantage of the better channel conditions by allocating more power. If

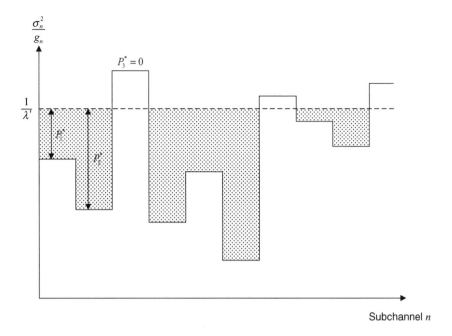

Figure 6.16   Water-filling power allocation over parallel subchannels

the channel gain of a subchannel is too low, when compared with the noise variance of the bottom, no transmission power is allocated to that subchannel. This means that that particular subchannel is too poor to transmit any data, so should be turned off.

6.2.1.2 Comparison with Equal-power Allocation

The water-filling allocation is the most effective power allocation for parallel subchannels in the single-user case. However, the allocation in (6.90) cannot be expressed in closed form, so we have no choice but to find the power allocation for each user by numerically searching for the height of the water surface, $1/\lambda'$. Due to this computational complexity, the water-filling allocation needs to be compared with other simpler methods to properly appreciate its performance.

The most straightforward way of allocating the shared transmission power is to distribute it equally to each subchannel. Figure 6.17 compares the spectral efficiencies achieved respectively by the water-filling and equal-power allocations over 10 subchannels with different channel gains.[19] The figure shows that the water-filling allocation achieves significant capacity gain over the equal-power allocation in a low-SNR regime, but the gain diminishes in a high-SNR regime. There may be several reasons for this. First, the capacity is a concave function of transmission power. The concavity implies that additional transmission power brings just marginal capacity improvement in the high-SNR regime. Thus, the capacity becomes relatively

---

[19] The channel gains are obtained from the CDF of a Rayleigh distribution. The gain of subchannel n is picked at the point where its CDF becomes $0.1n - 0.05$, the center of the nth interval among the 10 equally divided intervals in $[0, 1]$.

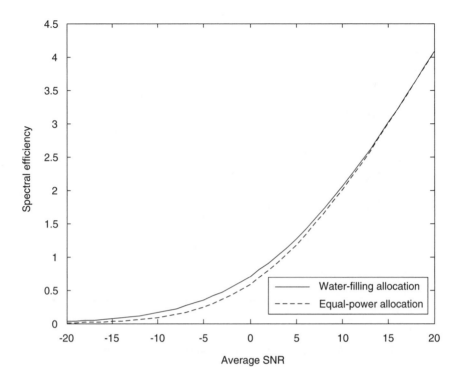

Figure 6.17   Comparison of spectral efficiency between water-filling and equal-power allocations

insensitive to the transmission power allocation in that regime. Second, the resulting allocation is not much different for the two allocation methods in the high-SNR regime.

Figure 6.18 illustrates a comparison of the water-filling allocation for low- and high-SNR cases. In the water-filling context, a high SNR implies that there is plenty of water to be poured into the vessel. So the fluctuation at the bottom of the vessel does not have much affect on the power allocation, thus yielding a similar result to the equal-power allocation.

Overall, the water-filling allocation brings considerable capacity improvement in a low-SNR regime. However, it is not an attractive solution in a high-SNR regime because its performance gain is negligible when compared with simple equal-power allocation.

6.2.2 Multiuser Case I: Throughput Maximization

Consider now the case where multiple users share the subchannels and transmission power. The resource management problem may generate various formulations according to the objectives and the constraints on allocation of resources. This subsection discusses the maximization of throughput.

First consider the issue of throughput maximization without any QoS constraint. The optimal allocation in this case is quite simple (Jang and Lee, 2003): the optimal subchannel allocation is to assign each subchannel to the user whose channel gain is the largest at that subchannel. It is obvious that any other allocation will decrease total throughput. Then, the optimal power allocation is to adopt the water-filling method according to the channel gain determined by the subchannel allocation. One important observation in (Jang and Lee, 2003) is that the

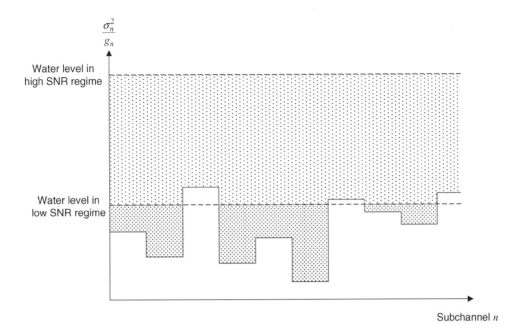

$$\frac{\sigma_n^2}{g_n}$$

Water level in high SNR regime

Water level in low SNR regime

Subchannel n

Figure 6.18 Comparison of water-filling power allocation in low- and high-SNR regimes

performance gap between water-filling and equal-power allocations is less in the multiuser case than in the single-user case. This is because the SNR of each subchannel improves through the subchannel allocation procedure: the allocation selects the best channel user, so the selected user has a much larger channel gain than the average. As discussed in the previous subsection, the gain from the water-filling allocation diminishes as the SNR increases.

Now consider the throughput maximization problem with a fairness constraint. The performance of a resource management algorithm is represented in the form of a rate vector which contains each user's transmission rate. In this rate vector, the ratio among the constituent users can be interpreted as a fairness criterion, which serves as a relative priority in providing the shared resources to each user. Under this proportional rate constraint, throughput maximization is formulated as

$$\underset{p_{k,n},\Omega_k}{\text{maximize}} \quad \sum_{k=1}^{K} R_k = \sum_{k=1}^{K} \sum_{n \in \Omega_k} \frac{W}{N} \log_2\left(1 + \frac{p_{k,n}g_{k,n}}{\sigma_{k,n}^2}\right)$$

$$\text{subject to} \quad \bigcup_{k=1}^{K} \Omega_k\{1,2,\dots,N\},$$

$$\Omega_k \cap \Omega_j = \phi \quad for \quad k \neq j, \tag{6.93}$$

$$p_{k,n} \geq 0,$$

$$\sum_{k=1}^{K} \sum_{n=1}^{N} p_{k,n} \leq P,$$

$$R_1 : R_2 : \dots : R_K = \rho_1 : \rho_2 : \dots : \rho_K,$$

where the transmission rate for user k is given by

$$R_k = \sum_{n \in \Omega_k} \frac{W}{N} \log_2 \left(1 + \frac{p_{k,n} g_{k,n}}{\sigma_{k,n}^2} \right). \tag{6.94}$$

In this formulation, Ω_k denotes the set of subchannels allocated to user k and the proportional rate constraint is given by $\rho_1 : \rho_2 : \ldots : \rho_K$. Since the ratio of each user's rate denotes the direction of the resultant rate vector in the achievable region, the largest achievable region can be obtained by solving (6.93) for various different rate constraints $\rho_1 : \rho_2 : \ldots : \rho_K$. In other words, the solution of (6.93) provides the Pareto-optimal rate vector for a given direction in the achievable region.

The formulation in (6.93) requires a simultaneous allocation of subchannel and transmission power. Unfortunately, however, this simultaneous approach is too complicated to solve and no explicit solution has been reported yet. The main reason is that the power allocation uses the channel gain of each subchannel but this channel gain information is unavailable before the subchannel allocation is completed. Therefore, it is reasonable to divide the whole problem into two subproblems – subchannel and transmission power allocations – and solve each separately. Note that this separation reduces the complexity burden but does not guarantee optimality.

6.2.2.1 Subchannel Allocation

Consider one of the most widely used subchannel allocation algorithms, which assumes a predetermined power distribution and iteratively allocates the best subchannel to the user with the lowest proportional capacity (Rhee and Cioffi, 2000).[20] If the transmission power of subchannel n is P_n, we can define $\gamma_{k,n} = g_{k,n} P_n / \sigma_{k,n}^2$ as the SNR of user k at that subchannel. Then, the subchannel allocation algorithm is described as follows:

1. Initialization

    (a) Set $R_k = 0$ and $\Omega_k = \phi$ for $k = 1, 2, \ldots, K$, and $A = \{1, 2, \ldots, N\}$.

2. For $k = 1$ to K

    (b) Find n satisfying $\gamma_{k,n} \geq \gamma_{k,j}$ for all $j \in A$.
    (c) Let $\Omega_k = \Omega_k \cup \{n\}$, $A = A - \{n\}$ and update R_k by (6.94).

3. While $A \neq \phi$

    (d) Find k satisfying $R_k / \rho_k \leq R_j / \rho_j$ for all j, $1 \leq j \leq K$.
    (e) For the found k, find n satisfying $\gamma_{k,n} \geq \gamma_{k,j}$ for all $j \in A$.
    (f) For the found k and n, put $\Omega_k = \Omega_k \cup \{n\}$, $A = A - \{n\}$ and update R_k by (6.94).

---

[20] The original algorithm in (Rhee and Cioffi, 2000) assumed an equal-power allocation $P_n = P/N$ and identical rate constraint $\rho_1 = \rho_2 = \ldots = \rho_K = 1$. It is extended in this subsection to encompass a more generalized power allocation and rate constraints.

The operation in line (2) is to allocate at least one subchannel to each user. As observed in line (3), this algorithm finds the user who has the least normalized transmission rate and allocates the best subchannel among the remaining users to that user. The algorithm repeats the iterative allocation until the set of the remaining subchannels becomes empty. This algorithm is heuristic and cannot achieve true optimality. However, it is known to provide an acceptable performance as a simple suboptimal subchannel allocation algorithm (Rhee and Cioffi, 2000; Shen *et al.*, 2005).

6.2.2.2 Transmission Power Allocation

Once Ω_k is determined for each user by subchannel allocation, the remaining problem is to allocate the transmission power between the subchannels. It is formulated as

$$
\begin{aligned}
\underset{p_{k,n}}{\text{maximize}} \quad & \sum_{k=1}^{K} R_k = \sum_{k=1}^{K} \sum_{n \in \Omega_k} \frac{W}{N} \log_2(1 + p_{k,n} G_{k,n}) \\
\text{subject to} \quad & p_{k,n} \geq 0 \\
& \sum_{k=1}^{K} \sum_{n=1}^{N} p_{k,n} \leq P, \\
& R_1 : R_2 : \ldots : R_K = \rho_1 : \rho_2 : \ldots : \rho_K,
\end{aligned}
\tag{6.95}
$$

where $G_{k,n} = g_{k,n}/\sigma_{k,n}^2$. The solution of this problem can be considered by a two-step approach. One is *inter-user allocation* which determines the amount of transmission power allocated to user k. We denote by $P_{k,\text{tot}}$ the total power allocated to user k. The other is *intra-user allocation* which distributes $P_{k,\text{tot}}$ determined by the inter-user allocation to the subchannels assigned to user k. As the intra-user allocation, it is optimal to adopt the water-filling method to each user. Therefore, we focus on inter-user allocation in the following.

As a result of the subchannel allocation, a set of subchannels, Ω_k, is assigned to user k. Depending on the channel condition and the result of inter-user allocation, each subchannel in Ω_k may or may not be used. This is because intra-user allocation (water-filling allocation) may allocate zero transmission power to some subchannels in poor channel condition. So, we denote by Ω'_k the set of subchannels to which positive transmission power is allocated and reassign the subchannel indices in Ω'_k by $G_{k,1} \leq G_{k,2} \leq \ldots \leq G_{k,N_k}$, where N_k is the number of subchannels in Ω'_k. Since all the subchannels in Ω'_k have positive transmission power, from (6.90) we get

$$
\frac{1 + G_{k,m} P_{k,m}}{G_{k,m}} = \frac{1 + G_{k,n} P_{k,n}}{G_{k,n}} \quad \text{for} \quad m, n \in \Omega'.
\tag{6.96}
$$

As we have sorted the subchannels into an increasing order, $P_{k,1}$ has the smallest value and the powers of the other subchannels are given by

$$
P_{k,n} = P_{k,1} + \frac{G_{k,n} - G_{k,1}}{G_{k,n} G_{k,1}}.
\tag{6.97}
$$

With this relation, we get the transmission power of the user k from

$$P_{k,\text{tot}} = \sum_{n=1}^{N_k} P_{k,n} = N_k P_{k,1} + \sum_{n=2}^{N_k} \frac{G_{k,n} - G_{k,1}}{G_{k,n} G_{k,1}}. \tag{6.98}$$

By applying (6.96) and (6.98) to the proportional capacity constraint $R_1/\rho_1 = R_k/\rho_k$, we get the $K-1$ equations for $P_{k,\text{tot}}$ by

$$\frac{1}{\rho_1} \frac{N_1}{N} \left(\log_2 \left(1 + G_{1,1} \frac{P_{1,\text{tot}} - U_1}{N_1} \right) + \log_2 V_1 \right) = \frac{1}{\rho_k} \frac{N_k}{N} \left(\log_2 \left(1 + G_{k,1} \frac{P_{k,\text{tot}} - U_k}{N_k} \right) + \log_2 V_k \right) \tag{6.99}$$

for

$$U_k = \sum_{n=2}^{N_k} \frac{G_{k,n} - G_{k,1}}{G_{k,n} G_{k,1}}, \tag{6.100}$$

$$V_k = \left(\prod_{n=2}^{N_k} \frac{G_{k,n}}{G_{k,1}} \right)^{1/N_k}. \tag{6.101}$$

We have an additional equation from the total power constraint:

$$\sum_{k=1}^{K} P_{k,\text{tot}} = P. \tag{6.102}$$

By solving these K equations, we obtain $P_{k,\text{tot}}$, the result of the inter-user allocation. Since these equations are nonlinear in general, we have no choice but to employ some numerical methods to find the solution.[21]

The above inter-user allocation is started for Ω'_k, the set of subchannels which has positive transmission power. However, it is not guaranteed that the result of the inter-user allocation, $P_{k,\text{tot}}$, is distributed over all the subchannels in the result of the subchannel allocation Ω_k. No transmission power may be allocated to the worst subchannel in Ω_k when $U_k > P_{k,\text{tot}}$.[22] When this situation happens, the set of Ω'_k, as well as the corresponding values of N_k, U_k, and V_k, need to be updated by excluding the worst subchannel from Ω'_k and the power allocation algorithm should be executed again. Figure 6.19 shows the overall allocation procedure.

The result of transmission power allocation in the multiuser case may be interpreted as *multilevel water-filling*, as depicted in Figure 6.20. This is because each user performs the water-filling allocation with the allocated subchannels and transmission power. Since the allocated

---

[21] Refer to (Shen *et al.*, 2005) for a detailed discussion on how to find the solution.

[22] We observe in (6.98) that $P_{k,1}$ should be a negative value if $U_k > P_{k,tot}$. Obviously, this violates the constraint of the original problem and thus power allocation should be performed again with a modified subchannel set Ω'_k.

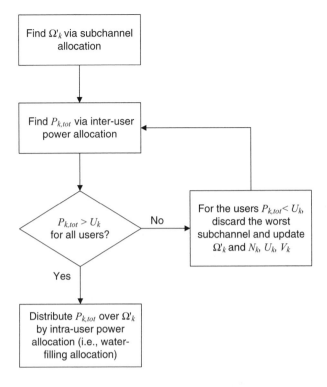

Figure 6.19 Flow chart of the subchannel and transmission power allocation algorithm

transmission power (i.e., the amount of pouring water) is different for each user, the vessel in the figure has multiple water levels.

6.2.3 Multiuser Case II: Utility Maximization

The problem of utility maximization over multiple parallel subchannels may be formulated as (Song and Li, 2005a, 2005b)

$$\underset{p_{k,n}, \Omega_k}{\text{maximize}} \quad \sum_{k=1}^{K} U_k(R_k)$$

$$\text{subject to} \quad \bigcup_{k=1}^{K} \Omega_k \{1, 2, \ldots, N\},$$

$$\Omega_k \cap \Omega_j = \phi \quad \text{for} \quad k \neq j,$$

$$p_{k,n} \geq 0,$$

$$\sum_{k=1}^{K} \sum_{n=1}^{N} p_{k,n} \leq P,$$

(6.103)

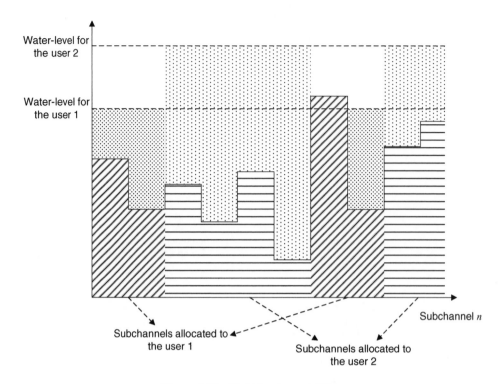

Figure 6.20   Multi-level water-filling

where $U_k(R_k)$ is the utility of user k which is a function of the transmission rate R_k. We assume that the utility function is concave, as it is suitable for elastic applications. The concavity assumption enables us to achieve the global optimum easily. As with the throughput maximization problem, we divide the utility maximization problem in (6.103) into two subproblems: subchannel and transmission power allocations.

6.2.3.1 Subchannel Allocation

First consider the subchannel allocation problem while assuming that a transmission power allocation is given. Since the transmission power at each subchannel is determined, the transmission rate available for each user is also determined at each subchannel. Denote by $r_{k,n}$ the transmission rate at which user k would be served if subchannel n is allocated to that user. Then, the problem can be rewritten as

$$
\begin{aligned}
\underset{x}{\text{maximize}} \quad & U(\mathbf{x}) = \sum_{k=1}^{K} U_k \left(\sum_{n=1}^{N} r_{k,n} x_{k,n} \right) \\
\text{subject to} \quad & \sum_{k=1}^{K} x_{k,n} = 1 \quad \text{for} \quad k = 1, 2, \ldots, K, \\
& x_{k,n} \in \{0, 1\} \quad \text{for} \quad k = 1, \quad 2, \ldots, K, n = 1, 2, \ldots, N,
\end{aligned}
\tag{6.104}
$$

where $x_{k,n}$ indicates whether or not the subchannel n is assigned to user k, and $\mathbf{x} = [x_{1,1}, \ldots, x_{1,N}, x_{2,1}, \ldots, x_{2,N}, \ldots, x_{K,1}, \ldots, x_{K,N}]^{\mathrm{T}}$ is the associate vector. The objective function $U(\mathbf{x})$ is concave as each $U_k(R_k)$ is concave. Then, for any two feasible vectors $\mathbf{x}$ and $\mathbf{y}$, we have

$$U(\mathbf{x}) - U(\mathbf{y}) \geq \nabla U(\mathbf{x})^T (\mathbf{x} - \mathbf{y}) \tag{6.105}$$

where the gradient of $U(\mathbf{x})$ is defined by

$$\nabla U(\mathbf{x}) = \begin{bmatrix} U_1'(R_1)r_{1,1} \\ \vdots \\ U_1'(R_1)r_{1,N} \\ \vdots \\ U_K'(R_K)r_{K,1} \\ \vdots \\ U_K'(R_K)r_{K,N} \end{bmatrix} \tag{6.106}$$

for $U_k'(R_k) = \mathrm{d}U_k(R_k)/\mathrm{d}R_k$. Then, if a vector $\mathbf{x}^*$ satisfies

$$\nabla U(\mathbf{x}^*)^T (\mathbf{x}^* - \mathbf{x}) \geq 0 \tag{6.107}$$

for an arbitrary feasible vector $\mathbf{x}$, we get $U(\mathbf{x}^*) - U(\mathbf{x}) \geq 0$ by (6.105), which means that $\mathbf{x}^*$ is the optimal solution of (6.104). The condition in (6.107) is equivalent to letting $x^*_{k,n} = 1$ for the k with

$$U_k'(R_k^*)r_{k,n} \geq U_j'(R_j^*)r_{j,n}, \quad \text{for all} \quad j \neq k, \tag{6.108}$$

where $R_k^* = \sum_{n=1}^{N} r_{k,n} x^*_{k,n}$ denotes the service rate of user k with the optimal allocation.

The condition on the optimality in (6.108) may be interpreted as follows. If the service rate of user k is determined by R_k^*, the marginal increment of the utility for the additional rate is given by $U_k'(R_k^*)$. Then, by a linear approximation of the utility function around R_k^* depicted in Figure 6.21, we can claim that subchannel n can increase the utility of user k by $U_k'(R_k^*)r_{k,n}$. In order to maximize the sum of utilities, it is optimal to allocate each subchannel to the user whose utility can be increased more than any other user at that subchannel. Under this linear approximation, the condition in (6.108) implies that any subchannel allocation other than the optimal one causes some loss in the total utility.

It is noteworthy that the condition in (6.108) is sufficient for the optimality but its necessity is lost due to the non-convexity of the achievable region. Actually, it may happen that there exists no vector $\mathbf{x}^*$ which satisfies the condition (6.108).[23] However, it is shown that this condition becomes equivalent to the optimal condition when the bandwidth of each subchannel is infinitesimal; that is, when the number of subchannels increases to infinity (Song and Li, 2005a, 2005b). This is because the increment of the service rate by an additional subchannel allocation

---

[23] For example, suppose two users share a single channel by $r_{1,1} = r_{2,1} = 1$ with the utility function $U_k = \ln(1 + R_k)$. Then, both of the two feasible indicator vectors $[1, 0]$ and $[0, 1]$ cannot satisfy the condition in (6.108).

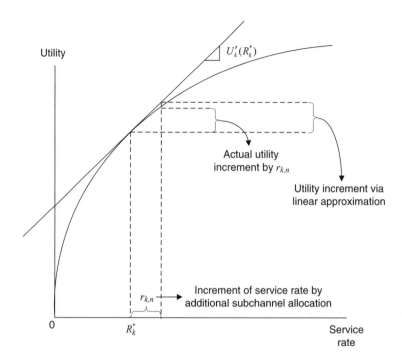

Figure 6.21 Optimal subchannel allocation for utility maximization

approaches zero and the utility increment obtained by the linear approximation gets closer to the actual utility increment in this case. Thus, the subchannel allocation criterion in (6.108) renders a near-optimal solution in a system where the bandwidth of each subchannel is narrow enough; that is, where the number of subchannels making up the whole bandwidth, N, is large enough.

6.2.3.2 Transmission Power Allocation

Now consider the transmission power allocation for a given subchannel allocation which is formulated as

$$
\begin{aligned}
\underset{p_{k,n}}{\text{maximize}} \quad & \sum_{k=1}^{K} U_k(R_k) = \sum_{k=1}^{K} U_k\left(\sum_{n \in \Omega_k} \frac{W}{N} \log_2\left(1 + \frac{g_{k,n} p_{k,n}}{\sigma_{k,n}^2} \right) \right) \\
\text{subject to} \quad & p_{k,n} \geq 0, \\
& \sum_{k=1}^{K} \sum_{n=1}^{N} p_{k,n} \leq P.
\end{aligned}
\tag{6.109}
$$

Since the objective is a concave function, the problem can be solved by the Lagrange multiplier method. Following a procedure similar to the case of (6.88), we can obtain the optimal power allocation

$$P_{k,n}^* = \left[\frac{U_k'(R_k)}{\lambda'} - \frac{\sigma_{k,n}^2}{g_{k,n}} \right]^+, \quad \text{for} \quad n \in \Omega_k. \tag{6.110}$$

The Lagrange multiplier λ' is determined by the total power constraint. The power allocation in (6.110) is nothing but the multi-level water-filling depicted in Figure 6.20 in the previous subsection. One important observation is that the water level of each user is proportional to the marginal utility increment $U_k'(R_k)$. Since each user's utility increment is proportional to $U_k'(R_k)$ under the linear approximation, it implies that the optimal power allocation pours a relatively larger amount of power to a user to whom an additional service rate brings forth more utility increment. Note that this allocation meets the necessary and sufficient condition for optimality.

Based on the above optimal conditions of subchannel and transmission power allocations, we can obtain the solution to (6.103) by iteratively repeating the two allocations. We start by performing the subchannel allocation under equal-power allocation, and then perform the transmission power allocation using the resulting subchannel allocation. In this iteration process, the total utility does not decrease after each allocation as long as the bandwidth of each subchannel is infinitesimal. Therefore, the iteration will eventually converge to the global optimal, as it is the only local maximum (Song and Li, 2005a, 2005b).

6.2.4 Multiuser Case III: With Time Diversity

So far we have discussed subchannel and transmission power allocations based on the instantaneous service rate. This implies that the performance metric and the QoS constraint are formulated in terms of the service rate for a given channel realization. For example, the formulation in (6.93) constrains the relative ratio among each user's service rate to be preserved at a given level regardless of the current channel condition.

Resource allocation based on the instantaneous service rate cannot exploit multiuser diversity in the time domain but utilizes only the subchannel domain diversity.[24] This means that, in order to satisfy the QoS constraint, wireless resources should be allocated to a user on a deep faded channel over the entire bandwidth. For example, consider the case of three channels and two users. Let the transmission rates of the three channels at the first time slot be $\{4, 4, 1\}$ if allocated to user 1, and $\{1, 2, 2\}$ if allocated to user 2; and let these numbers change to $\{1, 2, 2\}$ to user 1 and $\{4, 4, 1\}$ to user 2 at the second time slot. If the fairness constraint were to equalize each user's instantaneous service rate as formulated in (6.93), the optimal allocation would assign the first subchannel at the first time slot and the second and the third subchannels at the second time slot to user 1, rendering the time-average throughput of 4 for each user.

For delay-tolerable data traffics, however, this instantaneous QoS support is not necessary provided the long-term throughput is satisfactorily. Thus, performance can be improved if an opportunistic resource allocation strategy is applied to data traffics, since the resources may be allocated to each user only when the channel condition is relatively good. In the above example, the time-average throughput increases to 5 if the first and second subchannels at the first slot and the third subchannel at the second slot are allocated to user 1. Obviously, this throughput improvement results from the exploitation of time diversity. The resource allocation with time

---

[24] Note that the subchannel domain is equivalent to the frequency domain in OFDMA systems.

diversity does not now necessarily support the QoS constraint in terms of the instantaneous service rate, but it formulates the QoS constraint in terms of the time-averaged service rate.

Resource allocation with time diversity is a challenging issue because its formulation becomes very complicated due to the time-varying nature of the channel state. We approach the problem by dividing it into three subproblems: subchannel allocation, inter-user power allocation, and intra-user power allocation. Intra-user power allocation is straightforward in the case of time diversity, as water-filling is the optimal allocation. So we focus on the subchannel and inter-user power allocations in the following, based on the discussions given in (Seo and Lee, 2006).

6.2.4.1 Subchannel Allocation

Subchannel allocation with time diversity is equivalent to multichannel opportunistic scheduling. Hence, it suffices to apply one of the opportunistic scheduling algorithms discussed in Chapter 5 for each subchannel. Each algorithm selects the user who has the largest scheduling metric which is calculated from the channel gain and the weighting factor of the user. The fairness among the different users can be adjusted by properly controlling the weighting factors of the employed algorithm.

The channel gain of the channel selected by subchannel allocation with time diversity to a user is a random variable since the selected channel itself is random. However, it is known that the maximum of multiple independent random variables does not deviate much. (In other words, if we pick the maximum among several random variables, the picked maximum value is random but renders almost the same value at each trial.) This tendency becomes more apparent when more random variables contend to be the maximum. Since opportunistic scheduling is to take the maximum among multiple random variables, we can approximate the selected channel gain to a *constant* in this case.

Consider a user having a Rayleigh fading channel with the average channel gain $\bar{g}$ and assume that this user is scheduled by the CDF-based scheduling algorithm with selection probability p. It is shown in Seo and Lee (2006) that the channel gain when the user is selected can be approximated by $\bar{g}(\bar{X}-\ln p)$ for the mean of the Gumbel random variable, $\bar{X} \approx 0.577215$, because the approximation error converges to zero as the selection probability p approaches zero. This implies that we can approximate the channel gain of the scheduled user to a constant if there are many users contending to be scheduled for each subchannel.

6.2.4.2 Transmission Power Allocation

The above approximation is very useful to derive a guideline for efficient power allocation with time diversity. If we additionally assume that the number of independent subchannels is large enough, the number of allocated subchannels also becomes a constant for each user by the law of large numbers. Under these assumptions, we can obtain an approximation of the time-average throughput of each user in the form

$$T_k = \frac{W}{N} n_k \log_2 \left(1 + \frac{\tilde{g}_k P_k}{n_k \sigma_k^2} \right), \tag{6.111}$$

where n_k and P_k denote the number of subchannels and the transmission power allocated to user k, respectively, and $\tilde{g}_k$ is the channel gain selected by the subchannel allocation algorithm. This approximation is obtained as follows. We assume that n_k subchannels with an identical channel gain $\tilde{g}_k$ are assigned to user k after the subchannel allocation. Then, the allocated transmission power P_k is equally distributed over the n_k subchannels, making the SNR $\tilde{g}_k P_k/(n_k \sigma_k^2)$ for each subchannel. Since user k has n_k subchannels with this SNR, the throughput can be expressed in (6.111). We may call this approximation the *asymptotic throughput* as it is obtained under the asymptotic assumption that the numbers of contending users and subchannels increase to infinity.

With this asymptotic throughput, we can formulate the resource allocation problem as

$$\begin{aligned} \underset{p_k, n_k}{\text{maximize}} \quad & \sum_{k=1}^{K} T_k \\ \text{subject to} \quad & T_1 : T_2 : \ldots : T_K = \rho_1 : \rho_2 : \ldots : \rho_K, \\ & \sum_{k=1}^{K} P_k \leq P, \sum_{k=1}^{K} n_k \leq N, P_k, n_k \geq 0. \end{aligned} \qquad (6.112)$$

This problem has the same fairness criterion $\rho_1 : \rho_2 : \ldots : \rho_K$ as in (6.93). By introducing an intermediate variable t, we can rewrite the formulation in the form

$$\begin{aligned} \underset{p_k, n_k}{\text{maximize}} \quad & t \\ \text{subject to} \quad & t \leq \frac{1}{\rho_k} T_k, \sum_{k=1}^{K} P_k \leq P, \quad \sum_{k=1}^{K} n_k \leq N, \quad P_k \geq 0, \quad n_k \geq 0. \end{aligned} \qquad (6.113)$$

This formulation is a convex problem, so can be solved by the Lagrange multiplier method. Define the Lagrangian by

$$L = t - \sum_{k=1}^{K} \lambda_k \left\{ t - \frac{W n_k}{N \rho_k} \log_2 \left(1 + \frac{\tilde{G}_k P_k}{n_k} \right) \right\} - \nu \left(\sum_{k=1}^{K} n_k - N \right) - \upsilon \left(\sum_{k=1}^{K} P_k - P \right), \quad (6.114)$$

where $\tilde{G}_k = \tilde{g}_k / \sigma_k^2$. On differentiation we get

$$\frac{\partial L}{\partial P_k} = \frac{\lambda_k}{\rho_k} \frac{\partial T_k}{\partial P_k} - \upsilon = 0, \qquad (6.115)$$

$$\frac{\partial L}{\partial n_k} = \frac{\lambda_k}{\rho_k} \frac{\partial T_k}{\partial n_k} - \nu = \frac{W \lambda_k}{N \rho_k} \left\{ \log_2 \left(1 + \frac{\tilde{G}_k P_k}{n_k} \right) - \frac{\tilde{G}_k P_k}{\ln 2 (\tilde{G}_k P_k + n_k)} \right\} - \nu = 0. \qquad (6.116)$$

Noting that $\tilde{G}_k$, the channel gain after subchannel allocation, increases to a very large value under the asymptotic assumption, (6.116) can be approximated to

$$\log_2 \left(1 + \frac{\tilde{G}_k P_k}{n_k} \right) - \frac{\tilde{G}_k P_k}{\ln 2 (\tilde{G}_k P_k + n_k)} \approx \log_2 \left(1 + \frac{\tilde{G}_k P_k}{n_k} \right) \qquad (6.117)$$

as the first term increases to infinity as $\tilde{G}_k$ increases while the second term is always less than a finite number, $1/\ln 2$. Then, by (6.111) we have

$$\log_2\left(1 + \frac{\tilde{G}_k P_k}{n_k}\right) = \frac{NT_k}{W n_k} \tag{6.118}$$

and, by applying this approximation and equating (6.115) and (6.116), we get

$$\tilde{\phi}_k\left(\frac{P_k}{n_k}\right) \equiv \frac{n_k}{T_k}\frac{\partial T_k}{\partial P_k} = \frac{1}{\log_2\left(1 + \dfrac{\tilde{G}_k P_k}{n_k}\right)} \cdot \frac{\tilde{G}_k}{\ln 2\left(1 + \dfrac{\tilde{G}_k P_k}{n_k}\right)} = \frac{\upsilon}{\nu}. \tag{6.119}$$

This relation provides a guideline for efficient power allocation. The left side of the relation, $(n_k/T_k)(\partial T_k/\partial P_k)$, can be interpreted as the normalized throughput increment contributed by the increment of transmission power. If we increase the transmission power P_k marginally, the asymptotic throughput increases at the rate

$$\frac{\partial T_k}{\partial (P_k/n_k)} = n_k \frac{\partial T_k}{\partial P_k}$$

as the transmission power is equally distributed over the n_k subchannels. Then, the term $(n_k/T_k)(\partial T_k/\partial P_k)$ is the throughput increment normalized by the current throughput. The relation in (6.119) implies that the transmission power should be allocated in such a way that the resulting throughput increment becomes identical for all the users. We may call the power allocation in (6.119) the *asymptotically optimal power allocation* as it provides the optimal solution for the asymptotic formulation in (6.112).

Now we consider how to apply this asymptotically optimal power allocation to practical cases. Suppose that the subchannel n is assigned to user $k(n)$. Then, for each subchannel, we can define the metric

$$\phi_n(P_{k,n}) \equiv \frac{1}{r_{k,n}}\frac{\partial r_{k,n}}{\partial P_{k,n}}\bigg|_{k=k(n)} = \frac{G_{k,n}}{\ln 2(1 + G_{k,n}P_{k,n})\log_2(1 + G_{k,n}P_{k,n})}\bigg|_{k=k(n)} \tag{6.120}$$

for $G_{k,n} = g_{k,n}/\sigma_{k,n}^2$. By comparing (6.119) and (6.120), one can easily find that the metric $\tilde{\phi}_k(P_k/n_k)$ is the asymptotic version of $\phi_n(P_{k,n})$ because P_k/n_k can be interpreted as the transmission power allocated to each subchannel for user k, or $P_{k,n}$ in (6.120). Therefore, the asymptotically optimal allocation rule in (6.119) can be implemented in practical cases by allocating the transmission power such that the resulting normalized throughput increment $\phi_n(P_{k,n})$ becomes identical for each subchannel. This power allocation rule may be called the *proportional–fair power allocation* (PFPA) because it maximizes $\sum_{n=1}^{N} \ln r_{k(n),n}$ and thus distributes total power in a proportional–fair manner. One can easily show this property by the Lagrange multiplier method. Define the Lagrangian by

$$L = \sum_{n=1}^{N} \ln r_{k(n),n} - \lambda \sum_{n=1}^{N}(P_{k(n),n} - P), \tag{6.121}$$

and differentiate it to get

$$\frac{\partial L}{\partial P_{k(n),n}} = \frac{1}{r_{k(n),n}} \frac{\partial r_{k(n),n}}{\partial P_{k(n),n}} - \lambda = \phi_n(P_{k,n}) - \lambda = 0. \tag{6.122}$$

This implies that the resulting normalized throughput increment becomes identical for each subchannel. Since the proportional–fair power allocation is developed as an inter-user allocation, its purpose is to determine the transmission power allocated to each user, $P_{k,\mathrm{tot}} = \sum_{n=1}^{N} P_{k,n}$, which is distributed over the assigned subchannels by the employed intra-user allocation.

One interesting observation is that the operation of PFPA is independent of the fairness criterion $\rho_1 : \rho_2 : \cdots : \rho_K$ as opposed to the inter-user power allocation in (6.99) in Section 6.2.2. Thus, we can perform the asymptotically optimal allocation simply by applying the PFPA rule for each resultant subchannel allocation regardless of the fairness criterion. This property enables the PFPA to be employed as a power allocation with time diversity since it does not need to consider the fairness criterion at each time slot. Actually, the fairness criterion is achieved by adjusting the weighting factors of the subchannel allocation algorithm.

Another important property of PFPA is that it has the tendency of allocating more transmission power to a user with a lower channel gain. This is because the normalized throughput increment defined in (6.120) is a decreasing function of the channel gain and more transmission power should be allocated to a low channel user to equalize the resulting normalized throughput increment. By manipulating (6.119), we get the asymptotic optimal power allocation for each subchannel as

$$\frac{P_k}{n_k} = \frac{\nu}{\upsilon} \frac{Wn_k}{NT_k} - \frac{1}{G_k}. \tag{6.123}$$

This allocation is similar to the water-filling method, but the water level is inversely proportional to T_k/n_k, the throughput achieved per subchannel. This means that PFPA tends to allocate more power to a weaker user whose throughput is relatively low.

The above property appears to contradict the water-filling allocation which allocates more power to a better channel. However, it should be noted that the water-filling algorithm is not a good inter-user allocation method as it becomes difficult to support the fairness criterion with the water-filling approach. One may try to allocate many subchannels to the users with low average channel gains to compensate for the power allocation, but this severely degrades the throughput of the users in a good channel condition, thereby deteriorating the overall performance. On the other hand, the operation of PFPA is more desirable in terms of fairness because it allocates more power to the low channel users to whom an additional power allocation can contribute more than to the users with high channel gain due to the concavity of the capacity–power relation. PFPA is able to achieve the fairness criterion for low channel users by using fewer subchannels and thereby enhances the overall throughput performance.

6.3 Transmission Power Adaptation to Time-varying Environments

Since any communication environment inherently has randomness, the efficiency of wireless resource use can be improved by adjusting transmission power allocations according to the

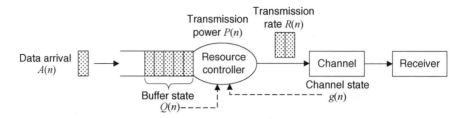

Figure 6.22   System model for power adaptation

state of the environment. The limiting factor is the finiteness of the energy the transmitter has available. The objective of transmission power management is to determine the most energy-efficient method that can yield maximal performance under a given constraint on average power consumption. In this context, the term *power adaptation* is used.

One main source of randomness is the fading process of wireless channels. In order to provide an effective communication service in such time-varying channels, the transmission power needs to be adjusted according to the channel state. Specifically, when the channel undergoes deep fading there are two main choices. One choice is to wait until the channel has recovered from deep fading, while exploiting opportunistic transmission. This choice saves transmission power when in a poor channel state and uses it to achieve a higher spectral efficiency when in a good channel state. The other choice is to compensate for the bad channel condition by using more transmission power. This approach may lead to poor spectral efficiency but is very helpful in providing a certain level of QoS even when in a bad channel state. As will be discussed later, the choice can be made differently depending on traffic characteristics and the coherent time of the wireless channel.

Another main source of randomness is the arrival pattern of user traffic. The amount of information generated by the applications may vary in time, and this time-varying nature of traffic arrival is salient for data traffic as packets are generated in bursts. In other words, a burst of data packets may arrive at a time slot, with no traffic arriving at the next time slot. Such randomness may be mitigated by employing a buffer at the transmitter. However, that cannot resolve the problem fundamentally because excessive buffering may cause a serious delay problem. In order to prevent such a problem it is necessary to increase the transmission rate by using more transmission power.

Figure 6.22 depicts a system model for power adaptation. It assumes that a dedicated feedback channel exists between the transmitter and receiver. The wireless channel and the traffic arrival pattern both vary randomly. Depending on the channel and the buffer states, the transmitter determines the transmission power and the transmission rate at each time slot. The subsections that follow first discuss power adaptation techniques by separating out the two considerations (i.e., the channel and the buffer states) and then introduce power adaptation techniques which take both of them into account.

6.3.1 Capacity of Time-varying Channels

Consider, first, power adaptation to time-varying channels, assuming static traffic arrival. Assume a discrete-time frequency-flat fading channel with stationary and ergodic channel gain $\{g(n)\}$, where the index n denotes the time slot. Assume that the transmitter knows the channel information before transmission and adjusts the transmission power and the corresponding transmission rate based on the channel information. In this case, the transmission power is a function of the channel gain. Denote by $P(g(n))$ the transmission power for channel gain $g(n)$.

Power adaptation in fading channels may use different transmission rates for each time slot. Thus, the concept of "capacity," the most widely used performance metric, is not as simple as in the case of time-invariant channels. In the following we will introduce two different definitions of the capacity for fading channels, namely outage capacity and ergodic capacity, which strongly depend on the rate of channel state change. The two definitions are derived under extreme assumptions of fading speed. Then the effective capacity that compromises the two different definitions is considered.

6.3.1.1 Slowly Fading Channel: Outage Capacity

Consider the case where the channel state varies very slowly. More specifically, the channel gain is constant over a transmission burst and then changes to a new value based on the fading distribution. This models the *slow-fading* situation where the coherence time of the channel is longer than the delay requirement of the traffic. In this case, an arrival of data traffic cannot be buffered until the channel state changes to a new level but should be transmitted under the current channel condition. The entire codeword (i.e., the transmission burst) is transmitted under the same channel condition with the same transmission rate. Since the delay requirement will be violated if the transmitter waits until the channel has recovered from a bad condition, a deeply faded channel should be compensated.

If the traffic arrival rate is R bits/sec/Hz, the transmitter sends data at the rate R with transmission power $P(g(n))$ for channel gain $g(n)$. The maximum rate of reliable communication supported by this transmission is $\log_2(1 + g(n)P(g(n))/\sigma^2)$ for the noise variance σ^2. Suppose that this rate is smaller than R. Then the decoding error probability of the transmission burst cannot be made arbitrarily small. So the system may go into outage with probability

$$p_{\text{out}}(R) = \Pr\{\log_2(1 + g(n)P(g(n))/\sigma^2) < R\}. \tag{6.124}$$

The transmitted traffic may be assumed reliably decoded unless the system goes into outage. Thus the acceptable level of the outage probability is a QoS parameter that is determined by the requirement on the reliability of the transmission.

The capacity in this slowly fading channel case is defined by the maximum transmission rate R which is supportable under the constraints on the outage probability and the average power consumption. This definition is called the *outage capacity* and is mathematically formulated as

$$\begin{aligned} \underset{P(g(n))}{\text{maximize}} \quad & R \\ \text{subject to} \quad & P_{\text{out}}(R) \leq \varepsilon, \quad E[P(g(n))] \leq P, \end{aligned} \tag{6.125}$$

for the maximum acceptable outage probability ε.

As to using the transmission power, we can make the following observations. First, if the system is in outage at a channel state, it is optimal to use zero transmission power at that state. Second, at the channel state where the outage does not occur, it is optimal to set the transmission power such that the resulting supportable transmission rate becomes identical, as any excessive transmission power would be wasted. This means that the received power, $g(n)P(g(n))$, is kept at a constant level if the system is not in outage.

The above observations lead us to the optimal power adaptation

$$P^*(g(n)) = \begin{cases} 0, & \text{if the system is in outage} \\ P_{\text{rec}}/g(n), & \text{otherwise} \end{cases} \qquad (6.126)$$

for a constant level of the received power P_{rec}. In practice, since much transmission power is required to maintain the reliability at a smaller channel gain, the optimal power adaptation allows outage when the channel gain becomes smaller than a threshold. Thus we may rewrite (6.126) by

$$P^*(g(n)) = \begin{cases} 0, & \text{if } g(n) < g_0 \\ P_{\text{rec}}/g(n), & \text{otherwise} \end{cases} \qquad (6.127)$$

for a channel gain threshold g_0, which is obtained from the outage probability requirement by $\Pr\{g(n) < g_0\} = \varepsilon$. The received power level P_{rec} is obtained from the average power constraint by $\mathbb{E}[P^*(g(n))] = P$. The optimal power adaptation in (6.127) is called *truncated channel inversion* as the transmission power is inversely proportional to the channel gain and is truncated to zero if the channel gain is smaller than the threshold. Figure 6.23 depicts this power adaptation graphically.

Truncated channel inversion tends to allocate more transmission power to a worse channel. This is because no buffering is allowed in the slowly fading model and the transmitter has no choice but to compensate the bad channel condition by utilizing a larger transmission power. An interpretation of this operation is that, when the delay requirement is stringent, we have to use more transmission power to save the channels in poor state. However, when the channel state becomes

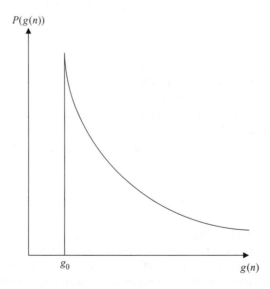

Figure 6.23   Graphical illustration of truncated channel inversion

extremely poor, requiring a significant level of power for compensation, it is desirable to abandon the transmission and declare outage, thereby saving the energy for use at a better channel state.

6.3.1.2 Fast-fading Channel: Ergodic Capacity

Consider the case where the channel state varies very fast. More specifically, traffic has a long delay constraint such that different channel states can exist during a transmission burst. In this case, the entire arrival is fragmented into multiple codewords, with different codewords undergoing different channel states. The encoding rate of each codeword is determined by its transmission power and the channel gain it experiences. This models the *fast-fading* situation where the delay requirement is much longer than the coherence time of the wireless channel. In this fast-fading case, the transmitter is allowed to delay the transmission of data until the channel condition becomes favorable.

Consider a transmission burst consisting of N different channel states. Assume that the channel remains constant over a coherence period for each channel realization. Then, the channel over N such coherence periods may be regarded as a parallel channel with N subchannels that have independent channel states. If the channel condition for the N periods is given by $g(1), \ldots, g(N)$, the maximum reliable transmission rate of a transmission burst is formulated as

$$\begin{aligned} \underset{P(g(n))}{\text{maximize}} \quad & \frac{1}{N} \sum_{n=1}^{N} \log_2 \left(1 + \frac{g(n)P(g(n))}{\sigma^2} \right) \\ \text{subject to} \quad & \frac{1}{N} \sum_{n=1}^{N} P(g(n)) \leq P. \end{aligned} \tag{6.128}$$

This problem is equivalent to the power allocation problem over parallel subchannels in Section 6.2. Thus, the optimal solution is the water-filling method, given by

$$P^*(g(n)) = \left[\frac{1}{\lambda} - \frac{\sigma^2}{g(n)} \right]^+, \tag{6.129}$$

where λ is determined by the power constraint

$$\frac{1}{N} \sum_{n=1}^{N} \left[\frac{1}{\lambda} - \frac{\sigma^2}{g(n)} \right]^+ = P. \tag{6.130}$$

Since the water-filling is done over the time slot, we can call it *time-domain water-filling*.

The optimal power allocation in (6.129) depends on the water level $1/\lambda$, which in turn depends on all the other channel gains by the power constraint in (6.130). Thus, implementation of the solution in (6.129) requires knowledge of the future channel states. That is, the transmitter needs to know the channel gains $g(1), \ldots, g(N)$ even for the first transmission slot. This non-causality requirement of the time-domain water-filling would be resolved if the traffic is delay-unconstrained so that a transmission burst consists of infinite coherence periods;

that is, $N \to \infty$. Since the fading process is assumed to be ergodic, almost all the possible realizations are supposed to appear during a transmission burst and the frequency of appearance of each state is proportional to its probability density. Then, as $N \to \infty$, (6.130) converges to

$$\mathbb{E}\left[\left[\frac{1}{\lambda} - \frac{\sigma^2}{g(n)}\right]^+\right] = P \qquad (6.131)$$

by the law of large numbers. Since the expectation in (6.131) is taken with respect to the marginal distribution of the channel gain, the water-level is independent of the specific realization of the fading process and thus can be obtained from the statistical information of $\{g(n)\}$. Hence, the optimal power adaptation in (6.129) depends only on the channel gain at each time slot. Figure 6.24 illustrates the time-domain water-filling method.

The capacity of this fast-fading channel model takes the expression

$$C_{\text{erg}} = \mathbb{E}\left[\log_2\left(1 + \frac{g(n)P^*(g(n))}{\sigma^2}\right)\right], \qquad (6.132)$$

where the optimal power adaptation $P^*(g(n))$ is determined by (6.129) and (6.131). This is called the *ergodic capacity* as it means the maximum achievable rate averaged over all fading states.

In contrast to the truncated channel inversion of the slowly fading model, time-domain water-filling tends to allocate more transmission power to a better channel in this fast-fading model. This is because the transmitter is allowed to buffer the data arrival until the channel condition becomes favorable. Thus, when the delay requirement is relatively loose compared with the coherent time of the fading process, it is more desirable to deliver a large amount of information with high transmission power on a favorable channel.

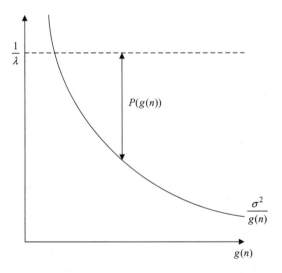

Figure 6.24 Time-domain water-filling

6.3.1.3 Compromise Model: Effective Capacity

Both definitions of capacity – outage and ergodic – are based on extreme assumptions on the fading speed of the wireless channel and the delay requirement of arriving traffic. In the model for outage capacity, the delay requirement is assumed to be extremely stringent. Arriving traffic is not allowed to be buffered because any buffering will violate the stringent delay requirement. Thus, at every slot time, incoming traffic is forced to be transmitted immediately if the channel condition is acceptable or to be dropped off otherwise. In this sense, the truncated channel inversion method that achieves outage capacity may be regarded as if it were developed for bufferless wireless links. On the other hand, no delay requirement is assumed in the model for ergodic capacity. Any arriving traffic is allowed to be buffered for an arbitrary time and no restriction exists on the amount of data transmitted in each slot time. In this sense, the water-filling method that achieves the ergodic capacity may be regarded as if it were developed for wireless links with infinite buffer size.

Due to such extreme assumptions, neither of the two capacity models can be practically applied to a wireless link that has a moderate delay requirement. The channel inversion method would consume too much transmission power for providing unnecessarily low delay. Meanwhile, the water-filling method would not provide any means to meet the delay requirement.

One way to compromise between the two capacity models is to introduce the concept of *effective capacity* discussed in Section 5.5. The effective capacity is defined as the maximum constant arrival rate that a given service process $R(n)$ can support in order to guarantee a statistical delay requirement. More specifically, the steady-state probability that the queue length process exceeds a threshold x is approximated by

$$\Pr\{Q(\infty) \geq x\} \approx e^{-\theta x} \tag{6.133}$$

for a given QoS parameter θ (>0). As noted in the previous chapter, the parameter θ determines how fast the buffer overflow probability decays as the queue threshold x increases. A higher θ implies a more stringent delay requirement since it imposes a lower buffer overflow probability. Then, the effective capacity that can support the overflow requirement in (6.133) is given by

$$C_{\text{eff}} = -\frac{1}{\theta}\ln(\mathbb{E}\left[e^{-\theta R(n)}\right]), \tag{6.134}$$

in a block fading channel where the channel remains constant within each time slot and independently changes to another state at the end of each slot. Therefore, we can adjust the definition of the "capacity" by controlling the QoS parameter θ according to the delay requirement of the arriving traffic. The transmission rate at time slot n is given by

$$R(n) = \log_2\left(1 + \frac{g(n)P(g(n))}{\sigma^2}\right) \tag{6.135}$$

for a power adaptation method $P(g(n))$.

With the above definitions, we can formulate a power adaptation method that maximizes the effective capacity in (6.134) under a constraint on average power consumption (Tang and Zhang, 2007). The formulation is given by

$$
\underset{P(g(n))}{\text{maximize}} \quad -\frac{1}{\theta}\ln\left\{\int_0^\infty \exp\left(-\theta\log_2\left(1+\frac{g(n)P(g(n))}{\sigma^2}\right)\right)dF_G(g(n))\right\}
$$

$$
\text{subject to} \quad \int_0^\infty P(g(n))dF_G(g(n)) \leq P,
$$

(6.136)

for a properly given QoS parameter θ. The objective of this problem is equivalent to

$$
\underset{P(g(n))}{\text{minimize}} \quad \left\{\int_0^\infty \exp\left(-\theta\log_2\left(1+\frac{g(n)P(g(n))}{\sigma^2}\right)\right)dF_G(g(n))\right\}
$$

(6.137)

as θ is a given parameter, and the overall problem becomes a convex optimization problem. Then, we can form the Lagrangian function

$$
L = \int_0^\infty \exp\left(-\theta\log_2\left(1+\frac{g(n)P(g(n))}{\sigma^2}\right)\right)dF_G(g(n)) + \lambda(\theta)\left(\int_0^\infty P(g(n))dF_G(g(n)) - P\right)
$$

$$
= (\ln 2)^\theta \int_0^\infty \left(1+\frac{g(n)P(g(n))}{\sigma^2}\right)^{-\theta} dF_G(g(n)) + \lambda(\theta)\left(\int_0^\infty P(g(n))dF_G(g(n)) - P\right),
$$

(6.138)

for a Lagrange multiplier $\lambda(\theta)$. On differentiating (6.138) we get

$$
\frac{\partial L}{\partial P(g(n))} = (\ln 2)^\theta \left\{\lambda'(\theta) - \theta\frac{g(n)}{\sigma^2}\left(1+\frac{g(n)P(g(n))}{\sigma^2}\right)^{-\theta-1}\right\}dF_G(g(n)) = 0
$$

(6.139)

for $\lambda'(\theta) = (\ln 2)^{-\theta}\lambda(\theta)$. On rearranging this relation, we derive the optimal power adaptation that maximizes the effective capacity as

$$
P^*(g(n)) = \left[\left(\frac{\theta}{\lambda'(\theta)}\right)^{\frac{1}{\theta+1}}\left(\frac{\sigma^2}{g(n)}\right)^{\frac{\theta}{\theta+1}} - \frac{\sigma^2}{g(n)}\right]^+,
$$

(6.140)

where $\lambda'\theta$ is determined such that the average power constraint is met with equality for the given QoS parameter θ. Observe in this power allocation that the transmission power is set to a positive level only if the current channel gain normalized by the noise variance is higher than a given threshold $\lambda'(\theta)/\theta$. More specifically, a positive transmission power is used only when

$$
\left(\frac{\theta}{\lambda'(\theta)}\right)^{\frac{1}{\theta+1}}\left(\frac{\sigma^2}{g(n)}\right)^{\frac{\theta}{\theta+1}} > \frac{\sigma^2}{g(n)} \Rightarrow \frac{g(n)}{\sigma^2} > \frac{\lambda'(\theta)}{\theta}.
$$

(6.141)

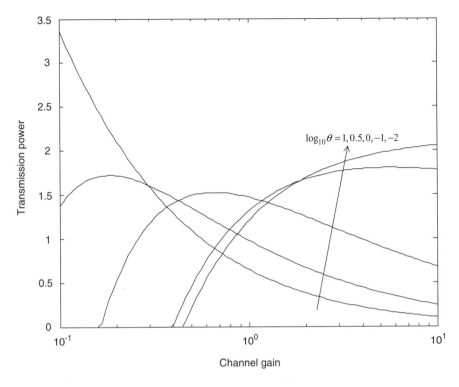

Figure 6.25   Optimal power adaptation for effective capacity maximization

The optimal power adaptation that maximizes the effective capacity heavily depends on the QoS parameter θ. Figure 6.25 illustrates the optimal power allocation in (6.140) for various QoS parameters. Based on this figure, we can state that the power allocation in (6.140) is able to compromise between the water-filling and channel-inversion algorithms according to the delay requirement. For a small θ, the power adaptation assigns more power to a better channel and less power to a worse channel, while exploiting the opportunistic transmission of a time-varying channel with loose delay requirement. In contrast, for a large θ, the power adaptation assigns more power to a worse channel in order to compensate for the destructive fading.

As the QoS parameter θ varies between 0 and ∞, reflecting different delay requirements, the corresponding power adaptation swings between the water-filling and the channel-inversion algorithms as follows. When the delay requirement is so loose that $\theta \to 0$, the transmission power in (6.140) becomes

$$\lim_{\theta \to 0} P^*(g(n)) = \left[\frac{1}{\lambda_{WF}} - \frac{\sigma^2}{g(n)} \right]^+ \tag{6.142}$$

for a positive threshold $1/\lambda_{WF} = \lim_{\theta \to 0} (\theta/\lambda'(\theta))^{1/(\theta+1)}$ which determines the minimum channel gain to which a positive transmission power is allocated. In this extremely loose delay requirement, the optimal transmission power in (6.140) reduces to the water-filling algorithm which is known to be the optimal strategy without delay constraint. On the other hand, when the delay requirement is so stringent that $\theta \to \infty$, (6.140) becomes

$$\lim_{\theta \to \infty} P^*(g(n)) = \frac{1}{\lambda_{CI}} \frac{\sigma^2}{g(n)} \tag{6.143}$$

for a positive value $1/\lambda_{CI} = \lim_{\theta \to \infty} (\theta/\lambda(\theta))^{1/(\theta+1)}$. This implies that the optimal transmission power in (6.140) converges to the simple channel-inversion algorithm when the delay constraint is extremely stringent.

For a moderate delay constraint, the optimal power adaptation is a mixture of the water-filling and channel-inversion methods. When the channel gain is relatively low, it operates like the water-filling method by using more power at a better channel condition and transmitting no data when the channel condition is below a threshold. This operation achieves the energy efficiency with an increased average delay, but that is not a big problem as the moderate delay requirement admits this delay increment. On the other hand, the power adaptation begins to decrease the transmission power as in the channel-inversion case as the channel gain grows above a given level. This is to compensate for the delay increment caused at the poor-condition channel; but this compensation can be done without increasing the power consumption greatly, since not much power is required in favorable channel states.

6.3.2 Transmission Time and Energy Efficiency

Consider now the power adaptation to time-varying traffic arrivals. In order to concentrate on the effect of the arrival pattern, assume a time-invariant channel; that is, $g(n) = g$. Then, the energy efficiency of a transmission power management scheme is determined wholly by the relation between the instantaneous rate and the transmission power, which is

$$R = \log_2 \left(1 + \frac{gP}{\sigma^2} \right) \Leftrightarrow P = \frac{\sigma^2(2^R - 1)}{g}. \tag{6.144}$$

According to this relation, a unit increment in the transmission rate requires an exponential increase in the transmission power. Thus, if it is permissible to reduce the transmission rate, the power consumption can be reduced at the cost of increased delay. This means that more energy-efficient transmission can be done by increasing the data transmission time. Denote by $E(\tau)$ the energy consumed in transmitting a bit, for the number of transmissions per bit, $\tau = 1/R$. Then, when we transmit B bits in transmission time τB, the energy consumption becomes

$$E(\tau) = P \times \tau B \times \frac{1}{B} = P\tau = \frac{\tau\sigma^2(2^{1/\tau} - 1)}{g}, \tag{6.145}$$

which is monotonically decreasing and convex in τ. Figure 6.26 shows the energy per bit with respect to transmission time.

The above properties of $E(\tau)$ give some insights on achieving energy-efficient transmissions. First, since the energy consumption decreases monotonically as the transmission time

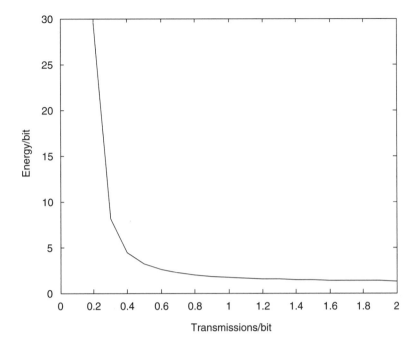

Figure 6.26 Plots of the relation in (6.145) for $\sigma^2/g = 1$

increases, it is beneficial to increase the transmission time as much as possible. Second, when transmitting two data packets within a given time, it is most energy-efficient to equalize the transmission times of the packets. For example, if the packet transmission times are τ_1 and τ_2 and the time limit is given by $T = \tau_1 + \tau_2$, the energy consumption decreases as the difference $|\tau_1 - \tau_2|$ decreases and it reaches the minimum when $\tau_1 = \tau_2 = T/2$. This property can be easily proved by applying Jensen's inequality to the convexity of $E(\tau)$.

However, there are several limiting factors in determining the transmission time. The delay requirement of the application constrains the transmission time not to grow arbitrarily. In addition, sometimes, the transmission time of each packet cannot be equalized. In the above example, if the second packet arrives after time $T/2$ has elapsed since the beginning of the first packet transmission, the maximum possible transmission time of the packet becomes smaller than $T/2$. Therefore, for an energy-efficient transmission, the above properties of $E(\tau)$ should be utilized with a careful consideration of those limiting factors.

Two different power adaptation methods are discussed in the following, each of which renders the highest energy-efficiency under its own formulation.

6.3.2.1 Under a Finite Time Limit

Consider a transmitter–receiver pair whose transmission time is limited by a finite value T. Assume that M packets with the same length arrive at the transmitter in the time interval $[0, T)$ and they must be transmitted before time T. Denote by t_m the arrival time of the mth packet and by d_m the inter-arrival time between the mth and the $(m + 1)$th packets. The transmission

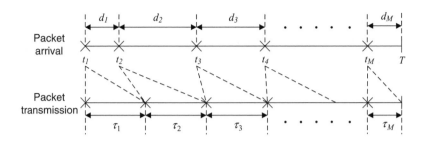

Figure 6.27   Packet arrival and transmission in $[0, T)$

scheduler determines the transmission time $\{\tau_m\}$ (equivalently, the transmission rate $\{R_m\}$ and the transmission power $\{P_m\}$) of each packet. The objective is to find the transmission schedule which minimizes the total energy consumption $\sum_{m=1}^{M} E(\tau_m)$ while transmitting all the M packets by the time limit T. We assume that the energy function $E(\tau)$ is a monotonically decreasing and convex function of the transmission time τ. Figure 6.27 illustrates the packet arrivals and transmissions. In this formulation, the time limit T may be interpreted as the delay constraint. In addition, a packet cannot be transmitted before it arrives at the transmitter, which constraint may be represented by

$$\sum_{m=1}^{k} \tau_m \geq \sum_{m=1}^{k} d_m \tag{6.146}$$

for all $k = 1, 2, \ldots, M-1$ and $\sum_{m=1}^{M} \tau_m = \sum_{m=1}^{M} d_m.$[25] We can call this the *feasibility constraint*.

The above problem is solved in Uysal-Biyikoglu *et al.* (2002). First consider the offline algorithm that is assumed to know the whole arrival process at the start time. This offline algorithm provides a lower bound on the energy consumption and also gives an insight into the design of an efficient power-adaptation method. Denote by $\{\tau_m^*\}$ the offline optimal transmission schedule.

One property of the offline optimal schedule is that the transmission time of a packet is longer than or equal to that of a packet that arrives later; that is, $\tau_m^* \geq \tau_{m+1}^*$ for $m = 1, 2, \ldots, M-1$. We can prove this by contradiction. Suppose $\tau_m^* < \tau_{m+1}^*$ for any m. Then it is feasible to equalize the transmission times of the two packets since the longer transmission time of the $(m+1)$th packet means that it arrives before a half of the sum of the two transmission times (i.e., before the time $\sum_{i=1}^{m-1} \tau_i^* + (\tau_m^* + \tau_{m+1}^*)/2$). Then, from the convexity of $E(\tau)$, the equalized time schedule consumes less energy, which contradicts the optimality of $\{\tau_m^*\}$. This property comes from the fact that equalizing the transmission time renders minimum energy consumption. The inequality of packet transmission times is caused by the feasibility constraint which may not

---

[25] The whole transmission may terminate before the time limit T but this case cannot achieve the optimal energy consumption due to the assumption on the energy consumption. Hence, we can only consider the "non-idling" scheduling which always has a packet to transmit in the time duration $[0, T]$.

allow transmission time equalization for some packet arrival patterns. For example, if the last packet arrives just before the time limit, it may not be possible to equalize its transmission time with the others.

Now consider the following transmission schedule. For given packet inter-arrival times, define

$$u_1 = \max_{k \in \{1,2,\dots,M\}} \left\{ \frac{1}{k} \sum_{i=1}^{k} d_i \right\}, \tag{6.147}$$

$$k_1 = \max_{k \in \{1,2,\dots,M\}} \left\{ k : \frac{1}{k} \sum_{i=1}^{k} d_i = u_1 \right\}, \tag{6.148}$$

and for $j \geq 1$ define

$$u_{j+1} = \max_{k \in \{1,2,\dots,M-k_j\}} \left\{ \frac{1}{k} \sum_{i=1}^{k} d_{k_j+i} \right\}, \tag{6.149}$$

$$k_{j+1} = k_j + \max_{k \in \{1,2,\dots,M-k_j\}} \left\{ k : \frac{1}{k} \sum_{i=1}^{k} d_{k_j+i} = u_{j+1} \right\}. \tag{6.150}$$

The process is repeated to obtain the pairs (u_j, k_j) until $k_j = M$. Let $J = \min\{j : k_j = M\}$. Define a transmission schedule $\{\hat{\tau}_m\}$ by

$$\hat{\tau}_m = u_j \quad \text{if} \quad k_{j-1} < m \leq k_j \text{ with } k_0 = 0. \tag{6.151}$$

This schedule can be interpreted as follows. It categorizes all the packets into J groups. The packets $k_{j-1} + 1$, $k_{j-1} + 2$, $\dots$, and k_j belong to group j and all of them have the same transmission time u_j. We have from (6.150) that

$$\sum_{m=1}^{k_j} \hat{\tau}_m = \sum_{i=1}^{j} (k_i - k_{i-1}) u_i = \sum_{i=1}^{j} \sum_{k=1}^{k_i - k_{i-1}} d_{k_{i-1}+k} = \sum_{m=1}^{k_j} d_m, \tag{6.152}$$

and this implies that the transmission of the k_jth packet (i.e., the last packet in group j) finishes when the $(k_j + 1)$th packet (i.e., the first packet in group $(j + 1)$) arrives. This property plays the role of "separation" since the transmission of the first packet in each group begins at the time of its arrival, thereby not being affected by the transmission times of the previous groups. The constituent packets of group j are determined such that the common transmission time u_j can be maximized while satisfying the group separation property in (6.152).

Figure 6.28 is an illustration of the transmission schedule $\{\hat{\tau}_m\}$ for a given packet arrival pattern (Uysal-Biyikoglu et al., 2002).

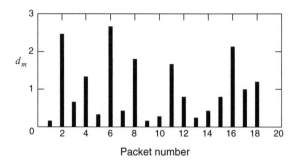

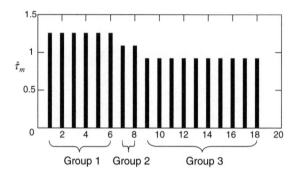

Figure 6.28 An example of the transmission schedule $\{\hat{\tau}_m\}$ (Reproduced with permission from E. Uysal-Biyikoglu, B. Prabhakar, and A. El Gamal, "Energy-efficient packet transmission over a wireless link," *IEEE/ACM Transactions on Networking*, **10**, August, 487–499, 2002. © 2002 IEEE)

We next consider the feasibility of the above transmission schedule $\{\hat{\tau}_m\}$. For a packet in group j (i.e., the kth packet with $k_{j-1} < k \le k_j$), we get

$$\sum_{m=1}^{k} \hat{\tau}_m = \sum_{m=1}^{k_{j-1}} d_m + \sum_{m=k_{j-1}+1}^{k_j} \hat{\tau}_m, \tag{6.153}$$

where the equality is due to (6.152). Then, by the definition of k_j, we get

$$\sum_{m=k_{j-1}+1}^{k_j} \hat{\tau}_m = (k_j - k_{j-1})u_j \ge \sum_{m=k_{j-1}+1}^{k_j} d_m, \tag{6.154}$$

and the feasibility can be shown by substitution into (6.153).

Now consider the energy efficiency of the time schedule $\{\hat{\tau}_m\}$. Assume another feasible schedule $\{\tau_m\}$. Define an index a to be the first packet number whose transmission time of the two schedules do not coincide; that is, $a = \arg\min\{m : \hat{\tau}_m \ne \tau_m\}$. This means that $\tau_m = \hat{\tau}_m$ for $m < a$. Then, there are two possibilities to consider as follows.

One possibility is the case $\hat{\tau}_a < \tau_a$. Since the time schedule $\{\hat{\tau}_m\}$ is non-idle, there must be at least one $b > a$ for which $\hat{\tau}_b > \tau_b$, as otherwise the schedule $\{\tau_m\}$ cannot complete the whole transmission by the time limit T. Let b be the index of the first packet with $\hat{\tau}_b > \tau_b$. Then, we consider the time schedule $\{\tilde{\tau}_m\}$ with

$$
\begin{aligned}
\tilde{\tau}_a &= \tau_a - \Delta, \\
\tilde{\tau}_b &= \tau_b + \Delta, \\
\tilde{\tau}_m &= \tau_m \quad \text{for all} \quad m \neq a, b,
\end{aligned}
\tag{6.155}
$$

where $\Delta = \min\{(\tau_a - \hat{\tau}_a), (\hat{\tau}_b - \tau_b)\}$. The schedule $\{\tilde{\tau}_m\}$ is obtained by making the transmission time of the ath or bth packet equal to that of $\{\hat{\tau}_m\}$, while using the transmission time of $\{\tau_m\}$ for the other packets. From the definition of $\{\tilde{\tau}_m\}$, we get

$$
\begin{aligned}
\tilde{\tau}_m &= \hat{\tau}_m \quad \text{for} \quad 1 \leq m \leq a-1, \\
\tilde{\tau}_m &\geq \hat{\tau}_m \quad \text{for} \quad a \leq m \leq b-1,
\end{aligned}
\tag{6.156}
$$

and by applying the feasibility of $\{\hat{\tau}_m\}$ we get

$$
\sum_{m=1}^{k} \tilde{\tau}_m \geq \sum_{m=1}^{k} \hat{\tau}_m \geq \sum_{m=1}^{k} d_m, \quad \text{for} \quad 1 \leq k \leq b-1.
\tag{6.157}
$$

In addition, from the feasibility of $\{\hat{\tau}_m\}$, we obtain

$$
\sum_{m=1}^{k} \tilde{\tau}_m = \sum_{m=1}^{k} \tau_m \geq \sum_{m=1}^{k} d_m, \quad \text{for} \quad b \leq k \leq M.
\tag{6.158}
$$

The inequalities in (6.157) and (6.158) altogether imply that the time schedule is feasible. When comparing the energy consumption of the two feasible schedules $\{\tau_m\}$ and $\{\tilde{\tau}_m\}$, it is easy to verify that the schedule $\{\tilde{\tau}_m\}$ consumes less energy since the transmission times of the ath and bth packets are better equalized.

The other possibility is the case $\hat{\tau}_a > \tau_a$. Suppose that the time schedule $\{\tau_m\}$ does not satisfy the relation $\tau_m \geq \tau_{m+1}$. Then, the schedule $\{\tau_m\}$ cannot be the offline optimal and we can easily reduce the energy consumption. Thus, we assume that $\tau_m \geq \tau_{m+1}$ for $m = 1, 2, \ldots, M-1$. We additionally assume that the ath packet is in group j (i.e., $k_{j-1} < a \leq k_j$). Then

$$
\sum_{m=1}^{k_j} \tau_m = \sum_{m=1}^{a-1} \hat{\tau}_m + \sum_{m=a}^{k_j} \tau_m \leq \sum_{m=1}^{a-1} \hat{\tau}_m + (k_j - a + 1)\tau_a < \sum_{m=1}^{a-1} \hat{\tau}_m + (k_j - a + 1)\hat{\tau}_a = \sum_{m=1}^{k_j} \hat{\tau}_m,
\tag{6.159}
$$

where the last equality holds due to the fact that all the packets in a group have the same transmission time for the time schedule $\{\hat{\tau}_m\}$. But, by definition of $\{\hat{\tau}_m\}$, we get $\sum_{m=1}^{k_j} \hat{\tau}_m = \sum_{m=1}^{k_j} d_m$ and, by inserting this into (6.159), we get $\sum_{m=1}^{k_j} \tau_m < \sum_{m=1}^{k_j} d_m$. This implies that the time schedule $\{\tau_m\}$ is not feasible, which contradicts the earlier feasibility assumption of $\{\tau_m\}$.

From the above discussion, it is verified that any feasible time schedule other than $\{\hat{\tau}_m\}$ cannot achieve optimality. If a time schedule other than $\{\hat{\tau}_m\}$ is employed, there exists at least one time schedule that is feasible and consumes less energy. Consequently, the time schedule $\{\hat{\tau}_m\}$ is equivalent to the offline optimal schedule $\{\tau_m^*\}$ and transmits all the packets by the time limit with the least energy consumption. This optimality can be easily extended to the case of variable-length packets (Uysal-Biyikoglu *et al.*, 2002) and to the case of multiple users (El Gamal *et al.*, 2002).

Observe in Figure 6.28 how this offline optimal schedule trades delay for energy efficiency. In the figure, the inter-arrival time highly fluctuates from packet to packet, but the inter-departure time (i.e., the transmission time) does not deviate much for each packet. For this tradeoff, the transmitter should necessarily buffer the arriving packets. In other words, the buffer absorbs the high randomness of the arrival process.

Now consider the offline optimal schedule in relation to the buffer state. The mth packet starts its transmission at time T_m^* which is given by

$$T_m^* = \sum_{i=1}^{m-1} \tau_i^*. \tag{6.160}$$

When the mth packet starts its transmission there are q_m packets in the buffer, where

$$q_m = \max\left\{ k : \sum_{i=1}^{k-1} d_i \le T_m^* \right\} - m, \tag{6.161}$$

which may be called the *backlog* for the mth packet. This relation is nothing but the difference between the number of packets that have arrived before time T_m^* and the number of packets departed until that time. Note that this backlog does not include the mth packet. That is, if $q_m = 1$, there is precisely one packet, the $(m + 1)$th packet, in the buffer when the mth packet starts transmitting. In addition, we denote by $c_i(T_m^*)$ the inter-arrival time of the packet that arrives for the ith time after T_m^*. The above notations imply that, when the mth packet starts transmitting (i.e., at time T_m^*), there are q_m packets backlogged and $M - m - q_m$ packets are yet to arrive with the arrival times

$$T_m^* + c_1(T_m^*), T_m^* + c_1(T_m^*) + c_2(T_m^*), \ldots, T_m^* + \sum_{i=1}^{M-m-q_m} c_i(T_m^*).$$

Then, from the properties of the offline optimal schedule $\{\tau_m^*\}$ discussed above, the optimal transmission time can be represented by

$$\tau_m^* = \max_{k \in \{1, \ldots, M-m-b_m\}} \left\{ \frac{1}{k + q_m} \sum_{i=1}^{k} c_i(T_m^*) \right\}. \tag{6.162}$$

The relation in (6.162) may be interpreted as follows. Suppose that the mth packet belongs to group j and the scheduler decides to include $(k - 1)$ additional arrivals in the group. Then, at time T_m^*, the residual transmission time for group j is equal to $\sum_{i=1}^{k} c_i(T_m^*)$. This residual transmission time is equally shared by $k + q_m$ packets (i.e., q_m in the buffer, $(k - 1)$ yet to arrive, and one being transmitted) that are included in group j but have not completed their transmission. Therefore, $\sum_{i=1}^{k} c_i(T_m^*)/(k + q_m)$ becomes the transmission time of each packet in the group and the optimal schedule is determined such that this transmission time can be maximized for each packet in the group. It is noteworthy that the transmission time in (6.162) is just an alternative representation of the offline optimal schedule defined in (6.147)–(6.150) and gives exactly the same schedule. It schedules packets one at a time, taking into account the current backlog and future arrivals.

The alternative representation of the offline optimal schedule in (6.162) suggests a transmission schedule that is adaptive to the current buffer state. Since the future arrival is not available actually, an online schedule should be proposed for practical use. For online scheduling algorithms proposed in (Uysal-Biyikoglu, 2002), the transmission time in (6.162) is a random variable as the inter-arrival times $c_i(T_m^*)$ are random. The proposed online algorithm simply sets the transmission time equal to the expectation of (6.162). This means that the transmission time of a packet which starts transmission at time $t < T$ with q backlogs is

$$\tau^*(q, t) = \mathbb{E}\left[\max_{k \in \{1,\dots,M\}} \left\{\frac{1}{k+q} \sum_{i=1}^{k} c_i(t)\right\}\right], \qquad (6.163)$$

where $c_i(t)$ is a random variable denoting the inter-arrival time of packets that will arrive in (t, T). This online algorithm can be a practical alternative to the offline optimal schedule, as it performs close to the optimum in terms of energy consumption.

6.3.2.2 Under an Average Delay Constraint

Now consider transmission power adaptation under constrained average packet delay. Assume a slotted system with the average packet generation rate of $\bar{A}$ packets per time slot. $A(n)$ packets with the same size arrive at the transmitter in time slot n. $A(n)$ has a distribution $\Pr\{A(n) = a\}$ with a finite support $[0, \dots, M]$ (i.e., the largest number of packets that arrive in a time slot is M), and the arrival process $\{A(n)\}$ is i.i.d. from slot to slot. Assume that the buffer size at the transmitter is infinitely large to make the system lossless,[26] and the transmitter sends $U(n)$ packets in time slot n by adjusting the transmission rate and the corresponding transmission power. Since the transmission rate should be determined before the transmission of each time slot, all the packets that arrive in time slot n can be transmitted only in time slot $(n + 1)$ or later. Thus, a natural constraint on $U(n)$ is that $0 \leq U(n) \leq Q(n)$ for the number of packets in the buffer, $Q(n)$. Assume that transmission energy $E(U(n))$ is required to transmit $U(n)$ packets in a time slot. Since the energy consumption is calculated for each time slot of the same length, minimizing the average energy consumption is equivalent to minimizing the average power consumption. Thus, in this subsection, we consider the power consumption $P(U(n))$ which is required to transmit $U(n)$ packets in a time slot.

The relation between the number of transmitted packets and the power consumption is obtained by a simple modification of (6.144), as $U(n)$ is proportional to the transmission rate. Under this system model, the buffer state is updated by

$$Q(n+1) = Q(n) + A(n) - U(n). \qquad (6.164)$$

---

[26] In a finite buffer system, packet loss may occur when the remaining buffer space is smaller than the number of arrivals: If $A(n) > L - Q(n) + U(n)$ for the buffer size L, $A(n) - (L - Q(n) + U(n))$ packets are dropped and cannot be stored in the buffer. Since the dropped packets are not transmitted, one can easily reduce the energy consumption by allowing a large number of packet drops, but at a cost of unreliability of packet delivery. This is why we restrict the discussion on the lossless schedulers which do not allow any packet drop at the buffer and all the arrived packets are delivered to the receiver.

Then, by Little's theorem, the average packet delay is related to the average buffer length by

$$D_{avg} = \frac{1}{A} \mathbb{E}[Q(n)]. \tag{6.165}$$

In this formulation, a scheduler is defined by a mapping from the current buffer state $Q(n)$ to the number of transmitted packets, $U(n)$. The objective is to devise the optimal scheduler that minimizes the energy consumption while satisfying the delay constraint D_{req}. This may be formulated by

$$P^*(D_{req}) = \min_{D_{avg} \leq D_{req}} \left\{ \lim_{n \to \infty} \mathbb{E}[P(U(n))] \right\} \tag{6.166}$$

for the optimal power consumption $P^*(D_{req})$.

Since a scheduler is a mapping from $Q(n)$ to $U(n)$, each scheduler can be characterized by the probability $\alpha_{j,i} = \Pr\{U(n) = j | Q(n) = i\}$, meaning the probability that j packets are transmitted when i packets are in the buffer. Then, every scheduler α can be represented by a matrix with entries $\alpha_{j,i}$. Since any scheduler cannot transmit more packets than available in the buffer, we get

$$\alpha_{j,i} = 0 \quad \text{for} \quad j > i. \tag{6.167}$$

In addition, since $\alpha_{j,i}$ are probabilities, we have

$$\sum_{j=0}^{i} \alpha_{j,i} = 1. \tag{6.168}$$

A scheduler is said to be *deterministic* if $\alpha_{j,i} \in \{0, 1\}$; that is, the scheduler takes the same action for each buffer state all the time. Otherwise, it is labeled *randomized*. Note that every scheduler is memoryless as the number of transmitted packets is independent of the previous buffer state.

Since the source is assumed to be i.i.d. and the schedulers are assumed to be memoryless, the buffer state forms a first-order Markov chain. Then, by combining the probabilities of $\alpha_{j,i}$ and the distribution of $A(n)$, we have the buffer state transition matrix $\mathbf{C} = [C_{j,i}]$ where $C_{j,i}$ is the probability of transition from the buffer state $Q(n) = i$ to the state $Q(n + 1) = j$. This transition probability is expressed by

$$C_{j,i} = \sum_{k=0}^{i} \Pr\{A(n) = j - i + k\} \Pr\{U(n) = k | Q(n) = i\} = \sum_{k=0}^{i} \Pr\{A(n) = j - i + k\} \alpha_{k,i}. \tag{6.169}$$

Define by s_i the stationary probability of buffer state i, and define by $\mathbf{s} = [s_0, s_1, \ldots]^{\mathrm{T}}$ the steady-state probability vector. Then, from the definitions of $\mathbf{C}$ and $\mathbf{s}$, we get

$$\mathbf{Cs} = \mathbf{s}. \tag{6.170}$$

For the average packet delay and average transmission power, we get the expressions

$$D_{avg}(\alpha) = \frac{1}{A}\mathbb{E}\{Q(n)\} = \frac{1}{A}\left(\sum_{i=0}^{\infty} is_i\right), \tag{6.171}$$

$$P_{avg}(\alpha) = \mathbb{E}[P(U(n))] = \sum_{i=0}^{\infty}\sum_{j=0}^{i} s_i\alpha_{j,i}P(j). \tag{6.172}$$

Based on the above formulations, some analytical results have been made available about the optimal scheduler that minimizes the energy consumption while satisfying the average delay requirement (Rajan *et al.*, 2004). The main result is that deterministic schedulers form a basis of the set of randomized schedulers. The average delay achieved by any randomized scheduler is given by a convex combination of the average packet delays achieved by all possible zero-outage deterministic schedulers and, in addition, the same convex combination of the average powers of the deterministic schedulers gives the average power of that randomized policy.

Consider now two lossless schedulers, α_A and α_B, which are identical in all columns except for the kth column. A third scheduler, α, is given by a convex combination of the two schedulers; that is, $\alpha = \theta\alpha_A + (1-\theta)\alpha_B$ for $\theta \in [0, 1]$. Then, we get the state transition probability matrix $\mathbf{C}_\alpha = \theta\mathbf{C}_{\alpha_A} + (1-\theta)\mathbf{C}_{\alpha_B}$. If we assume that the steady-state probability vector $\mathbf{s}_\alpha$ is given by

$$\mathbf{s}_\alpha = \theta'\mathbf{s}_{\alpha_A} + (1-\theta')\mathbf{s}_{\alpha_B} \tag{6.173}$$

for the steady-state probability vector $\mathbf{s}_{\alpha_A}$ and $\mathbf{s}_{\alpha_B}$ respectively of the schedulers α_A and α_B, then, by applying (6.170) to $\mathbf{s}_\alpha$, we get

$$[\theta\mathbf{C}_{\alpha_A} + (1-\theta)\mathbf{C}_{\alpha_B}][\theta'\mathbf{s}_{\alpha_A} + (1-\theta')\mathbf{s}_{\alpha_B}] = \theta'\mathbf{s}_{\alpha_A} + (1-\theta')\mathbf{s}_{\alpha_B}, \tag{6.174}$$

which reduces to

$$[\mathbf{C}_{\alpha_A} - \mathbf{C}_{\alpha_B}][\theta(1-\theta')\mathbf{s}_{\alpha_B} - \theta'(1-\theta)\mathbf{s}_{\alpha_A}] = \mathbf{0}. \tag{6.175}$$

Since $\mathbf{C}_{\alpha_A}$ and $\mathbf{C}_{\alpha_B}$ differ only in the kth column, $[\mathbf{C}_{\alpha_A} - \mathbf{C}_{\alpha_B}]$ is a matrix with 0 elements in all columns other than the kth column. Thus, the relation in (6.175) is satisfied if and only if the kth element of the vector $[\theta(1-\theta')\mathbf{s}_{\alpha_B} - \theta'(1-\theta)\mathbf{s}_{\alpha_A}]$ becomes zero. This implies that

$$\theta(1-\theta')s_{\alpha_B,k} - \theta'(1-\theta)s_{\alpha_A,k} = 0 \tag{6.176}$$

which may be rearranged to

$$\theta' = \frac{\theta s_{\alpha_B,k}}{\theta s_{\alpha_B,k} + (1-\theta)s_{\alpha_A,k}}. \tag{6.177}$$

This implies that the steady-state probability vector $\mathbf{s}_\alpha$ is represented by a convex combination of $\mathbf{s}_{\alpha_A}$ and $\mathbf{s}_{\alpha_B}$; that is, the relation (6.173) holds.

The average delay of α in (6.171) can be rewritten as

$$D_{\mathrm{avg}}(\alpha) = \frac{1}{A}\left(\sum_{i=0}^{\infty} i\{\theta' s_{\alpha_A,i} + (1-\theta') s_{\alpha_B,i}\}\right) = \theta' D_{\mathrm{avg}}(\alpha_A) + (1-\theta') D_{\mathrm{avg}}(\alpha_B), \quad (6.178)$$

which implies that the average delay of the scheduler α is a convex combination of the average delay achieved by the schedulers α_A and α_B. Likewise, the average power in (6.172) can be rewritten as

$$P_{\mathrm{avg}}(\alpha) = \sum_{i=0}^{\infty}\sum_{j=0}^{i} s_{\alpha,i}\alpha_{j,i}P(j) = \sum_{i=0}^{\infty}\sum_{j=0}^{i}(\theta' s_{\alpha_A,i} + (1-\theta') s_{\alpha_B,i})(\theta\alpha_{A,j,i} + (1-\theta)\alpha_{B,j,i})P(j)$$

$$= \theta\theta' P_{\mathrm{avg}}(\alpha_A) + (1-\theta)(1-\theta') P_{\mathrm{avg}}(\alpha_B) + \sum_{i=0}^{\infty}\sum_{j=0}^{i}(\theta'(1-\theta) s_{\alpha_A,i}\alpha_{B,j,i}$$

$$+ \theta(1-\theta') s_{\alpha_B,i}\alpha_{A,j,i})P(j).$$

$$(6.179)$$

By (6.176) and the fact that $\alpha_{A,j,i} = \alpha_{B,j,i}$ for all $i \neq k$, we have

$$0 = (\alpha_{A,j,i} - \alpha_{B,j,i})(\theta(1-\theta') s_{\alpha_B,i} - \theta'(1-\theta) s_{\alpha_A,i}), \quad (6.180)$$

which can be rearranged as

$$\theta'(1-\theta) s_{\alpha_A,i}\alpha_{B,j,i} + \theta(1-\theta') s_{\alpha_B,i}\alpha_{A,j,i} = \theta'(1-\theta) s_{\alpha_A,i}\alpha_{A,j,i} + \theta(1-\theta') s_{\alpha_B,i}\alpha_{B,j,i}. \quad (6.181)$$

Finally, we insert (6.181) into (6.179) to get

$$P_{\mathrm{avg}}(\alpha) = \theta\theta' P_{\mathrm{avg}}(\alpha_A) + (1-\theta)(1-\theta') P_{\mathrm{avg}}(\alpha_B)$$

$$+ \sum_{i=0}^{\infty}\sum_{j=0}^{i}(\theta'(1-\theta) s_{\alpha_A,i}\alpha_{A,j,i} + \theta(1-\theta') s_{\alpha_B,i}\alpha_{B,j,i})P(j)$$

$$= \theta\theta' P_{\mathrm{avg}}(\alpha_A) + (1-\theta)(1-\theta') P_{\mathrm{avg}}(\alpha_B) + \theta'(1-\theta) P_{\mathrm{avg}}(\alpha_A) + \theta(1-\theta') P_{\mathrm{avg}}(\alpha_B)$$

$$= \theta' P_{\mathrm{avg}}(\alpha_A) + (1-\theta') P_{\mathrm{avg}}(\alpha_B),$$

$$(6.182)$$

which implies that the average power consumption also has the same convex combination relation as for the average delay.

As an example, consider the schedulers[27]

$$\alpha_A = \begin{bmatrix} 1 & 0 & 0 & 0 \\ 0 & 1 & 0 & 0 \\ 0 & 0 & 1 & 0 \\ 0 & 0 & 0 & 1 \end{bmatrix}, \quad \alpha_B = \begin{bmatrix} 1 & 1 & 0 & 0 \\ 0 & 0 & 0 & 0 \\ 0 & 0 & 1 & 0 \\ 0 & 0 & 0 & 1 \end{bmatrix}, \quad (6.183)$$

---

[27] These schedulers empty the buffer when they store four packets in the buffer. Thus, the probability that the buffer state grows over 4 becomes zero and hence it suffices to consider only the buffer states less than or equal to 4.

and $\alpha = 0.6\alpha_A + 0.4\alpha_B$ (i.e., $\theta = 0.6$) for a uniform arrival distribution with $M = 2$. Then, by (6.170) we get the steady-state probabilities

$$s_{\alpha_A} = \begin{bmatrix} \dfrac{1}{3} & \dfrac{1}{3} & \dfrac{1}{3} & 0 \end{bmatrix}, \quad s_{\alpha_B} = \begin{bmatrix} \dfrac{2}{3} & \dfrac{1}{3} & \dfrac{1}{3} & \dfrac{1}{9} \end{bmatrix}. \tag{6.184}$$

By (6.171) and (6.172), we get the scheduling results $D_{avg}(\alpha_A) = 1, P_{avg}(\alpha_A) = 2.7, D_{avg}(\alpha_B) = 1.33$ and $P_{avg}(\alpha_A) = 4.25$. Since $\theta' = 0.6$ by (6.177), we finally get, by (6.178) and (6.182), $D_{avg}(\alpha) = 1.13$ and $P_{avg}(\alpha) = 3.32$.

The above discussion reveals that, for a scheduler obtained by a convex combination of two schedulers which are identical except for a column, the average delay and the power consumption are a convex combination of those of the two individual schedulers. Conversely, it is also possible to decompose a randomized scheduler into a convex combination of multiple schedulers which differ only in one column and, if we repeat the decomposition, it eventually leads to a convex combination of deterministic schedulers. For example:

$$
\begin{bmatrix} 1 & 0.2 & 0.1 \\ 0 & 0.8 & 0.3 \\ 0 & 0 & 0.6 \end{bmatrix} = 0.2 \begin{bmatrix} 1 & 1 & 0.1 \\ 0 & 0 & 0.3 \\ 0 & 0 & 0.6 \end{bmatrix} + 0.8 \begin{bmatrix} 1 & 0 & 0.1 \\ 0 & 1 & 0.3 \\ 0 & 0 & 0.6 \end{bmatrix}
$$

$$
= 0.2 \left\{ 0.1 \begin{bmatrix} 1 & 1 & 1 \\ 0 & 0 & 0 \\ 0 & 0 & 0 \end{bmatrix} + 0.3 \begin{bmatrix} 1 & 1 & 0 \\ 0 & 0 & 1 \\ 0 & 0 & 0 \end{bmatrix} + 0.6 \begin{bmatrix} 1 & 1 & 0 \\ 0 & 0 & 0 \\ 0 & 0 & 1 \end{bmatrix} \right\}
$$

$$
+ 0.8 \left\{ 0.1 \begin{bmatrix} 1 & 0 & 1 \\ 0 & 1 & 0 \\ 0 & 0 & 0 \end{bmatrix} + 0.3 \begin{bmatrix} 1 & 0 & 0 \\ 0 & 1 & 1 \\ 0 & 0 & 0 \end{bmatrix} + 0.6 \begin{bmatrix} 1 & 0 & 0 \\ 0 & 1 & 0 \\ 0 & 0 & 1 \end{bmatrix} \right\}. \tag{6.185}
$$

In the first step of the decomposition, the scheduler has been decomposed into two schedulers that are different only in the second column. In the second decomposition, each scheduler has been decomposed into three schedulers that are different only in the third column. Note that all the schedulers after the second decomposition are deterministic schedulers.

By applying the above property repeatedly we can decompose a randomized scheduler into a convex combination of multiple deterministic schedulers. Accordingly the average delay and power consumption of a randomized scheduler is a convex combination of those of the deterministic schedulers. More rigorously, for any randomized scheduler α, there exists $\theta'_i \in [0, 1]$ with $\sum_{i=1}^{F'} \theta'_i = 1$ such that

$$
D_{avg}(\alpha) = \sum_{i=1}^{F'} \theta'_i D_{avg}(\alpha_i) \quad \text{and} \quad P_{avg}(\alpha) = \sum_{i=1}^{F'} \theta'_i P_{avg}(\alpha_i) \tag{6.186}
$$

for all the deterministic schedulers α_i with $i = 1, 2, \ldots, F'$. Here, F' denotes the number of deterministic schedulers.[28] This discussion establishes that the optimal scheduler can be constructed by a proper convex combination of optimal deterministic schedulers. The optimal delay–power curve becomes a piecewise-linear curve with the vertices achieved by optimal

---

[28] Refer to (Rajan *et al.*, 2004) for a detailed proof.

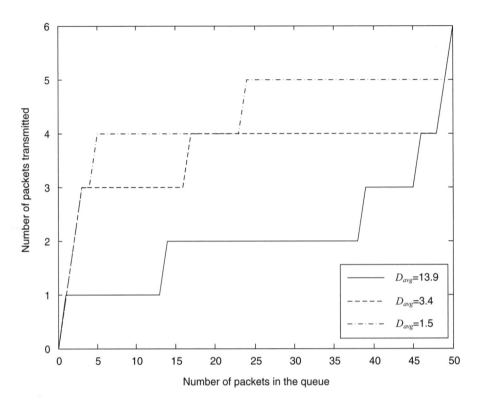

Figure 6.29 Number of packets transmitted at each queue state for the optimal scheduling policy with various delay requirements

deterministic schedulers. These optimal schedulers can be determined by dynamic programming methods. This method is discussed in the next subsection with a more generalized channel model including the time-varying channels.

Figure 6.29 shows an example of the optimal scheduler and the number of transmitted packets for each buffer state. The number of arriving packets is assumed to be an i.i.d. random variable distributed in the range [0, 6]. As the buffer state increases, the optimal policy transmits more packets to satisfy the delay constraint. However, the number of packets transmitted does not increase as fast as the buffer state; it increases slowly while still storing some packets in the buffer. For example, suppose that six packets arrive at an empty buffer at a time slot. Then, the optimal scheduler for the delay requirement of 3.4 slots transmits only three packets while storing the other packets in the buffer. This implies that a burst of packet arrivals is split over several time slots with differently delayed transmissions. Thus, the deviation of inter-departure time of the packets is much reduced when compared with the highly fluctuating inter-arrival time, and this operation is equivalent to the absorption of high randomness of the arrival process that was observed in Figure 6.28. We also observe that more packets need to be transmitted with a higher transmission power to achieve a lower delay. For the delay requirement of 1.5 slots, for example, almost all the arriving packets (in the range [0, 6]) are transmitted immediately to prevent the buffer from growing fast. Consequently, the delay and the energy efficiency are traded in this average delay requirement case.

6.3.3 Power Adaptation Based on Buffer and Channel States

Here we extend the power adaptation process to the case where both traffic arrival and channel state vary in time. The channel gain and traffic arrivals are given by $g(n)$ and $A(n)$, respectively. A scheduling policy determines the number of transmitted packets, $U(n)$, according to the current channel state $g(n)$ and the buffer state $Q(n)$ which is updated by

$$Q(n+1) = Q(n) + A(n) - U(n). \tag{6.187}$$

We assume an infinite buffer size for a simple formulation.

The objective of the formulation is minimization of the average transmission power under a constraint on the average packet delay; that is

$$P^*(D_{\text{req}}) = \min_{D_{\text{avg}} \leq D_{\text{req}}} \left\{ \lim_{n \to \infty} \mathbb{E}[P(g(n), U(n))] \right\}, \tag{6.188}$$

where

$$D_{\text{avg}} = \frac{1}{A} \mathbb{E}[Q(n)]. \tag{6.189}$$

The differences from the formulation in the previous subsection are that the power consumption is also affected by the channel state $g(n)$, and the transmission power is denoted by a function of two input parameters, $P(g(n)$ and $U(n))$.

Before considering how to obtain the optimal policy, we investigate the tradeoff relation between the average delay and the power consumption. It is obvious that $P^*(D_{\text{req}})$ is non-increasing with D_{req}. One observation is that, even in this time-varying communication environment, the tradeoff relation is convex, similarly to the relation between transmission time and power consumption (Berry and Gallager, 2002). Let D^1_{req} and D^2_{req} be two different delay requirements with the corresponding power consumptions $P^*(D^1_{\text{req}})$ and $P^*(D^2_{\text{req}})$. Then, let $\{g(n,w)\}_{n=1}^{\infty}$ and $\{A(n,w)\}_{n=1}^{\infty}$ be the given sample paths (corresponding to the sample point w) of the channel states and arrival states. Denote by $\{U^i(n,w)\}$ and $\{Q^i(n,w)\}$ the sequence of the optimal policy which attains $P^*(D^i_{\text{req}})$ and the resulting buffer state, respectively. Now consider a new policy $\{U^\theta(n,w)\}$ in the form

$$U^\theta(n,w) = \theta U^1(n,w) + (1-\theta)U^2(n,w) \quad \text{for} \quad 0 \leq \theta \leq 1, \tag{6.190}$$

and denote by $Q^\theta(n,w)$ the buffer state of this policy.[29] Since the initial buffer state is the same for all the policies – that is, $Q^1(0,w) = Q^2(0,w) = Q^\theta(0,w)$ – we have

$$Q^\theta(n,w) = \theta Q^1(n,w) + (1-\theta)Q^2(n,w) \quad \text{for all} \quad n, \tag{6.191}$$

---

[29] It should be noted that this kind of combination of scheduling policies is different from the combination discussed in the previous subsection. The one in Section 6.3.2 is the combination in the buffer state domain while the combination in (6.190) is done in the time domain for each sample path. For example, consider two different policies as follows. One policy sends four packets at the buffer state 4 and stays idle at the other states; and the other sends two and four packets at the buffet states 2 and 4, respectively. Then, if we combine these two policies with an equal weighting, the resulting policy sends two packets with probability 1/2 at the buffer state 2, and four packets at the buffer state 4; and it stays idle at the other states. However, if we consider the arrival process $A(n) = \{2,0,2,0,\ldots\}$, the combination in the time domain renders the scheduling sequence $U^\theta(n) = \{0,1,0,3,0,1,0,3,\ldots\}$, which is different from the one combined in the buffer domain.

by applying the mathematical induction to (6.187). By taking the expectation over all sample paths, we get the average delay of the new policy $\{U^{\theta}(n, w)\}$ as

$$D_{\text{req}}^{\theta} = \theta D_{\text{req}}^{1} + (1-\theta)D_{\text{req}}^{2}, \tag{6.192}$$

which implies that the average delay attained by a convex combination of two policies is expressed by the same convex combination of the delay of each combined policy. Then, by applying the convexity of $P(g(n)$ and $U(n))$ to (6.190), we get

$$P(g(n), U^{\theta}(n, w)) \leq \theta P(g(n), U^{1}(n, w)) + (1-\theta)P(g(n), U^{2}(n, w)), \tag{6.193}$$

and, by taking the expectations again, we get

$$P(g(n), U^{\theta}(n)) \leq \theta P^{*}(D_{\text{req}}^{1}) + (1-\theta)P^{*}(D_{\text{req}}^{2}). \tag{6.194}$$

The optimal policy with the delay requirement D_{req}^{θ} will consume less power than $\{U^{\theta}(n, w)\}$ does, so we have

$$P^{*}(\theta D_{\text{req}}^{1} + (1-\theta)D_{\text{req}}^{2}) \leq \theta P^{*}(D_{\text{req}}^{1}) + (1-\theta)P^{*}(D_{\text{req}}^{2}), \tag{6.195}$$

which implies that the optimal delay–power relation is convex.

Figure 6.30 shows an example of this optimal delay–power tradeoff. The packet arrival is assumed to be an i.i.d random variable in the range [0, 1]. The channel is modeled as a Markov chain with five states assuming a Doppler frequency of 20 Hz and an average SNR of 0 dB. As

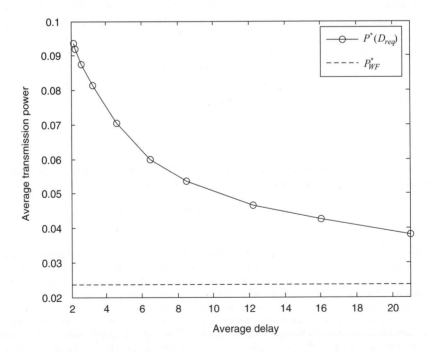

Figure 6.30 Example of the optimal power consumption for various delay requirements

discussed above, the power consumption convexly decreases as the delay requirement increases. The power consumption approaches the line P_{WF}^* which denotes the power consumption required to achieve the ergodic capacity of the average arrival rate $\bar{A}$ by the water-filling method. Note that this corresponds to the case with infinite delay constraint and the minimum power consumption for the buffer to be stabilized.

The optimal scheduling policy can be obtained by the dynamic programming method. Under the assumption that the traffic arrival $A(n)$ is an i.i.d process and the channel state $g(n)$ is a stationary Markov chain independent of the arrival process, determination of the optimal policy becomes a *Markov decision process* (MDP).[30] The state of this problem is described by the concatenation of the channel and buffer states, $[g(n), Q(n)]$. The action at each state is given by the number of transmitted packets $U(n)|_{[g(n),Q(n)]}$ and a policy is defined as a set of the actions determined over the whole state space. The transition probability for a given action is expressed by

$$\Pr\{[g(n+1), Q(n+1)] = [g_1, Q_1] | [g(n), Q(n)] = [g_0, Q_0], U(n)\}$$
$$= \Pr\{g(n+1) = g_1 | g(n) = g_0\} \times \Pr\{A(n) = Q_1 - Q_0 + U(n)\}. \tag{6.196}$$

The objective of the MDP problem is to minimize the weighted combination of the average power consumption and the average delay; that is

$$\text{minimize} \lim_{m \to \infty} \frac{1}{m} \sum_{n=1}^{m} \mathbb{E}\left[P(g(n), U(n)) + \beta \frac{Q(n)}{\bar{A}}\right], \tag{6.197}$$

where β is a positive parameter that indicates the relative importance of the average delay over the power consumption: a larger value of β corresponds to placing more importance on delay, thereby attaining a lower average delay with a higher power consumption. From the power–rate relation in (6.144), we get

$$P(g(n), U(n)) = \frac{\sigma^2 (2^{cU(n)} - 1)}{g(n)} \tag{6.198}$$

for a constant c that relates the transmission rate and the number of packet transmissions. Figure 6.31 illustrates the overall problem formulation as an MDP.[31] In the figure, the transition probabilities are determined by the distributions of the arrival and channel processes and the applied scheduling policy. We can consider a related discounted-cost problem where future costs are discounted by a factor of λ; that is

$$\text{minimize} \lim_{m \to \infty} \sum_{n=1}^{m} \mathbb{E}\left[\lambda^n \left(P(g(n), U(n)) + \beta \frac{Q(n)}{\bar{A}}\right)\right], \tag{6.199}$$

---

[30] It is just for simplicity of problem formulation that we assume the traffic arrival is an i.i.d process. Note that it is also possible to assume that traffic arrival is a Markov chain if the state of traffic generation is added when defining the system state.

[31] The optimal scheduler with a static channel (i.e., the one discussed in Section 6.3.2) is a special case of this formulation. There exists only one channel state (that is, $g(n) = g_1$ for all n), and the state diagram reduces to a one-dimensional one.

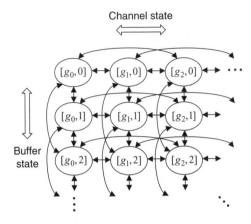

Figure 6.31 State diagram for problem formulation as a Markov decision process

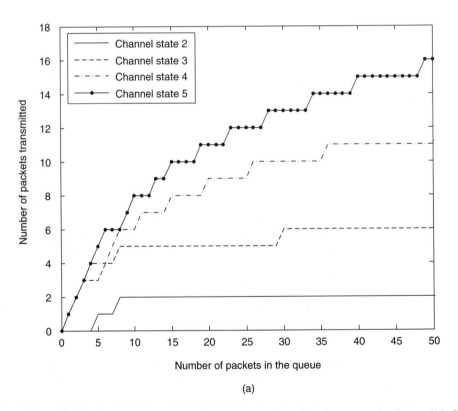

(a)

Figure 6.32 The optimal power-adaptation policy: (a) number of packets transmitted at each buffer state, and (b) transmission power at each channel state

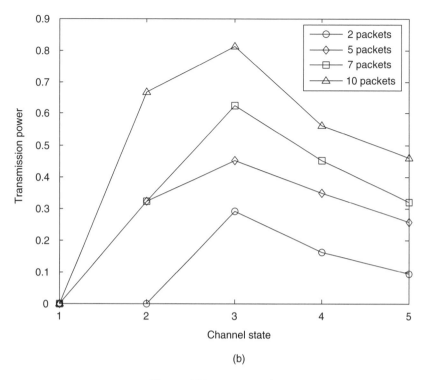

(b)

Figure 6.32 (*Continued*)

instead of the average-cost problem in (6.197). Then, as discussed in Section 4.2, it is sufficient to find an optimal policy for a discounted-cost problem for a large enough discounting factor. The optimal policy for (6.199) can be determined by a dynamic programming method such as value iteration and policy iteration discussed in Section 4.2.

Figure 6.32 illustrates the performance of the optimal scheduling policy. The system model is the same as that in Figure 6.30 and the average delay requirement is set to 4.6 slots. From Figure 6.32(a), which plots the number of transmitted packets with respect to the number of packets in the buffer, observe that the number of packet transmission increases as the buffer state increases but it does not increase as fast as the buffer state while storing several packets in the buffer. This is consistent with Figure 6.29 which was obtained for a static channel model. For different channel states, the optimal policy sends more packets at a better channel condition to exploit the favorable channel condition. From Figure 6.32(b), which plots the transmission power with respect to the channel states, observe that the transmission power curve is similar to the optimal transmission power with a moderate delay requirement in Figure 6.25: it operates like the water-filling method by using more power at a better channel condition but begins to decrease the transmission power like the channel inversion as the channel gain grows above a given level. Note also that the optimal policy tends to use more transmission power at each channel state as the buffer state increases. This is because an increased buffer state is equivalent to a more strict delay requirement. As a result, the optimal policy transmits more packets in a worse channel condition, so more transmission power is required to support this transmission.

References

Bambos, N., Chen, S.C. and Pottie, G.J. (2000) Channel access algorithms with active link protection for wireless communication networks with power control. *IEEE/ACM Transactions on Networking*, **8**, 583–597.

Berggren, F. and Kim, S.-L. (2004) Energy-efficient control of rate and power in DS-CDMA systems. *IEEE Transactions on Wireless Communications*, **3**, 725–733.

Berggren, F., Jantti, F.R. and Kim, S.-L. (2001) A generalized algorithm for constrained power control with capability of temporary removal. *IEEE Transactions on Vehicular Technology*, **50**, 1604–1612.

Berry, R.A. and Gallager, R.G. (2002) Communication over fading channels with delay constraints. *IEEE Transactions on Information Theory*, **48**, 1135–1149.

Bigloo, A.M.Y., Gulliver, T.A. and Bhargava, V.K. (1997) Maximum-likelihood decoding and code combining for DS/SSMA slotted ALOHA. *IEEE Transactions on Communications*, **45**, 1602–1612.

Cai, X., Sun, Y. and Akansy, A.N. (2003) Performance of CDMA random access systems with packet combining in fading channels. *IEEE Transactions on Communications*, **51**, 413–419.

El Gamal, A., Nair, C., Prabhakar, B., Uysal-Biyikoglu, E. and Zahedi, S. (2002) Energy-efficient scheduling of packet transmissions over wireless networks. In: *Proceedings IEEE Infocomm*, New York, 1773–1782.

Foschini, G.J. and Miljanic, Z. (1993) A simple distributed autonomous power control algorithm and its convergence. *IEEE Transactions on Vehicular Technology*, **42**, 641–646.

Grandhi, S.A. and Zander, J. (1994) Constrained power control in cellular radio systems. In: *Proceedings IEEE Vehicular Technology Conference*, Tokyo.

Hanly, S. (1995) An algorithm for combined cell-site selection and power control to maximize cellular spread spectrum capacity. *IEEE Journal of Selected Areas in Communications*, **13**, 1332–1340.

Hanly, S. (1996) Capacity and power control in a spread spectrum macrodiversity radio networks. *IEEE Transactions on Communications*, **44**, 247–256.

Jafar, S.A. and Goldsmith, A. (2003) Adaptive multirate CDMA for uplink throughput maximization. *IEEE Transactions on Wireless Communications*, **2**, 218–228.

Jang, J. and Lee, K.B. (2003) Transmit power adaptation for multiuser OFDM systems. *IEEE Journal of Selected Areas in Communications*, **21**, 171–178.

Kwon, S., Kim, K., Yun, Y., Kim, S.G. and Yi, B.K. (2005) Power controlled HARQ in CDMA2000 1xEV-DV. *IEEE Communications Magazine*, **43**, 77–81.

Leung, K.K., Sung, C.W., Wong, W.S. and Lok, T.M. (2004) Convergence theorem for a general class of power-control algorithms. *IEEE Transactions on Communications*, **52**, 1566–1574.

Oh, S.-J. and Soong, A.C.K. (2003) QoS-constrained information-theoretic sum capacity of reverse link CDMA systems. In: *Proceedings IEEE Globecomm*, San Francisco, CA.

Oh, S.-J. and Soong, A.C.K. (2006) QoS-constrained information-theoretic sum capacity of reverse link CDMA systems. *IEEE Transactions on Wireless Communications*, **5**, 3–7.

Proakis, J.G. (1995) *'Digital Communications'*, McGraw-Hill.

Rajan, D., Sabharwal, A. and Aazhang, B. (2004) Delay-bounded packet scheduling of bursty traffic over wireless channels. *IEEE Transactions on Information Theory*, **50**, 125–144.

Rhee, W. and Cioffi, J.M. (2000) Increasing in capacity of multiuser OFDM system using dynamic subchannel allocation. In: *Proceedings IEEE Vehicular Technology Conference*, Tokyo, Japan.

Sampath, A., Kumar, P.S. and Holzman, J.M. (1995) Power control and resource management for a multimedia CDMA wireless system. In: *Proceedings IEEE International Symposium on Personal, Indoor and Mobile Radio Communications*, vol. 1, 21–25.

Seo, H. and Lee, B.G. (2006) Proportional-fair power allocation with CDF-based scheduling for fair and efficient multiuser OFDM systems. *IEEE Transactions on Wireless Communications*, **5**, 978–983.

Seo, H., Park, S. and Lee, B.G. (2005) A retransmission power adjustment scheme for performance enhancement in DS/SSMA ALOHA with packet combining. *Journal of Communications and Networks*, **7**, 36–44.

Shen, Z., Andrews, J.G. and Evans, B.L. (2005) Adaptive resource allocation in multiuser OFDM systems with proportional rate constraints. *IEEE Transactions on Wireless Communications*, **4**, 2726–2737.

Song, G. and Li, Y.G. (2005a) Cross-layer optimization for OFDM wireless network. Part I: Theoretical framework. *IEEE Transactions on Wireless Communications*, **4**, 614–624.

Song, G. and Li, Y.G. (2005b) Cross-layer optimization for OFDM wireless network. Part II: Algorithm development. *IEEE Transactions on Wireless Communications*, **4**, 625–634.

Sung, C.W. and Leung, K.-K. (2005) A generalized framework for distribute power control in wireless networks. *IEEE Transactions on Information Theory*, **51**, 2625–2635.

Sung, C.W. and Wong, W.S. (2003) A noncooperative power control game for multirate CDMA data networks. *IEEE Transactions on Wireless Communications*, **2**, 186–194.

Tang, J. and Zhang, X. (2007) Quality-of-service driven power and rate adaptation over wireless links. *IEEE Transactions on Wireless Communications*, **6**, 3058–3068.

Tuninetti, D. and Caire, G. (2002) The throughput of some wireless multiacess systems. *IEEE Transactions on Information Theory*, **48**, 2773–2785.

Uysal-Biyikoglu, E., Prabhakar, B. and El Gamal, A. (2002) Energy-efficient packet transmission over a wireless link. *IEEE/ACM Transactions on Networking*, **10**, 487–499.

Xiao, M., Shroff, N.B. and Chong, E.K.P. (2003) A utility-based power-control scheme in wireless cellular systems. *IEEE/ACM Transactions on Networking*, **11**, 210–221.

Yates, R.D. (1995) A framework for uplink power control in cellular radio systems. *IEEE Journal on Selected Areas in Communications*, **13**, 1341–1347.

Yates, R. and Huang, C.Y. (1995) Integrated power control and base station assignment. *IEEE Transactions on Vehicular Technology*, **44**, 638–644.

Zhang, J. and Chong, E.K.P. (2000) CDMA systems in fading channels: Admissibility, network capacity, and power control. *IEEE Transactions on Information Theory*, **46**, 962–981.

Zhang, Q., Wong, T.F. and Lehnert, J.S. (1999) Performance of a type-II hybrid ARQ protocol in slotted DS-SSMA packet radio systems. *IEEE Transactions on Communications*, **47**, 281–290.

7

Antenna Management

Multiple-antenna technology has emerged as a key technology for next-generation wireless communication systems, as multiple antennas installed at transmitter and receiver facilitate a high rate of data transmission. The reason for this is that multiple antennas can contribute to increasing channel capacity without requiring additional bandwidth or transmission power. They can increase the data rate in proportion to the number of antennas.

However, since antennas necessitate circuitry operating at radio frequency in the transmitting and receiving devices, increasing their number involves increased device cost and processing complexity. In addition, the channel capacity relies heavily on the types of antenna usage, and optimal antenna usage varies depending on the wireless channel connecting the transmitter–receiver pair. So, for effective operation of multiple antennas, it is essential to adopt efficient antenna management methods that can optimize the usage of antennas to the given channel state.

Multiple-antenna technologies may be classified into "transmit diversity" and "spatial multiplexing" types. A *diversity transmission scheme* is a method to obtain spatial diversity on fading channels by sending the same (or slightly modified) data on different antennas. A *spatial-multiplexing scheme* transmits multiple streams of independent data from different antennas in order to maximize the data rate. It is also possible to achieve both diversity transmission and spatial multiplexing simultaneously, but there is a fundamental tradeoff: a high multiplexing gain can be achieved at the sacrifice of diversity gain. Therefore, it is desirable to choose the most appropriate transmission scheme for the given transmission environment.

This chapter discusses a formulation of the channel capacity of multiple-antenna systems for various channel conditions, yielding an ultimate upper bound to transmission performance. It investigates both the diversity-transmission and spatial-multiplexing modes to maximize multiple antenna gains, and finally extends the discussion to multiuser multiple-antenna systems.

7.1 Capacity of MIMO Channels

In multiple-antenna systems, there are several transmit and receive antennas. When the numbers of transmit and receive antennas are M and N, respectively, the resulting

Wireless Communications Resource Management   Byeong Gi Lee, Daeyoung Park, and Hanbyul Seo
© 2009 John Wiley & Sons (Asia) Pte Ltd

multiple-antenna channel is characterized by

$$\mathbf{r} = \mathbf{Hx} + \eta, \tag{7.1}$$

where $\mathbf{r}$ is an N-vector representing the received signal, $\mathbf{H}$ is an $N \times M$ channel matrix, $\mathbf{x}$ is an M-vector representing the transmitted symbol, and η is an additive zero-mean white Gaussian noise vector.

Since both the input and output in (7.1) are vectors with $M > 1$ and $N > 1$, the system is called a *multiple-input multiple-output* (MIMO) system. Likewise, systems with $M = 1$ and $N > 1$ are called *single-input multiple-output* (SIMO); those with $M > 1$ and $N = 1$ are *multiple-input single-output* (MISO); and those with $M = 1$ and $N = 1$ are *single-input single-output* (SISO). In this section we consider the capacity of the MIMO system under various different conditions of the channel matrix $\mathbf{H}$.

7.1.1 Capacity of a Deterministic Channel

Consider the MIMO system whose matrix channel $\mathbf{H}$ is fixed during the whole transmission period. Assume that each element of the noise vector η is i.i.d. with variance N_0 unless specified otherwise. The covariance matrix of this noise vector is an $N \times N$ identity matrix $\mathbb{E}[\eta\eta^H] = N_0\mathbf{I}_N$, where $\mathbf{I}_N$ is the $N \times N$ identity matrix and η^H is the Hermitian of η. Let $\mathbf{Q}$ be the covariance matrix of the transmitted signal x. Assume that the *channel state information* (CSI), which is a realization of the channel $\mathbf{H}$, is known at the receiver. In addition, if the channel is known at the transmitter (that is, *CSI available at transmitter*, "CSIT"), then the capacity of the system described in (7.1) is

$$C = \max_{\mathrm{Tr}(\mathbf{Q}) \leq P} \log_2 \det\left(\mathbf{I}_N + \frac{1}{N_0}\mathbf{HQH}^H \right) \quad \text{(bits/sec/Hz)}, \tag{7.2}$$

where $\mathrm{Tr}(\mathbf{Q})$ is the trace of matrix $\mathbf{Q}$ and the trace constraint $\mathrm{Tr}(\mathbf{Q}) \leq P$ implies that the total transmit power is constrained by P. In order to achieve the capacity, we have to allocate transmit power over each antenna under the total power constraint.

If we apply a *singular-value decomposition* (SVD) to $\mathbf{H}$, it is decomposed into

$$\mathbf{H} = \mathbf{U\Lambda V}^H, \tag{7.3}$$

where $\mathbf{U}$ and $\mathbf{V}$ are the unitary matrices and $\Lambda = \mathrm{diag}\,(\rho_1, \rho_2, \ldots, \rho_m, 0, 0, \ldots, 0)$ is a diagonal matrix for the rank m of the matrix $\mathbf{H}$ (i.e., the number of positive singular values). Note that the rank of $\mathbf{H}$ does not exceed $\min(M, N)$. Applying (7.3), we may rewrite (7.1) as

$$\tilde{\mathbf{r}} = \Lambda\tilde{\mathbf{x}} + \tilde{\eta}, \tag{7.4}$$

where $\tilde{\mathbf{r}} = \mathbf{U}^H\mathbf{r}$, $\tilde{\mathbf{x}} = \mathbf{V}^H\mathbf{x}$ and $\tilde{\eta} = \mathbf{U}^H\eta$. This equation represents m equivalent SISO channels with the channel gains $\rho_1, \rho_2, \ldots, \rho_m$, as shown in Figure 7.1.

We now consider how to optimize the covariance matrix $\mathbf{Q}$ for (7.2) under the trace constraint $\mathrm{Tr}(\mathbf{Q}) \leq P$. With the aid of the equivalent channel representation in Figure 7.1, the capacity may

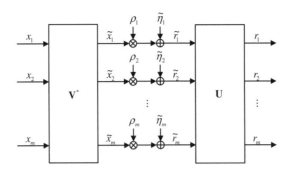

Figure 7.1   MIMO channel decomposed by SVD

be rewritten as

$$C = \max_{P_i} \sum_{i=1}^{m} \log_2\left(1 + \frac{P_i\rho_i^2}{N_0}\right), \tag{7.5}$$

where P_i is the power allocated to the ith equivalent SISO channel and $\sum_{i=1}^{m} P_i = P$. Since the optimal power allocation over the Gaussian parallel channels is water-filling (see Section 4.1.5), the optimal power allocation is given by

$$P_i^* = \max\left\{\frac{1}{\lambda} - \frac{N_0}{\rho_i^2}, 0\right\}, \tag{7.6}$$

where λ is chosen to satisfy the power constraint $\sum_{i=1}^{m} P_i^* = P$.

In the high-SNR case, the equal-power allocation is nearly optimal, and the capacity is approximated by

$$C \approx \sum_{i=1}^{m} \log_2\left(1 + \frac{P\rho_i^2}{mN_0}\right). \tag{7.7}$$

Applying Jensen's inequality to this concave log function, we get

$$\sum_{i=1}^{m} \log_2\left(1 + \frac{P\rho_i^2}{mN_0}\right) \leq m\log_2\left(1 + \frac{P}{m^2N_0}\sum_{i=1}^{m}\rho_i^2\right), \tag{7.8}$$

where the equality holds if and only if all the singular values are equal. Therefore, a higher capacity is obtained if the channel matrix is well-conditioned such that the singular values are very similar.

On the other hand, in the low-SNR case the optimal strategy is to allocate all power to the channel associated with the largest singular value. This yields the capacity

$$C \approx \log_2\left(1 + \frac{P}{N_0}\max_i \rho_i^2\right) \approx \frac{P}{N_0}\max_i \rho_i^2 \log_2 e. \tag{7.9}$$

If the transmitter has no knowledge about the channel information ("no CSIT"), the optimal covariance matrix is $\mathbf{Q} = (P/M)\mathbf{I}_M$. Note that this satisfies the trace constraint $\text{Tr}(\mathbf{Q}) \leq P$. In this case, the capacity is given by

$$
\begin{aligned}
C &= \log_2 \det\left(\mathbf{I}_N + \frac{P}{MN_0}\mathbf{H}\mathbf{H}^H \right) \\
&= \sum_{i=1}^{m} \log_2\left(1 + \frac{P}{MN_0}\rho_i^2 \right).
\end{aligned}
\tag{7.10}
$$

On comparing (7.10) with equal-power allocation for the CSIT case in (7.7), we find that P/M is replaced by P/m. This indicates that the transmit power is equally allocated into M transmit antennas in the case of "no CSIT", but in the case of "CSIT" it is equally allocated among the $m (\leq M)$ channels associated with the non-zero singular values. Thus, with a high signal-to-noise ratio, the power gain of the "CSIT" case over the "no CSIT" case is M/m.

7.1.2 Ergodic Capacity

Consider now the case where the channel is time-varying. Then the channel matrix $\mathbf{H}$ is a random matrix and each channel use corresponds to an independent realization of $\mathbf{H}$. Each entry of $\mathbf{H}$ is assumed to be i.i.d. complex Gaussian with zero mean and independent real and imaginary parts, having variance 1/2 each. Equivalently, each entry has Rayleigh magnitude and uniform phase.

Assume that the realization of $\mathbf{H}$ is known at the receiver, and the distribution of $\mathbf{H}$ is known at the transmitter. Additionally, assume that the channels vary from one time instant to the next time. Then, the average capacity over a long block can be achieved, as the channel $\mathbf{H}$ is generated by an ergodic process. For canonical i.i.d. Rayleigh distributed channels, the optimal covariance matrix is isotropic; that is, $\mathbf{Q} = (P/M)\mathbf{I}_M$.[1] Then the capacity is given by

$$
C = \mathbb{E}\left[\log_2 \det\left(\mathbf{I}_N + \frac{P}{MN_0}\mathbf{H}\mathbf{H}^H \right) \right] = \mathbb{E}\left[\sum_{i=1}^{m} \log_2\left(1 + \frac{P}{MN_0}\rho_i^2 \right) \right].
\tag{7.11}
$$

We define an $m \times m$ random non-negative definite matrix $\mathbf{W}$ by

$$
\mathbf{W} = \begin{cases} \mathbf{H}\mathbf{H}^H, & N < M \\ \mathbf{H}^H\mathbf{H}, & N \geq M \end{cases}
\tag{7.12}
$$

where $m = \min(M, N)$. Then the square of each singular value, ρ_i^2, is an eigenvalue of $\mathbf{W}$. The matrix $\mathbf{W}$ is Wishart-distributed with the parameters m and n, where $n \equiv \max(M, N)$, and the joint p.d.f. of its eigenvalues is given by (Telatar, 1999)

$$
f(\rho_1^2, \rho_2^2, \cdots, \rho_m^2) = \frac{1}{m! K_{m,n}} e^{-\sum_{i=1}^{m}\rho_i^2} \prod_{i=1}^{m} (\rho_i^2)^{n-m} \prod_{i<j} (\rho_i^2 - \rho_j^2)^2,
\tag{7.13}
$$

---

[1] Note that this isotropic input is not necessarily optimal for non-canonical channels (Lozano *et al.*, 2006).

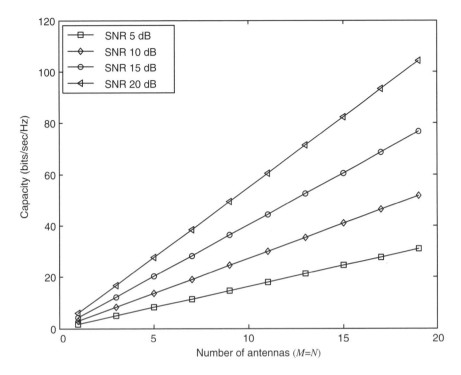

Figure 7.2 Capacity of MIMO system for Rayleigh fading channel with respect to the number of antennas ($M = N$)

where $K_{m,n}$ is a normalizing factor. So, the ergodic capacity becomes

$$C_{erg} = \sum_{i=1}^{m} \mathbb{E}\left[\log_2\left(1 + \frac{P}{MN_0}\rho_i^2\right)\right] = m\mathbb{E}\left[\log_2\left(1 + \frac{P}{MN_0}\rho_1^2\right)\right] \qquad (7.14)$$

and the expectation is taken over one eigenvalue.

Figure 7.2 plots the capacity of the system with respect to the number of antennas for the case $M = N$. Observe that the capacity is approximately a linear function of M. More specifically, as expressed in (7.14), the capacity increases linearly with the minimum of the transmit and receive antennas; that is, min (M, N). So, it is desirable to make the numbers of transmit and receive antennas the same when the total number of antennas is fixed.

7.1.3 Outage Capacity

When the channel **H** is ergodic, transmission rates arbitrarily close to the average mutual information can be achieved. On the other hand, when **H** is chosen randomly at the beginning of all transmission time slots and is held fixed during the whole transmission, the maximum achievable rate is random and the average capacity is not an adequate metric during this transmission. Then, it is more meaningful to consider the *outage probability*, which is the probability that the mutual information drops below the transmission rate R. For example, consider a SISO fading channel in which the mutual information is given by

$$C = \log_2\left(1 + |h|^2 \frac{P}{N_0}\right). \tag{7.15}$$

where channel h is random but remains unchanged during the transmission period. Note that the mutual information in (7.15) is a random variable that depends on the channel realization h and remains constant during the transmission. Since the channel gain may fluctuate, possibly dropping close to zero, it does not make sense to guarantee a data rate. So, the outage probability P_{out} is defined to be

$$P_{\text{out}} \equiv \Pr(C(h) < R). \tag{7.16}$$

The outage probability indicates the probability that a channel falls in a deep fade and thus fails to support the transmission rate R. The maximum rate that the channel can support with the given outage probability is called the *outage capacity*. For real-time applications, the outage capacity gives a more realistic performance bound than does the ergodic capacity, because the relevant delay constraint limits the manageable size of the block length of error-correction codes.

For Rayleigh fading, the outage probability is given by

$$P_{\text{out}} = \Pr\left(|h|^2 < (2^R - 1)\frac{N_0}{P}\right) = 1 - \exp\left(-(2^R - 1)\frac{N_0}{P}\right). \tag{7.17}$$

Conversely, the outage capacity C_{out} for a given outage probability P_{out} takes the expression

$$C_{\text{out}} = \log_2\left(1 - \frac{P}{N_0}\ln(1 - P_{\text{out}})\right) \tag{7.18}$$

Since $\exp(x) = 1 + x + O(x^2)$ for small x, at high SNR, the outage probability in (7.17) may be approximated to

$$P_{\text{out}} \approx (2^R - 1)\frac{N_0}{P}. \tag{7.19}$$

So, the outage probability falls off inversely proportional to the SNR P/N_0. As we will see in more detail in Section 7.2.1, the exponent of SNR is called the *diversity order*, and the diversity order for this case is 1.

The concepts of "outage probability" and "outage capacity" can be extended to MIMO fading channels. Then the outage probability is given by

$$P_{\text{out}} \equiv \Pr(C(\mathbf{H}) < R) = \Pr\left(\log_2 \det\left(\mathbf{I}_N + \frac{P}{MN_0}\mathbf{H}\mathbf{H}^{\text{H}}\right) < R\right). \tag{7.20}$$

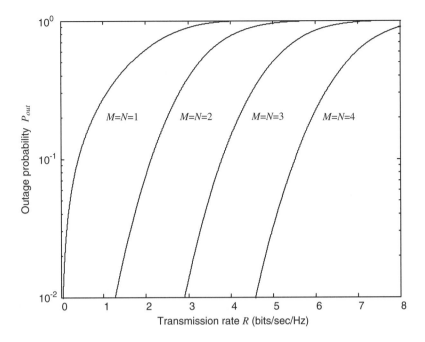

Figure 7.3 Outage probability of a MIMO system with Rayleigh fading channels (5 dB SNR)

If $M = N = 1$, the outage probability in (7.20) reduces to (7.17). Unfortunately, it is difficult to make a closed-form expression of the outage probability for MIMO systems, so we need to perform a Monte Carlo simulation to evaluate it. Figure 7.3 shows the outage probability of a MIMO system when the channel is Rayleigh distributed and the number of transmit antennas is equal to the number of receive antennas. Observe that the outage probability reduces significantly as the number of antennas increases.

7.2 MIMO Transmission

The information-theoretic analysis of the previous section showed that multiple antennas can possibly yield considerable gains in terms of capacity enhancement and outage probability reduction. However, such gains can be fully exploited only when the practical transmission and reception schemes are designed properly. The multiple antennas may be designed to achieve the diversity effect, or the spatial multiplexing effect, or a combination of the two, depending on the channel environment and the design goal. This section deals with the diversity-transmission and spatial-multiplexing schemes individually and then discusses diversity/multiplexing tradeoffs.

7.2.1 Diversity Transmission

When the channel falls in deep fade, the received signal is likely to be too weak to detect correctly. If multiple components of the same transmitted signal that have experienced independent fading channels are received, it is possible to improve the error performance substantially by combining them intelligently since the probability that all the signal components fall in deep fade simultaneously is very low. This is the *diversity effect*, which plays an important role in wireless communication systems. We can take advantage of the diversity effect in MIMO transmission because the signals transmitted over multiple antennas experience different fading channels. So, we consider how to design MIMO transmission systems in order to exploit the full diversity effect the channel can provide.

7.2.1.1 Design Guidelines

Consider a MIMO transmission with M transmit and N receive antennas in the Rayleigh fading channel environment. Assume that the channel is random and remains constant during the transmission duration T. Then the received signal takes the expression

$$\mathbf{R} = \mathbf{H}\mathbf{X} + \eta, \tag{7.21}$$

where $\mathbf{R}$ is an $N \times T$ received signal matrix, $\mathbf{H}$ is an $N \times M$ channel matrix, $\mathbf{X}$ is an $M \times T$ transmitted symbol matrix, and η is an $N \times T$ additive zero-mean white Gaussian noise matrix with variance N_0. Note that transmitted symbols occupy M transmit antennas and T transmission times, for which we assume $T \geq M$. The noise matrix η is spatially and temporally independent so that $\mathbb{E}[\eta\eta^H] = TN_0\mathbf{I}_N$.

The *maximum-likelihood* (ML) detection is intended to find the transmit codeword $\mathbf{X}$ such that the squared metric

$$\min_{X} \|\mathbf{R} - \mathbf{H}\mathbf{X}\|^2 \tag{7.22}$$

is minimized, where $\|\bullet\|^2$ denotes the Frobenius norm which is the sum of the absolute squares of the elements. The pairwise error probability $P(\mathbf{X} \to \hat{\mathbf{X}})$ is the probability that the detector makes a decision of $\hat{\mathbf{X}}$ even though the transmitted vector was $\mathbf{X}$. It takes the expression

$$P(\mathbf{X} \to \hat{\mathbf{X}}) = \Pr\left(\|\mathbf{R} - \mathbf{H}\hat{\mathbf{X}}\|^2 < \|\mathbf{R} - \mathbf{H}\mathbf{X}\|^2\right) = \mathbb{E}\left[Q\left(\frac{d(\mathbf{X}, \hat{\mathbf{X}})}{\sqrt{2N_0}}\right)\right], \tag{7.23}$$

where $Q(x)$ is the Gaussian tail function and the squared distance $d^2(\mathbf{X}, \hat{\mathbf{X}})$ is defined by

$$d^2(\mathbf{X}, \hat{\mathbf{X}}) = \sum_{n=1}^{N}\sum_{t=1}^{T}\left|\sum_{m=1}^{M} h_{n,m}(x_{m,t} - \hat{x}_{m,t})\right|^2. \tag{7.24}$$

Since $Q(x) \leq \exp(-x^2/2)$ (Proakis, 2001), the pairwise error probability is bounded by

$$P(\mathbf{X} \to \hat{\mathbf{X}}) \leq \mathbb{E}\left[\exp\left(-\frac{d^2(\mathbf{X}, \hat{\mathbf{X}})}{4N_0}\right)\right]. \tag{7.25}$$

In the case of Rayleigh fading channels, it is bounded by (Tarokh, *et al.*, 1998)

$$P(\mathbf{X} \to \hat{\mathbf{X}}) \leq \left(\prod_{m=1}^{k} \lambda_m \right)^{-N} SNR^{-kN}, \tag{7.26}$$

where $SNR \equiv 1/4N_0$, and λ_1, λ_2, ..., λ_k are the non-zero eigenvalues of the codeword difference matrix $\mathbf{X} - \hat{\mathbf{X}}$. Note that the number of non-zero eigenvalues, k, is the rank of the matrix $\mathbf{X} - \hat{\mathbf{X}}$. The power of SNR in the pairwise error probability determines the slope of the curve in the logarithmic graph and it is called the *diversity order*. In this MIMO transmission, the diversity order is kN in general, which takes the maximum value MN when the codeword difference matrix is full rank; that is, $k = M$. The term $(\prod_{m=1}^{k} \lambda_m)^{-n}$ denotes the coding gain which indicates the gain over the uncoded system having the same diversity order. When the codeword difference matrix is full rank, the coding gain is nothing but the determinant of $(\mathbf{X} - \hat{\mathbf{X}})(\mathbf{X} - \hat{\mathbf{X}})^{\mathrm{H}}$. Therefore, the design guideline is given as follows:

1. *Rank criterion:* In order to extract the full diversity gain of MN, the codeword difference matrix $\mathbf{X} - \hat{\mathbf{X}}$ of any codeword pair $(\mathbf{X}, \hat{\mathbf{X}})$ has to be full rank.
2. *Determinant criterion:* In order to maximize the coding gain, the minimum determinant of $(\mathbf{X} - \hat{\mathbf{X}})(\mathbf{X} - \hat{\mathbf{X}})^{\mathrm{H}}$ over all possible codeword pairs $(\mathbf{X}, \hat{\mathbf{X}})$ has to be maximized.

7.2.1.2 Space–Time Block Codes

The joint encoding schemes over multiple transmit antennas are called *space–time codes*. One of the simple space–time codes is Alamouti's *space–time transmit-diversity* (STTD) scheme (Alamouti, 1998). Alamouti's code is designed for $M = 2$ and $T = 2$, assuming that the channel matrix remains fixed for two symbol periods. The codeword matrix has the form

$$\mathbf{X} = \begin{bmatrix} x_1 & -x_2^{\mathrm{H}} \\ x_2 & x_1^{\mathrm{H}} \end{bmatrix}, \tag{7.27}$$

where x_1 and x_2 are complex symbols. During the first symbol period, x_1 is transmitted from antenna 1, and x_2 is transmitted from antenna 2. During the second symbol period, their complex conjugate symbols $-x_2^{\mathrm{H}}$ and x_1^{H} are transmitted at antennas 1 and 2, respectively. So, the spatial transmission rate is one symbol per symbol period.

When there is one receive antenna, the signals received in two adjacent symbol periods are given by

$$\begin{aligned} r_1 &= h_1 x_1 + h_2 x_2 + \eta_1 \\ r_2 &= -h_1 x_2^{\mathrm{H}} + h_2 x_1^{\mathrm{H}} + \eta_2 \end{aligned} \tag{7.28}$$

or equivalently in matrix form:

$$\begin{bmatrix} r_1 \\ r_2^{\mathrm{H}} \end{bmatrix} = \begin{bmatrix} h_1 & h_2 \\ h_2^{\mathrm{H}} & -h_1^{\mathrm{H}} \end{bmatrix} \begin{bmatrix} x_1 \\ x_2 \end{bmatrix} + \begin{bmatrix} \eta_1 \\ \eta_2^{\mathrm{H}} \end{bmatrix} \tag{7.29}$$

We multiply the Hermitian of the effective matrix in (7.29) to each side to get

$$
\begin{bmatrix} h_1^H & h_2 \\ h_2^H & -h_1 \end{bmatrix} \begin{bmatrix} r_1 \\ r_2^H \end{bmatrix} = (|h_1|^2 + |h_2|^2) \begin{bmatrix} x_1 \\ x_2 \end{bmatrix} + \begin{bmatrix} h_1^H & h_2 \\ h_2^H & -h_1 \end{bmatrix} \begin{bmatrix} \eta_1 \\ \eta_2^H \end{bmatrix}. \tag{7.30}
$$

Since the multiplication by an orthogonal matrix does not change the probabilistic characteristics of an AWGN vector, except for the noise power, the noise term in (7.30) is another AWGN vector. As one can easily show, the ML detection, in effect, is to determine the transmit symbols x_i, $i = 1$, 2, such that

$$
\min_{x_i} |x_i - \tilde{x}_i|, \tag{7.31}
$$

where the combined symbols $\tilde{x}_i$, $i = 1$, 2, are given by

$$
\begin{bmatrix} \tilde{x}_1 \\ \tilde{x}_2 \end{bmatrix} = \frac{1}{|h_1|^2 + |h_2|^2} \begin{bmatrix} h_1^H & h_2 \\ h_2^H & -h_1 \end{bmatrix} \begin{bmatrix} r_1 \\ r_2^H \end{bmatrix}. \tag{7.32}
$$

This enables detection of each symbol individually, which greatly reduces the complexity of the ML detection in (7.22). This desirable property originates from the fact that the transmit matrix in (7.27) follows the orthogonal design (Tarokh, *et al.*, 1999). The structure of the transmit matrix makes the effective channel matrix orthogonal irrespective of the channel realization, thus decoupling the joint ML detection into a simple scalar detection. This detection rule can be easily extended to the case of N receive antennas.

Consider now the diversity order that Alamouti's space–time block code achieves. The determinant of the codeword difference matrix is given by

$$
\det(\mathbf{X} - \hat{\mathbf{X}}) = \det \left(\begin{bmatrix} x_1 - \hat{x}_1 & -x_2^H + x_2^H \\ x_2 - \hat{x}_2 & x_1^H - x_1^H \end{bmatrix} \right) = |x_1 - \hat{x}_1|^2 + |x_2 - \hat{x}_2|^2. \tag{7.33}
$$

If $x_1 \neq \hat{x}_1$ or $x_2 \neq \hat{x}_2$, $\mathbf{X} - \hat{\mathbf{X}}$ is non-singular and thus has full rank. Therefore, Alamouti's space–time block code achieves the diversity order of $2N$, which is the maximal diversity order that the MIMO system with two transmit and N receive antennas can achieve.

Figure 7.4 shows the bit error-rate performance of uncoded *binary phase-shift keying* (BPSK) for Alamouti's space–time code and other transmission schemes (Alamouti, 1998). Observe that the 1×1 SISO scheme exhibits very poor performance at high SNR because it has no diversity effect. However, both the 1×2 SIMO scheme and Alamouti's space–time code, having the diversity order of 2, exhibit much better performances. The performance of the 1×2 SIMO scheme with transmit power P is identical to the performance of Alamouti's code with transmit power $2P$, with each antenna radiating power P. So, in terms of transmitted power, Alamouti's code has a 3 dB SNR penalty when compared with the 1×2 SIMO.

7.2.1.3 Space–Time Trellis Codes

In order to achieve a higher coding gain, a conventional trellis structure can be applied to multiple antenna systems. Space–time trellis codes are also designed to achieve the maximal

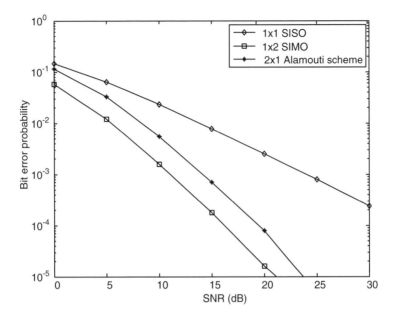

Figure 7.4   Performance comparison of 1×1 no-diversity scheme, 1×2 receive diversity scheme, and Alamouti's space–time codes

diversity and coding gains. Figure 7.5 shows a trellis diagram for a space–time trellis code for two transmit antennas and QPSK (*quadrature phase-shift keying*) constellation with rate 2 bits/sec/Hz (Tarokh *et al.*, 1998). This code has four states (0–3), with each state corresponding to each node in the trellis diagram. At the beginning the encoder is required to be in state 0. The pairs of numbers at each node denote the output symbols of the encoder that are transmitted through two transmit antennas. The output symbols are mapped at the QPSK constellation. At each time, two input bits determine the next state and the output symbols, depending on the

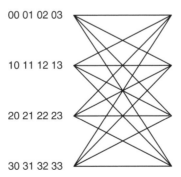

Figure 7.5   Trellis diagram for a four-state space–time trellis code for $M = 2$ and QPSK (2 bits/sec/Hz) (Reproduced with permission from V. Tarokh, N. Seshadri, and A.R. Calderbank, "Space-time codes for high data rate wireless communication: Performance criterion and code construction," *IEEE Transactions on Information Theory*, 44, no. 2, 744–765, March 1998. © 1998 IEEE)

00 01 02 03

10 11 12 13

20 21 22 23

30 31 32 33

22 23 20 21

32 33 30 31

02 03 00 01

12 13 10 11

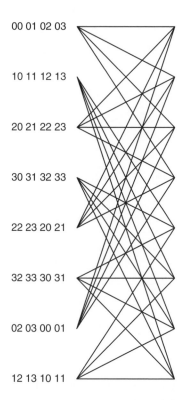

Figure 7.6 Trellis diagram for an eight-state space–time trellis code for $M = 2$ and QPSK (2 bits/sec/Hz) (Reproduced with permission from V. Tarokh, N. Seshadri, and A.R. Calderbank, "Space-time codes for high data rate wireless communication: Performance criterion and code construction," *IEEE Transactions on Information Theory*, 44, no. 2, 744–765, March 1998. © 1998 IEEE)

current state. For example, at state 2, if information bits are 00, then the output symbols are 2 for the first transmit antenna and 0 for the second transmit antenna and the state makes a transition to state 0. Similarly, at state 2, if information bits are 01, then the symbols 2 and 1 are transmitted through antennas 1 and 2, respectively, and the state 1 becomes the next state. This space–time trellis code achieves a diversity order of $2N$ when the number of receive antennas is N.

Figure 7.6 shows an eight-state space–time trellis code for two transmit antennas and QPSK constellation with rate 2 bits/sec/Hz (Tarokh, *et al.*, 1998). Increasing the number of states yields better performance in terms of the *frame error rate* (FER). The coding gain of this eight-state trellis code is higher than that of the four-state trellis code. However, the diversity order is the same.

Figure 7.7 shows the frame error rate with respect to SNR for the four-state and eight-state space–time codes with two transmit and two receive antennas. In the figure, the outage probability $\Pr(C(\mathbf{H}) < 2)$ in (7.20) for two transmit and two receive antennas is overlaid. Observe that both space–time trellis codes achieve the diversity order of 4. The eight-state code has a higher coding gain than the four-state code, but it also has a higher complexity. Both of them have room for improvements, because their frame error rates are still higher than the outage probability.

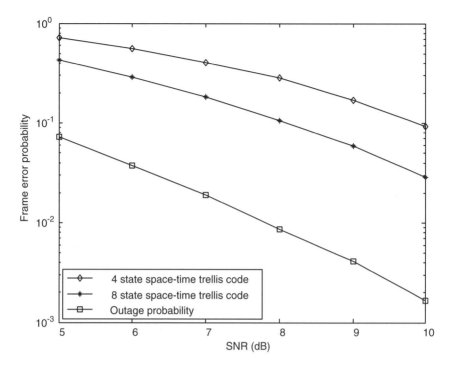

Figure 7.7 Outage probability for 2 bits/sec/Hz and the frame error-rate performances of four-state and eight-state space–time trellis codes

7.2.2 Spatial Multiplexing

Space–time codes can transmit up to one independent symbol per unit symbol period. Each antenna transmits a slightly modified copy of the original data stream so that the data stream experiences an independent fading while propagating and then combines with the other data stream at the receiver to yield higher resolution detection. In this way, the space–time codes achieve the spatial diversity gain. On the other hand, one may transmit multiple different data streams simultaneously through multiple antennas, thereby increasing the transmission rate at the expense of the diversity gain. In the following, we discuss the latter technique, called spatial multiplexing.

7.2.2.1 Detection Algorithms[2]

Consider a *spatial-multiplexing* (SM) system for M transmit and N receive antennas. The symbols $x_1, x_2, \ldots, x_M$ are transmitted through antennas 1 to M, respectively. The received signal is expressed by (7.1). Here we assume that the noise vector η is spatially independent, so

---

[2] Part of this subsection is reproduced by permission from Lee and Choi, *Broadband Wireless Access and Local Networks,* Norwood, MA: Artech House, Inc., 2008. © 2008 by Artech House, Inc.

that $\mathbb{E}[\eta\eta^H] = N_0\mathbf{I}_N$. Differently from the Alamouti's space–time code which transmitted two symbols during two symbol periods yielding the coding rate of 1, the spatial-multiplexing scheme transmits M symbols during one symbol period yielding the coding rate of M.

The optimal detector for the spatial-multiplexing scheme is the ML detector, which detects the transmitted signal vector by

$$\hat{x}_{ML} = \arg\min_x |r - \mathbf{H}x|^2. \tag{7.34}$$

The ML detection is performed through an exhaustive search over all candidate vector symbols. In the case of L *quadrature amplitude-modulation* (QAM) symbols, the ML detector has to compute L^M distance metrics, which grows to a large number for high-order modulations and large number of transmit antennas. For example, for a four-antenna system with 16 QAM, the number of searches that the ML receiver has to perform increases to $16^4 = 65\,536$.

Linear receivers such as *zero-forcing* (ZF) and *minimum mean-square-error* (MMSE) detectors can separate the transmit streams and decode each of them individually. They decrease the decoding complexity significantly even for a large number of transmit antennas and/or a high modulation order, but they usually do not achieve full diversity gain. As a consequence, the detection error rate is high when compared with that of the ML receiver, especially in the high-SNR region. In contrast, in the case of the *successive interference-cancellation* (SIC) detector which employs the ZF or MMSE detector for detecting the transmit signals in a sequential manner, the detection error is reduced at the cost of increased complexity.

In the case of the ZF detector, we can suppress the interference among the transmit streams by multiplying the received signal vector r with the Moore–Penrose pseudo-inverse of the channel matrix, $\mathbf{H}^+ = (\mathbf{H}^H\mathbf{H})^{-1}\mathbf{H}^H$, to get

$$\tilde{x}_{ZF} = \mathbf{H}^+ r = x + (\mathbf{H}^H\mathbf{H})^{-1}\mathbf{H}^H\eta. \tag{7.35}$$

Once $\tilde{x}_{ZF}$ is calculated, the demodulation is done by determining the M-vector $\hat{x}$ whose ith element is the constellation point closest to the ith element of $\tilde{x}_{ZF}$. So, the ZF detector decouples the matrix channel into parallel scalar channels at the cost of noise enhancement.

In the case of the MMSE detector, we minimize the mean square error

$$\mathbf{G}_{MMSE} = \arg\min_{\mathbf{G}} \mathbb{E}\left[||\mathbf{G}r - x||^2\right]. \tag{7.36}$$

By the orthogonal principle, the optimal matrix $\mathbf{G}_{MMSE}$ is given by $\mathbf{G}_{MMSE} = (\mathbf{H}^H\mathbf{H} + N_0\mathbf{I}_M)^{-1}\mathbf{H}^H$, which corresponds to a modified version of $\mathbf{H}^+ = (\mathbf{H}^H\mathbf{H})^{-1}\mathbf{H}^H$, with the noise power N_0 incorporated in it. The MMSE detection is given by

$$\tilde{x}_{MMSE} = (\mathbf{H}^H\mathbf{H} + N_0\mathbf{I}_M)^{-1}\mathbf{H}^H r. \tag{7.37}$$

Once $\tilde{x}_{MMSE}$ is calculated, the remaining demodulation process is the same as that for ZF detection. MMSE detection also decouples the matrix channel into parallel scalar channels. However, the performance degradation caused by noise is smaller in the MMSE case than in the ZF case.

Given received vector r and channel matrix $\mathbf{H}$

Initialize $i \leftarrow 1$ and $r^1 = r$ and $\mathbf{H}^1 = \mathbf{H}$

Repeat

$$\mathbf{G}^i = ((\mathbf{H}^i)^H \mathbf{H}^i + N_0\mathbf{I})^{-1}(\mathbf{H}^i)^H$$

$$k = \arg\max_j SINR_j^i$$

$$\tilde{x}_k = g_k^i r^i$$

$$\hat{x}_k = \Phi(\tilde{x}_k)$$

$$r^{i+1} = r^i - h_k^i \hat{x}_k$$

$$\mathbf{H}^{i+1} = [h_1^i, \cdots, h_{k-1}^i, 0, h_{k+1}^i, \cdots, h_M^i]$$

$$i \leftarrow i+1$$

Until $i > M$

Figure 7.8   Algorithm for OSIC detection

In general, MMSE detection is perceived as a better detection scheme than ZF detection as it reflects the noise effect in the matrix inversion process. Note that ZF detection becomes identical to ML detection when the channel matrix is orthogonal. The interference suppression process of the ZF detection scheme leads to noise amplification, which affects the detection performance significantly when the SNR is low. As the SNR increases, however, ZF detection approaches, in detection algorithm as well as in performance, MMSE detection. The maximum diversity order that ZF detection can achieve is $N - M + 1$.

SIC detection detects the transmit signal vector sequentially by detecting one signal at each iteration and eliminating the relevant interference in the received signal before continuing to the next iteration. The ZF and MMSE algorithms can be used for interference suppression. If the interference is suppressed by using ZF (or MMSE), it is called ZF-SIC (or ZF-MMSE). In order to enhance the error performance of SIC, we may arrange the signals in the order of SINR. In this *ordered-SIC* (OSIC) detection, the strongest signal is detected, at each iteration, among the remaining set of the transmit signals. By ordering in that way, it can minimize the accumulation of the detection errors in each iteration. The conventional OSIC algorithm based on the MMSE receiver is described in Figure 7.8. In the algorithm, $SINR_j^i$ indicates the SINR of the jth transmit signal at the ith iteration, g_k^i is the kth row vector of $\mathbf{G}^i$, and Φ is the demodulation function.

Figure 7.9 compares the performances of the above detection schemes for the SM system. Observe that ML detection outperforms the other schemes in terms of error probability and achieves a diversity order of 2. The performance of OSIC is quite close to that of ML even though its complexity is much lower. This testifies that the antenna ordering process in OSIC elicits significant performance improvement over SIC.

7.2.2.2 The BLAST Scheme

For spatial multiplexing, the *Bell Labs layered space–time* (BLAST) system was designed to layer the signals in space and time (Foschini, 1996). There are two types of BLAST system,

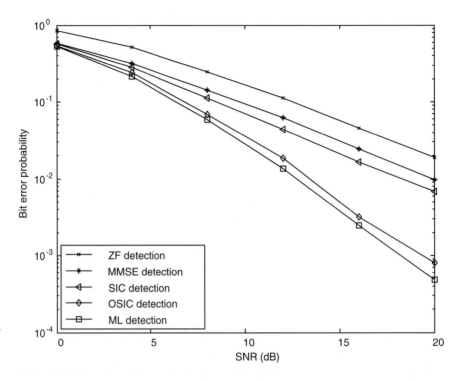

Figure 7.9   Performance of various detectors in a spatial multiplexing system ($M = N = 2$)

V-BLAST and *D-BLAST*, depending on how the data streams are allocated to antennas. Figure 7.10 shows the layering structure of the two systems for case with four transmit antennas. With V-BLAST, the data stream is demultiplexed into multiple substreams A, B, C, and D, and each substream is mapped into each antenna, as shown in Figure 7.10(a). In contrast, with D-BLAST, the substreams are arranged to the diagonal direction in the space–time plain, as shown in Figure 7.10(b). No symbols are transmitted through the transmit antenna associated with the shaded area in the figure. This boundary overhead becomes negligible as the block length increases and contributes to the reduction of the detection complexity.

To illustrate the operation of D-BLAST, consider the system with two transmit antennas, which has the transmit matrix

$$\mathbf{X} = \begin{bmatrix} x_{11} & x_{21} & x_{31} & x_{41} & \cdots \\ 0 & x_{12} & x_{22} & x_{32} & \cdots \end{bmatrix}, \tag{7.38}$$

where $\{x_{i1}, x_{i2}\}$ forms data stream i. At the receiver, we first estimate x_{11} which is not interfered with by any symbols. Then we estimate x_{12} by treating x_{21} as interference and nulling it by using the ZF or MMSE detections. Once x_{12} is estimated, $\{x_{11}, x_{12}\}$ are fed to the channel decoder to decode the first substream. After decoding it, we can subtract the first substream from the received signal to remove the interference from the first substream. The next

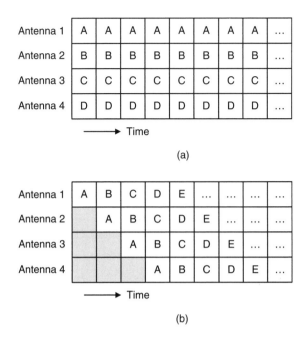

Figure 7.10 Space–time layering for (a) V-BLAST, and (b) D-BLAST

substream is decoded in the same manner. This demonstrates how helpful the initial overhead is for detecting the signal.

7.2.2.3 Double STTD

Since Alamouti's code is designed for two transmit antennas, we may use another space–time coding scheme when there are more antennas. The space–time codes are useful for diversity transmission, not for spatial multiplexing. The *double space–time transmit–diversity* (D-STTD) scheme is an extension of Alamouti's space–time block code to the case of four transmit antennas and two or more receive antennas (Onggosanusi, *et al.*, 2002). The D-STTD scheme achieves diversity and multiplexing gains simultaneously. At the transmitter, antennas 1 and 2 are paired and antennas 3 and 4 are paired. Alamouti's space–time block codes are applied to those two pairs separately. Figure 7.11 shows the block diagram of the D-STTD.

D-STTD is designed for $M = 4$ and $T = 2$ and the channel matrix is assumed to remain fixed for two consecutive symbol periods. The codeword matrix has the form

$$\mathbf{X} = \begin{bmatrix} x_1 & -x_2^H \\ x_2 & x_1^H \\ x_3 & -x_4^H \\ x_4 & x_3^H \end{bmatrix}. \tag{7.39}$$

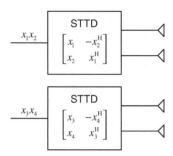

Figure 7.11   Block diagram of D-STTD

If there are two receive antennas, the received signal in (7.21) may be expressed by

$$\begin{bmatrix} r_{11} & r_{12} \\ r_{21} & r_{22} \end{bmatrix} = \begin{bmatrix} h_{11} & h_{12} & h_{13} & h_{14} \\ h_{21} & h_{22} & h_{23} & h_{24} \end{bmatrix} \begin{bmatrix} x_1 & -x_2^H \\ x_2 & x_1^H \\ x_3 & -x_4^H \\ x_4 & x_3^H \end{bmatrix} + \begin{bmatrix} \eta_{11} & \eta_{12} \\ \eta_{21} & \eta_{22} \end{bmatrix} \qquad (7.40)$$

where r_{ij} denotes the received signal at antenna i during time j. This can be rewritten in the form

$$\begin{bmatrix} r_{11} \\ r_{12}^H \\ r_{21} \\ r_{22}^H \end{bmatrix} = \mathbf{H} \begin{bmatrix} x_1 \\ x_2 \\ x_3 \\ x_4 \end{bmatrix} + \begin{bmatrix} \eta_{11} \\ \eta_{12}^H \\ \eta_{21} \\ \eta_{22}^H \end{bmatrix} \qquad (7.41)$$

for the effective channel matrix

$$\mathbf{H} = \begin{bmatrix} h_{11} & h_{12} & h_{13} & h_{14} \\ h_{12}^H & -h_{11}^H & h_{14}^H & -h_{13}^H \\ h_{21} & h_{22} & h_{23} & h_{24} \\ h_{22}^H & -h_{21}^H & h_{24}^H & -h_{23}^H \end{bmatrix} \qquad (7.42)$$

Then the first two rows in $\mathbf{H}$ are orthogonal and the second two rows are orthogonal too. For detection, any detection algorithms for spacial multiplexing may be applied to D-STTD. Since the signals are orthogonal within each group, the signal on each transmit antenna is affected by interference from the two antennas in different groups. Each data symbol achieves the diversity order of 2.

7.2.3 Diversity/Multiplexing Tradeoff

The error probability in a Rayleigh fading channel is given by $P_e \approx SNR^{-d}$ at high SNR for the diversity order d. The capacity of a MIMO system is of the order of min (M, N) log SNR, for the

M transmit and N receive antenna system as shown in (7.14). One may question if it is possible to get both full diversity and full multiplexing gains at the same time. The answer is that a fundamental tradeoff exists between diversity and multiplexing.

To see this, denote by $P_e(SNR)$ and $R(SNR)$ the error probability and the data rate at the given SNR level, respectively. A scheme is said to achieve the spatial multiplexing gain r and diversity gain d if the data rate is

$$\lim_{SNR \to \infty} \frac{R(SNR)}{\log SNR} = r \tag{7.43}$$

and the average error probability is

$$\lim_{SNR \to \infty} \frac{\log P_e(SNR)}{\log SNR} = -d. \tag{7.44}$$

For each r, we define by $d^*(r)$ the supremum of all possible diversity gains. In addition, we define the maximal diversity gain and maximal spatial multiplexing gain to be

$$d^*_{max} \equiv d^*(0) \tag{7.45}$$

$$r^*_{max} \equiv \sup\{r : d^*(r) > 0\}, \tag{7.46}$$

respectively. For the case of M transmit and N receive antennas, it is not difficult to show that

$$d^*_{max} = MN \tag{7.47}$$

$$r^*_{max} = \min\{M, N\}. \tag{7.48}$$

For a block length $l \geq M + N - 1$, the optimal tradeoff curve $d^*(r)$ is given by the piecewise-linear function connecting the points $(k, d^*(k))$, $k = 0, 1, \ldots, \min(M, N)$, where (Zheng and Tse, 2003)

$$d^*(r) = (M-r)(N-r). \tag{7.49}$$

Figure 7.12 illustrates the optimal diversity/multiplexing tradeoff relation (Zheng and Tse, 2003). The overall trend is as if r transmit and r receive antennas, for an integer r, were dedicated to multiplexing and the remainder of the antennas to diversity.

The diversity/multiplexing tradeoff is a useful measure for comparing various MIMO transmission schemes. The following discussion is of the tradeoff for a case with two transmit and one receive antenna.

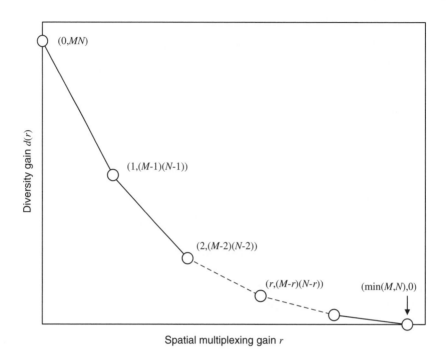

Figure 7.12 Optimal diversity/multiplexing tradeoff relation (Reproduced with permission from L. Zheng and D. Tse, "Diversity and multiplexing: A fundamental tradeoff in multiple-antenna channels," *IEEE Transactions on Information Theory*, 49, no. 5, 1073–1096, May 2003. © 2003 IEEE)

Consider some repetition coding, with a symbol repeated on the two transmit antennas in two consecutive symbol times. A symbol is transmitted through antenna 1 at the first symbol time, and the same symbol is transmitted through antenna 2 at the second symbol time. Then we get

$$\mathbf{R} = \mathbf{HX} + \eta \qquad (7.50)$$

for the transmitted symbol matrix

$$\mathbf{X} = \begin{bmatrix} x_1 & 0 \\ 0 & x_1 \end{bmatrix}. \qquad (7.51)$$

Note that it differs from Alamouti's space–time code in (7.27). The repetition code can achieve the maximal diversity order $d^*_{max} = 2$ with a multiplexing gain $r = 0$. If we increase the size of the constellation to support a data rate $R = r \log SNR$ for some $r > 0$, the distance between the constellation points shrinks with the SNR and the achievable diversity gain decreases (Zheng and Tse, 2003). In this case, the maximal spatial multiplexing gain achieved by this repetition code is 0.5, because only one symbol is transmitted in two symbol times.

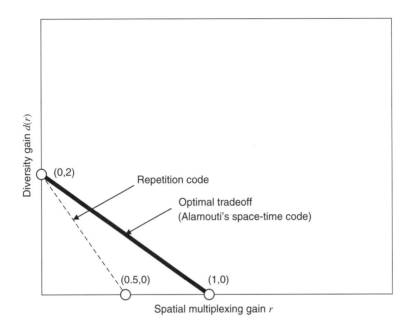

Figure 7.13 Diversity/multiplexing tradeoff for the case with two transmit and one receive antennas (Reproduced with permission from L. Zheng and D. Tse, "Diversity and multiplexing: A fundamental tradeoff in multiple-antenna channels," *IEEE Transactions on Information Theory*, 49, no. 5, 1073–1096, May 2003. © 2003 IEEE)

Next consider the Alamouti space–time code. It too achieves a diversity order of 2 with a multiplexing gain $r = 0$. However, since it transmits two independent symbols over two consecutive symbol times, the spatial multiplexing gain is 1. Figure 7.13 illustrates the diversity/multiplexing tradeoff of the repetition code and Alamouti's space–time code (Zheng and Tse, 2003). Note that Alamouti's space–time code outperforms the repetition code. In fact, Alamouti's code is optimal in terms of diversity/multiplexing tradeoff for the case with $M = 2$ and $N = 1$.

Now consider the case with two transmit and two receive antennas. The repetition code and Alamouti's space–time code achieve the maximal diversity order of 4. Adding one more receive antenna results in a higher diversity gain. However, the spatial multiplexing gains of these two systems remain the same as in the case with two transmit and one receive antennas. The repetition code and Alamouti's space–time code still achieve the multiplexing gains of 0.5 and 1, respectively. Since the maximal multiplexing gain given in (7.48) is 2, both of them are not optimal in terms of diversity/multiplexing tradeoff. If the number of receive antennas is no smaller than the number of transmit antennas, we can apply the spatial multiplexing schemes such as V-BLAST and D-BLAST. When V-BLAST is detected with ML detection, it achieves a diversity gain of $d(r) = 2(1 - r/2)$ for a given multiplexing gain r (Zheng and Tse, 2003). So, it is not optimal. However, D-BLAST with MMSE SIC detection is optimal in terms of diversity/multiplexing tradeoff, if we ignore the overhead for initial transmission. Figure 7.14 shows the diversity/multiplexing tradeoff relation for various 2×2 MIMO transmission schemes (Zheng and Tse, 2003).

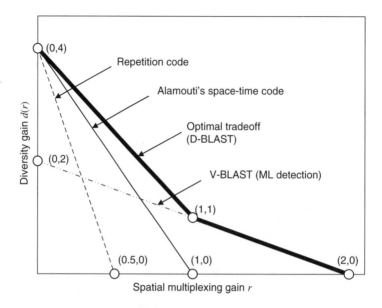

Figure 7.14 Diversity/multiplexing tradeoff for the case with two transmit and two receive antennas (Reproduced with permission from L. Zheng and D. Tse, "Diversity and multiplexing: A fundamental tradeoff in multiple-antenna channels," *IEEE Transactions on Information Theory*, 49, no. 5, 1073–1096, May 2003. © 2003 IEEE)

7.3 Multiuser MIMO

Previous sections have considered MIMO transmission in point-to-point communications, implicitly assuming a single-user environment. This section deals with point-to-multipoint or multipoint-to-point communications in a multiuser environment. Now each terminal may play the role of separate antennas, so that the communication link becomes a MIMO transmission link. In multiuser systems, multiple *mobile stations* (MS) communicate with one base station (BS) sharing the same bandwidth. The uplink and the downlink refer to the transmission paths from MS to BS, and from BS to MS, respectively. The uplink is often referred to as the *multiple-access channel* (MAC) because multiple mobile stations access the same channel to transmit to the BS in the uplink. Likewise, the downlink is often called the *broadcast channel* (BC) because the base station broadcasts messages to multiple mobile stations. This section of the chapter uses the two pairs of terms (i.e., uplink/MAC, and downlink/BC) interchangeably.

Mobile and base stations may be equipped with many antennas and the link may form a MIMO transmission link. The capacity of a MIMO channel grows in proportion to the minimum of the numbers of transmit and receive antennas. In general, mobile stations are much smaller than the BS, making it difficult to accommodate as many antennas. The capacity is thus limited by the number of MS antennas. In an environment where multiple mobile stations communicate with one BS, however, the antennas in the other mobile stations can be arranged to help to overcome this limitation. This is possible because multiuser data are transmitted through multiple antennas simultaneously, with each antenna transmitting a combination of the multiuser data. As a consequence, the multiuser MIMO technique provides a new method of increasing the capacity.

In the uplink, multiple receive antennas at the BS detect the signals transmitted by multiple mobile stations simultaneously. In the usual case where each MS has a single transmit antenna and the BS has multiple transmit antennas, each uplink data stream is independently coded and thus looks like the V-BLAST scheme. So, the receive antennas in the BS can detect the different data streams by using the same MIMO detection algorithm that was used for detecting multiple streams from a single user, such as the ML and MMSE-OSIC detectors.

In the downlink, the BS has to transmit independent data streams to multiple mobile stations simultaneously, but each MS cannot cooperate to decode its associated data stream. So, what is needed is some precoding scheme that subtracts the inter-user interference component at the BS using the channel information fed back from each MS. The optimal strategy for downlink precoding is *dirty-paper coding* which can achieve the sum capacity of the multiuser downlink, but it imposes a high computing burden and requires perfect CSI at the base station. There are other practical precoding approaches that require less computing and less information.

7.3.1 Uplink Channel

In the uplink, multiple mobile stations share the same wireless channel to transmit to the base station. This subsection first examines the capacity of the uplink channel in the single-antenna case and extends the result to multiple antennas. Then it discusses how multiple-access gain can be achieved in the uplink channel by applying the diversity/multiplexing tradeoff concept.

7.3.1.1 Uplink Capacity with Single Transmit Antenna

Consider the case where each user has a single transmit antenna and the base station has N receive antennas. The received signal at the BS is given by

$$\mathbf{r} = \sum_{k=1}^{K} \mathbf{h}_k \mathbf{x}_k + \eta, \tag{7.52}$$

where $\mathbf{r}$ is an N-vector representing the received signal, $\mathbf{h}_k$ is a channel vector from user k to the BS, $\mathbf{x}_k$ is a transmitted symbol from user k, and η is an additive zero-mean white Gaussian noise vector with variance N_0.

This uplink channel may be considered as a single-user MIMO transmission scheme. Since K user signals are received by N receive antennas at the BS, the uplink channel looks like a single-user MIMO transmission with K transmit and N receive antennas. This type of transmission operation is called *space-division multiple-access* (SDMA), or *collaborative spatial multiplexing* (CSM). In this case, the effective MIMO channel matrix is given by $\mathbf{H} = [h_1, h_2, \ldots, h_K]$ in (7.1). Consequently, all the detection schemes in the previous section are applicable to the uplink to yield spatial multiplexing gains. If the effective MIMO channel matrix $\mathbf{H}$ is full rank, the independent K data streams can be transmitted provided there are enough receive antennas; that is, if $N \geq K$. Such SDMA is advantageous over the single-user SM in that it does not require multiple antennas at the mobile stations while exhibiting multiuser diversity. When the number of users is larger than the antenna numbers in the BS, only some of the users can transmit through the common channel simultaneously and, in selecting those users to transmit, multiuser diversity can be obtained.

Consider now a two-user uplink channel in which channel vectors are static. If R_1 and R_2 denote the transmission rates of user 1 and user 2, respectively, the capacity region for the uplink is given by

$$R_1 \leq \log_2 \left(1 + \frac{|h_1|^2 P_1}{N_0} \right),$$

$$R_2 \leq \log_2 \left(1 + \frac{|h_2|^2 P_2}{N_0} \right), \tag{7.53}$$

$$R_1 + R_2 \leq \log_2 \det \left(\mathbf{I} + \frac{1}{N_0} \mathbf{HQH}^{\mathrm{H}} \right),$$

where P_1 and P_2 are the transmit powers of users 1 and 2, respectively, and the covariance matrix of the transmitted signal is represented by $\mathbf{Q} = \mathrm{diag}\,(P_1, P_2)$. The first and second constraints in (7.53) correspond to the capacity of a single-user SIMO transmission. The third constraint is the sum capacity of the two-user uplink channel and it is derived from the capacity of the MIMO transmission in (7.2). Since the signals transmitted by two users are independent, all the off-diagonal entries in $\mathbf{Q}$ are zero. Consequently, we can rewrite the sum capacity constraint as

$$R_1 + R_2 \leq \log_2 \det \left(\mathbf{I} + \frac{P_1}{N_0} h_1 h_1^{\mathrm{H}} + \frac{P_2}{N_0} h_2 h_2^{\mathrm{H}} \right). \tag{7.54}$$

Figure 7.15 shows the capacity region for this two-user uplink channel. The achievable rate region is pentagon-shaped. When user 1 transmits at the full rate $\log_2 (1 + |h_1|^2 P_1/N_0)$, and user

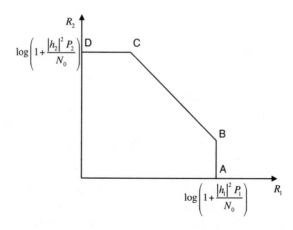

Figure 7.15   Capacity region of the two-user uplink channel

2 remains silent, then the channel becomes an interference-free Gaussian channel, so the rate point A in Figure 7.15 is achieved. Likewise, when user 2 transmits at the full rate $\log_2(1 + |h_2|^2 P_2/N_0)$ and user 1 keeps silent, then the rate point D is achieved. In order to achieve point B, user 1 transmits at the full rate $\log_2(1 + |h_1|^2 P_1/N_0)$, and user 2 makes a transmission regarding the signal from user 1 as interference. In this case, the receiver first detects the signal from user 2 and regards the signal from user 1 as an additive interference, and subtracts the contribution of user 2 from the received signal, and finally detects the signal from user 1. So, this interference cancellation scheme helps to achieve the rate point B. Likewise, the rate point C can be achieved by detecting the signal of user 1 first and then the signal of user 2. So, the *decoding order* determines which rate point can be achieved. All the points between B and C can be achieved by applying time-sharing to those two schemes.

If we extend (7.54) to the case of a K-user uplink channel, we get

$$\sum_{k=1}^{K} R_K \leq \log_2 \det\left(\mathbf{I} + \sum_{k=1}^{K} \frac{P_k}{N_0} h_k h_k^{\mathrm{H}}\right) \tag{7.55}$$

On rearranging the right-hand side of (7.55) we get

$$\sum_{k=1}^{K} R_K \leq \log_2 \det\left(\mathbf{I} + \sum_{k=1}^{K-1} \frac{P_k}{N_0} h_k h_k^{\mathrm{H}}\right) + \log_2 \det\left(\mathbf{I} + \left(\mathbf{I} + \sum_{k=1}^{K-1} \frac{P_k}{N_0} h_k h_k^{\mathrm{H}}\right)^{-1} \frac{P_K}{N_0} h_K h_K^{\mathrm{H}}\right)$$

$$= \log_2 \det\left(\mathbf{I} + \sum_{k=1}^{K-1} \frac{P_k}{N_0} h_k h_k^{\mathrm{H}}\right) + \log_2\left(1 + \frac{P_K}{N_0} h_K^{\mathrm{H}}\left(\mathbf{I} + \sum_{k=1}^{K-1} \frac{P_k}{N_0} h_k h_k^{\mathrm{H}}\right)^{-1} h_K\right), \tag{7.56}$$

where the equality holds due to the relation $\det(\mathbf{I} + \mathbf{AB}) = \det(\mathbf{I} + \mathbf{AB})$. If we repeat this process, we get (Varanasi and Guess, 1997)

$$\sum_{k=1}^{K} R_K \leq \sum_{k=1}^{K} \log_2\left(1 + \frac{P_k}{N_0} h_k^{\mathrm{H}}\left(\mathbf{I} + \sum_{j=1}^{k-1} \frac{P_j}{N_0} h_j h_j^{\mathrm{H}}\right)^{-1} h_k\right). \tag{7.57}$$

Now consider MMSE-SIC detection for the received signal (7.52). We first detect the signal from user K, treating the signals of all the other users as interference, and then subtract its contribution from the received signal. Then the detection process is repeated for user $K-1$ to user 1. In general, the SINR for user k may be expressed by

$$SINR_k = \frac{P_k}{N_0} h_k^{\mathrm{H}}\left(\mathbf{I} + \sum_{j=1}^{k-1} \frac{P_j}{N_0} h_j h_j^{\mathrm{H}}\right)^{-1} h_k \tag{7.58}$$

since the signals from user 1 to user $k-1$ are treated as interference and the signals from user $k+1$ to user K are already subtracted. So, the rate of user k is

$$R_k = \log_2\left(1 + \frac{P_k}{N_0}h_k^{\mathrm{H}}\left(\mathbf{I} + \sum_{j=1}^{k-1}\frac{P_j}{N_0}h_jh_j^{\mathrm{H}}\right)^{-1}h_k\right) \tag{7.59}$$

for $k = 1, 2, \ldots, K$. Adding all the rates in (7.59) from user 1 through to user K, the right-hand side of (7.57) is obtained. Therefore, MMSE-SIC detection achieves the sum capacity in the uplink channel (Varanasi and Guess, 1997). If the decoding order is changed, the resulting achievable rates may be different from that in (7.59). However, the sum rate is achieved to be the same as given in (7.57). In general, the capacity region of a K-user uplink channel is a polyhedron, and each vertex can be achieved by MMSE-SIC detection with each using a different detection order.

Now we return to the case of a two-user uplink channel. If the signal from user 2 is detected first and then the signal from user 1, the achievable rate is given by

$$\begin{aligned}
R_1 &= \log_2\left(1 + \frac{P_1}{N_0}|h_1|^2\right), \\
R_2 &= \log_2\left(1 + \frac{P_2}{N_0}h_2^{\mathrm{H}}\left(\mathbf{I} + \frac{P_1}{N_0}h_1h_1^{\mathrm{H}}\right)^{-1}h_2\right).
\end{aligned} \tag{7.60}$$

If we apply the matrix inversion lemma

$$(\mathbf{A}-\mathbf{BCD})^{-1} = \mathbf{A}^{-1} + \mathbf{A}^{-1}\mathbf{B}(\mathbf{C}^{-1}-\mathbf{DA}^{-1}\mathbf{B})^{-1}\mathbf{DA}^{-1} \tag{7.61}$$

to the second equation, we get

$$\begin{aligned}
R_2 &= \log_2\left(1 + \frac{P_2}{N_0}h_2^{\mathrm{H}}\left(\mathbf{I} - \left(\frac{N_0}{P_1} + |h_1|^2\right)^{-1}h_1h_1^{\mathrm{H}}\right)h_2\right) \\
&= \log_2\left(1 + \frac{P_2}{N_0}h_2^{\mathrm{H}}\left(\mathbf{I} - \left(\frac{N_0}{P_1|h_1|^2} + 1\right)^{-1}\frac{h_1h_1^{\mathrm{H}}}{|h_1|^2}\right)h_2\right).
\end{aligned} \tag{7.62}$$

If the two vectors h_1 and h_2 are orthogonal (i.e., $h_1^{\mathrm{H}}h_2 = 0$), the rate of user 2 is maximized. On the other hand, if $h_2 = ah_1$, for a non-zero scalar number a, then the rate is minimized. Thus, the upper and lower bounds of R_2 are given by

$$\log_2\left(1 + \frac{P_2|h_2|^2}{N_0 + P_1|h_1|^2}\right) \le R_2 \le \log_2\left(1 + \frac{P_2}{N_0}|h_2|^2\right). \tag{7.63}$$

With orthogonal channels, each user channel looks like an interference-free AWGN channel, so the maximum achievable rate of each user is the single-user AWGN capacity. However, when one user channel is aligned with the other user channel, then the user signals interfere with each other and the achievable rate of user 2 is reduced due to the interference from user 1. This relation can be extended to a K-user uplink channel.

7.3.1.2 Uplink Capacity with Multiple Transmit Antennas

Assume that user k has M_k antennas and the base station has N receive antennas. The received signal for the K-user uplink channel is given by

$$\mathbf{r} = \sum_{k=1}^{K} \mathbf{H}_k \mathbf{x}_k + \eta$$

where $\mathbf{r}$ is an N-vector representing the received signal, $\mathbf{H}_k$ is a channel matrix from user k to the BS, $\mathbf{x}_k$ is a transmitted symbol vector from user k, and η is an additive zero-mean white Gaussian noise vector with variance N_0. This uplink channel may also be considered as a single-user MIMO transmission scheme. The uplink channel looks like a single-user MIMO transmission with $\sum_{k=1}^{K} M_k$ transmit and N receive antennas. The sum capacity of the K-user uplink channel is

$$C = \max_{\mathrm{Tr}(\mathbf{Q}_k) \leq P_k} \log_2 \det \left(\mathbf{I} + \frac{1}{N_0} \sum_{k=1}^{K} \mathbf{H}_k \mathbf{Q}_k \mathbf{H}_k^{\mathrm{H}} \right). \tag{7.64}$$

The trace constraint $\mathrm{Tr}(\mathbf{Q}_k) \leq P_k$ corresponds to the total transmit power constraint for user k. In order to achieve the capacity, we have to find the optimal covariance matrices $\mathbf{Q}_k$. Since $\log \det (\bullet)$ is a concave function, this optimization becomes a convex optimization problem.

The covariance matrices for the problem in (7.64) are optimal if and only if $\mathbf{Q}_k$ is the single-user water-filling covariance matrix of channel $\mathbf{H}_k$ with the noise term

$$N_0 \mathbf{I} + \sum_{j=1, \, j \neq k}^{K} \mathbf{H}_k \mathbf{Q}_k \mathbf{H}_k^{\mathrm{H}}$$

for all $k = 1, 2, \ldots, K$ (Yu et al., 2004). This statement can easily be proved as follows. Suppose that, at the sum-rate optimal point, there exists a $\mathbf{Q}_k$ that does not satisfy the single-user water-filling condition. Then set $\mathbf{Q}_k$ to meet the single-user water-filling condition while fixing all the other covariance matrices. This update increases the sum-rate objective, which contradicts the optimality assumption of $\mathbf{Q}_k$. So, all the $\mathbf{Q}_k$ must meet the single-user water-filling condition. Conversely, since it is a convex optimization problem, we can apply the KKT condition. The Lagrangian is given by

$$L = \log_2 \det \left(\mathbf{I} + \frac{1}{N_0} \sum_{k=1}^{K} \mathbf{H}_k \mathbf{Q}_k \mathbf{H}_k^{\mathrm{H}} \right) - \sum_{k=1}^{K} \lambda_k (\mathrm{Tr}(\mathbf{Q}_k) - P_k) + \sum_{k=1}^{K} \mathrm{Tr}(\psi_k \mathbf{Q}_k), \tag{7.65}$$

where λ_k is the Lagrange multiplier associated with the power constraint, and Ψ_k is the Lagrange multiplier matrix associated with the positive-definite constraint for $\mathbf{Q}_k$. Since the gradient of $\log \det(\mathbf{X})$ is $\mathbf{X}^{-1}$, the KKT condition is given by (Yu *et al.*, 2004)

$$\lambda_k = \frac{1}{\ln 2} \mathbf{H}_k^{\mathrm{H}} \left(N_0 \mathbf{I} + \sum_{j=1}^{K} \mathbf{H}_j \mathbf{Q}_j \mathbf{H}_j^{\mathrm{H}} \right)^{-1} \mathbf{H}_k + \psi_k,$$
$$\mathrm{Tr}(\mathbf{Q}_k) = P_k, \tag{7.66}$$
$$\mathrm{Tr}(\psi_i \mathbf{Q}_k) = 0,$$
$$\psi_k, \mathbf{Q}_k, \lambda_k \geq 0,$$

for all $k = 1, 2, \ldots, K$. For each user k, the KKT condition corresponds to the single-user water-filling KKT condition if the noise is

$$N_0 \mathbf{I} + \sum_{j=1, j \neq k}^{K} \mathbf{H}_j \mathbf{Q}_j \mathbf{H}_j^{\mathrm{H}}.$$

This implies that each $\mathbf{Q}_k$ satisfies the single-user optimality condition while regarding the signals of other users as interference. We can construct an algorithm to obtain the optimal covariance matrices as given in Figure 7.16, which is called an *iterative water-filling algorithm* (Yu *et al.*, 2004).

7.3.1.3 Diversity/Multiplexing Tradeoff in Uplink Channels

The diversity/multiplexing tradeoff discussed in the previous section was for a point-to-point communication. The optimal diversity/multiplexing tradeoff, in effect, indicates how much diversity gain can be exploited for a given multiplexing gain. In the case of M transmit and N receive antennas, the maximal diversity gain for a given multiplexing gain $r \leq \min(M, N)$ is expressed by

$$d_{M,N}^*(r) = (M-r)(N-r). \tag{7.67}$$

Given   channel matrix $\mathbf{H}_k$ and a required tolerance $\varepsilon > 0$

Initialize $\mathbf{Q}_k = 0$ and $C_{new} = 0$

Repeat

  $C_{old} = C_{new}$

  For $k = 1$ to K

   $\mathbf{Q}_k = \arg \max_{\mathbf{Q}_k} \log_2 \det \left(\mathbf{H}_k \mathbf{Q}_k \mathbf{H}_k^{\mathrm{H}} + \sum_{j=1, j \neq k}^{K} \mathbf{H}_j \mathbf{Q}_j \mathbf{H}_j^{\mathrm{H}} + N_0 \mathbf{I} \right)$

  $C_{new} = \log_2 \det \left(\mathbf{I} + \frac{1}{N_0} \sum_{k=1}^{K} \mathbf{H}_k \mathbf{Q}_k \mathbf{H}_k^{\mathrm{H}} \right)$

Until $|C_{old} - C_{new}| < \varepsilon$

Figure 7.16   Algorithm for iterative water-filling

In the uplink channel, the receive antennas at the base station can separate out the signals from different users, and thus obtain a multiple-access gain. So, we can extend the diversity/multiplexing tradeoff in the single-user MIMO transmission to the case of uplink MIMO transmission.

Consider the i.i.d. Rayleigh fading uplink channel with K users. Each user has M transmit antennas and the BS has N receive antennas. Each user k has the multiplexing gain r_k, so that the rate of user k is given by $R_k = r_k \log SNR$. When maximum-likelihood (ML) detection is applied, the error probability decays with $P_e = SNR^{-d}$, so the diversity gain of each user is d. It is necessary to characterize the set of rates $(r_1, r_2, \ldots, r_k)$ that yields the diversity gain d. For simplicity, we consider the symmetric case where the multiplexing gains of all users are equal to r. Then the maximal multiplexing gain achieved by each user is given by min $(M, N/K)$, which is the degree of freedom per user.

Interestingly, the tradeoff performance differs in two regimes as follows. In a lightly loaded regime (i.e., with a low data rate), the maximal diversity order for a given multiplexing gain r is (Tse *et al.*, 2004)

$$d^*_{\text{sym}}(r) = d^*_{M,N}(r) \tag{7.68}$$

for $r \leq$ min $(M, N/(K + 1)$. In a heavily loaded regime, the maximal diversity gain is (Tse *et al.*, 2004)

$$d^*_{\text{sym}}(r) = d^*_{KM,N}(Kr) \tag{7.69}$$

for min $(M, N/(K + 1) \leq r \leq$ min $(M, N/K)$. This is because the typical error event for the lightly loaded regime is that only one of the users' packets is in error while the typical error event for the heavily loaded regime is that all the users' packets are erred simultaneously.

If the load of the system is sufficiently light, the single-user diversity multiplexing gain is achieved. Then min $(M, N/(K + 1)$ is the threshold on the multiplexing gain below which the error probability is the same as in the single-user MIMO case. If $M \leq N/(K + 1)$, then we have min $(M, N/(K + 1)) = M$, so the error probability is the same as the case when only one user exists in the system and the multiple-access gain is obtained at no cost (Tse *et al.*, 2004). In this case, the optimal diversity multiplexing tradeoff is as plotted in Figure 7.12. On the other hand, if $M \geq N/(K + 1)$, a single-user performance is achieved provided $r \leq N/(K + 1)$. In addition, adding another user does not degrade the error performance of the other users as long as the system stays in the lightly loaded region, which is very desirable in multiple access (Tse *et al.*, 2004). Figure 7.17 shows the optimal diversity/multiplexing tradeoff.

In the heavily loaded region, the symmetric diversity gain is $d^*_{KM,N}(Kr)$, which is the tradeoff for the system in which K users are pooled together into a single user with KM transmit antennas and multiplexing gain Kr (Tse *et al.*, 2004). In this regime, the error performance is affected by the other users. In the case of $M \geq N/(K + 1)$, the value $N/(K + 1)$ plays the role of the rate threshold beyond which the dominant error event is that all the users' packets are in error. Figure 7.17 shows that the diversity/multiplexing tradeoff behaves differently depending on the rate threshold min $(M, N/(K + 1))$.

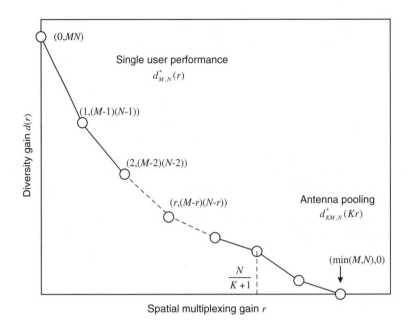

Figure 7.17 Symmetric diversity/multiplexing tradeoff for $M \geq N/(K + 1)$ (Reproduced with permission from N.C. Tse, P. Viswanath, and L. Zheng, " Diversity-multiplexing tradeoff in multiple-access channels," *IEEE Transactions on Information Theory*, 50, no. 9, 1859–1874, September 2004. © 2004 IEEE)

7.3.2 Dirty-paper Coding

In the downlink, the base station transmits independent data streams simultaneously to multiple users in order to increase the data rate. In this case, each signal may interfere with other signals, thus denying a high rate of transmission. In order to overcome this limitation, it is necessary to suppress the interference either at the receiver or at the transmitter. When there is one transmit antenna, it is optimal to decode the desired signal after cancelling out the other signals at the receiver, so that precoding is not necessary at the transmitter. However, when there are two or more transmit antennas, it is desirable to precancel the interference through precoding at the transmitter. The optimal precoding scheme at the transmitter is "dirty-paper coding" (DPC: a term explained later). If interference is known at the transmitter, then DPC allows encoding of the signal without additional power so that the receiver can decode it without being affected by the interference.

Consider the interference channel shown in Figure 7.18. The received signal is

$$Y = X + S + Z, \tag{7.70}$$

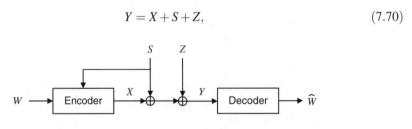

Figure 7.18 Block diagram of interference channel

where S is the interference, X is the transmitted signal after encoding based on the message W and the interference S, and Z is an additive Gaussian noise with variance P_Z. Note that the state of S is known to the transmitter but not to the receiver.

In this channel, if the transmitter encodes X without using information about S, then the capacity is given by

$$C = \frac{1}{2} \log_2\left(1 + \frac{P_X}{P_S + P_Z}\right), \tag{7.71}$$

where P_X, P_S, and P_Z are the powers of X, S, and Z, respectively. If P_X is very small compared with P_S, then the capacity is very small. When the interference S is known at the transmitter, the transmitter can presubtract the interference so that the receiver can be free from the interference. We can imagine a sheet of paper covered with dirt spots, on which a writer writes a message. The writer knows where dirt is located but the reader cannot distinguish them from the writer's ink marks. This problem is analogous to the communication problem in (7.70). So, the optimal transmission scheme is called *dirty-paper coding* (DPC) (Costa, 1983). The capacity of the DPC is given by

$$C = \frac{1}{2} \log_2\left(1 + \frac{P_X}{P_Z}\right), \tag{7.72}$$

which is nothing but the capacity of the AWGN channel in which no interference exists.

One simple precancelling approach is to subtract the interference from the message. In this case, $X = W - S$ is transmitted, so the received signal is

$$Y = (W - S) + S + Z = W + Z. \tag{7.73}$$

Even though the interference is perfectly cancelled, the transmit power may increase. The total power becomes $\mathbb{E}[X^2] = \mathbb{E}[W^2] + P_S$, which may be larger than P_X.

To avoid consuming additional power, we introduce a lattice modulo operation at the transmitter. One example of the modulo operation is (Erez, *et al.*, 2005)

$$\begin{aligned} X &= (W - \alpha S - U) \bmod \Lambda \\ &= W - \alpha S - U - Q_\Lambda(W - \alpha S - U), \end{aligned} \tag{7.74}$$

where $Q_\Lambda(\bullet)$ denotes a vector quantizer corresponding to the lattice Λ; that is, $Q_\Lambda(X)$ is the lattice point nearest to X

$$Q_\Lambda(X) = \arg \min_{\lambda \in \Lambda} |X - \lambda|^2, \tag{7.75}$$

and U is a uniform random variable which randomizes the quantization error and makes X and W independent. Due to the modulo operation in (7.74), the transmit power can be made smaller than the power constraint P_X, no matter how large the interference power may be. The random variable U is known both at the transmitter and the receiver. It makes the transmit signal uniform over the Voronoi region, irrespective of the interference S. Parameter α is chosen to minimize

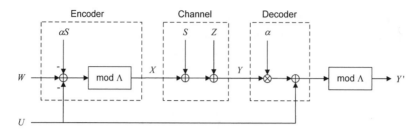

Figure 7.19   Encoding and decoding of the DPC scheme

the mean square error and is given by

$$\alpha = \frac{P_X}{P_X + P_Z}. \tag{7.76}$$

The receiver uses the modulo operation to remove the effect of interference; that is

$$Y' = (\alpha Y + U) \bmod \Lambda. \tag{7.77}$$

Then the channel defined by (7.70), (7.76), and (7.77) is equivalent to (Erez, *et al.*, 2005)

$$Y' = (W + Z') \bmod \Lambda, \tag{7.78}$$

where Z' is independent of W and Z, and is given by

$$Z' = [(1-\alpha)U + \alpha Z] \bmod \Lambda. \tag{7.79}$$

Figure 7.19 shows the encoding and the decoding schemes for DPC. The maximum rate achieved by this DPC scheme is (Erez, *et al.*, 2005)

$$R \geq \frac{1}{2}\log_2\left(1 + \frac{P_X}{P_Z}\right) - \frac{1}{2}\log_2 2\pi eG(\Lambda), \tag{7.80}$$

where $G(\Lambda)$ is the normalized second moment of the lattice Λ. With a proper choice of the lattice, the rate loss $(1/2)\log_2 2\pi eG(\Lambda)$ can be made arbitrarily small. Consequently, this DPC scheme conforms with the DPC capacity in (7.72).

7.3.3 Downlink Channel

In the downlink the base station transmits independent messages to multiple mobile stations. Since the receivers cannot cooperate to decode the messages, joint decoding is not possible. Instead, we can pursue interference presubtraction at the transmitter by adopting dirty-paper coding so that the capacity can be achieved in the downlink channel. In the following we examine the capacity of the MIMO downlink channel.

7.3.3.1 Degraded Broadcast Channel

Consider the two-user case in which the BS has a single transmit antenna and each user also has a single receive antenna. The received signal at user k is

$$r_k = x + \eta_k, \tag{7.81}$$

where x is a transmit symbol, and r_k and η_k are the received signal and the additive zero-mean white Gaussian noise of user k, respectively. The variance of the noise of user k is given by $\mathbb{E}[\eta_k^2] = \sigma_k^2$ for $k = 1,2$ and we may assume $\sigma_1^2 \leq \sigma_2^2$ without loss of generality. Due to the Gaussian distribution of the noise, the distribution of the received signal for a given transmit symbol x is given by

$$P(r_k|x) = \frac{1}{\sqrt{2\pi}\sigma_k} \exp\left(-\frac{(r_k-x)^2}{2\sigma_k^2}\right) \tag{7.82}$$

for $k = 1,2$. Thus, the joint distribution of the downlink channel may be expressed by

$$P(r_1, r_2|x) = P(r_1|x)P(r_2|r_1). \tag{7.83}$$

If this relation holds, the channel is called a *degraded channel* (Cover and Thomas, 1991). The terminology means that the received signal at user 2 is a degraded version of the received signal at user 1. The received signals can be expressed in the following equivalent form:

$$\begin{aligned} r_1 &= x + \eta_1 \\ r_2 &= r_1 + \eta'_2 \end{aligned} \tag{7.84}$$

where η'_2 is the additive zero-mean white Gaussian noise with variance $\mathbb{E}\left[\eta_2'^2\right] = \sigma_2^2 - \sigma_1^2$. Figure 7.20 depicts the original broadcast channel and its equivalent degraded channel.

The equivalence above helps to determine the capacity region of the downlink channel easily. In the degraded channel, user 1 experiences additive noise η_1, but user 2 experiences additive noise η_2 and the interference P_1 that is the power of the signal destined to user 1. Since the

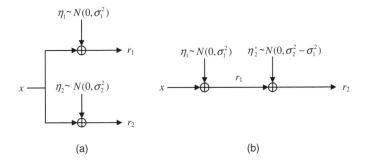

(a)   (b)

Figure 7.20   Two-user downlink channel: (a) broadcast channel, and (b) equivalent degraded channel

downlink channel is equivalent to the degraded channel, the downlink channel capacity region is given by

$$R_1 \leq \frac{1}{2} \log_2 \left(1 + \frac{P_1}{\sigma_1^2} \right)$$

$$R_2 \leq \frac{1}{2} \log_2 \left(1 + \frac{P_2}{P_1 + \sigma_2^2} \right),$$

(7.85)

where the factor 1/2 is included because the signal is generated from a real signal constellation. We have the total power constraint at the BS; that is, $P_1 + P_2 \leq P$. Figure 7.21 shows the capacity region determined according to (7.85) for the case $P = 100$, $\sigma_1^2 = 1$, $\sigma_2^2 = 5$. Observe that the sum capacity is achieved when $P_1 = P$ and $P_2 = 0$, which is equivalent to the single-user capacity of user 1's channel; that is

$$R_1 + R_2 \leq \frac{1}{2} \log_2 \left(1 + \frac{P}{\sigma_1^2} \right).$$

(7.86)

Therefore, in order to obtain the sum capacity, it is necessary to allocate all the power to the user whose channel gain is the best.

In order to achieve the capacity, it is necessary to encode and decode the message in a successive manner. If the codeword length is n, the codeword of user 2 is generated from 2^{nR_2} codewords independently of the message of user 1. This codeword serves as a cloud center

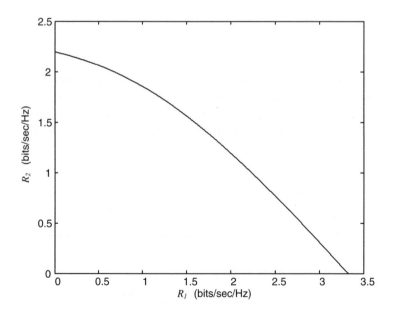

Figure 7.21   Capacity region for two-user downlink channel with $P = 100$, $\eta_1 \sim N(0,1)$, and $\eta_2 \sim N(0,5)$

which is detectable by both users. Then, the codeword of user 1 is generated and superimposed on the codeword of user 2 to form a composite codeword. Each cloud consists of 2^{nR_1} codewords which is detectable by user 1. Consequently, the weaker receiver (user 2) can see only the clouds, while the stronger receiver (user 1) can see each individual codeword within the cloud. This coding scheme is called *superposition coding* (Cover and Thomas, 1991). User 2 detects only the message destined to itself, treating the signal power of user 1 as interference. User 1 detects the message for user 2 first and subtracts its contribution to the received signal and finally detects the message for user 1. Thus, the stronger user always detects the weaker user's message first before decoding its own message. This process is *successive decoding*.

Now consider the downlink channel in which each user has N antennas and the base station has a single antenna. The received signal at user k ($k = 1, 2, \ldots, K$) is given by

$$r_k = h_k x + \eta_k \qquad (7.87)$$

where r_k is a received signal, h_k is a channel from the BS to user k, x_k is a transmitted symbol, and η is an additive zero-mean white Gaussian noise vector with variance N_0. In this case, the downlink channel becomes a degraded broadcast channel after rating users by their channel strength. Let π be the permutation of $\{1, 2, \ldots, K\}$ that sorts the channel gains in descending order; that is

$$|h_{\pi(1)}|^2 \geq |h_{\pi(2)}|^2 \geq \cdots \geq |h_{\pi(K)}|^2. \qquad (7.88)$$

Then, the encoding order for the superposition coding is $\pi(K), \pi(K-1), \ldots, \pi(1)$. We can extend (7.85) to obtain the capacity region

$$R_{\pi(k)} \leq \log_2 \left(1 + \frac{|h_{\pi(k)}|^2 P_{\pi(k)}}{|h_{\pi(k)}|^2 \sum_{j=1}^{k-1} P_{\pi(j)} + N_0} \right). \qquad (7.89)$$

However, if the BS has multiple antennas, then unfortunately this channel is not a physically degraded channel. So, superposition encoding or successive decoding no longer achieve the capacity.

7.3.3.2 MIMO Broadcast Channel

If the base station has more than one antenna, the downlink channel is not a degraded channel any longer because each user receives different strength signals from different transmit antennas. In addition, the channel matrices do not necessarily have the same eigenvectors, so it is not possible to diagonalize each channel matrix simultaneously. Nevertheless, it is possible to determine the capacity region by applying dirty-paper coding to the matrix downlink channel.

Consider the downlink channel in which each user has N antennas and the BS has M antennas. Then the received signal at each user is given by

$$\mathbf{r}_k = \mathbf{H}_k \mathbf{x} + \eta_k, \tag{7.90}$$

where $\mathbf{r}_k$ is an N-vector representing the received signal, $\mathbf{H}$ is an $N \times M$ channel matrix from the BS to user k, $\mathbf{x}$ is an M-vector representing the transmitted symbol, and η is an additive zero-mean white Gaussian noise vector with variance N_0. In this case, the downlink channel is not a degraded broadcast channel. Assume that the channel information (i.e., the channel matrix) is known at the BS. So, the interference from the signals destined for other users may be known non-causally at the transmitter. Thus users can be ordered and encoded successively. The DPC presubtracts the interference at the transmitter without increasing the transmit power so that the capacity remains the same as if there were no interference at the receiver.

If there are two users ($K=2$), each having one antenna ($N=1$), then the sum capacity can be obtained by maximizing the dirt-paper lower bound (Caire and Shamai, 2003)

$$R_1 + R_2 \leq \begin{cases} \log_2(1 + |\mathbf{H}_1|^2 A), & A \leq A_1, \\ \log_2 \dfrac{(A\det(\mathbf{HH}^H) + \mathrm{Tr}(\mathbf{HH}^H))^2 - 4|\mathbf{H}_2\mathbf{H}_1^H|^2}{4\det(\mathbf{HH}^H)}, & A > A_1, \end{cases} \tag{7.91}$$

where $\mathbf{H} = [\mathbf{H}_1^T \mathbf{H}_2^T]^T$. Without loss of generality, we may assume $|\mathbf{H}_1|^2 \geq |\mathbf{H}_2|^2$ and define

$$A = \frac{|\mathbf{H}_1|^2 - |\mathbf{H}_2|^2}{\det(\mathbf{HH}^H)}. \tag{7.92}$$

This approach is hard to extend to general cases with $K > 2$ or $N > 1$.

In the MIMO downlink, dirty-paper coding is applied to encode messages for different users. Assume there are K users. First, the codeword for user K is generated independently of the other users. The codeword for user K is known before encoding the message of user $K-1$. The DPC scheme can generate the codeword for user $K-1$ to presubtract the interference from user K so that the signal for user K does not interfere with the signal for user $K-1$. This process is repeated until encoding the message for user 1. If user $\pi(K)$ is encoded first, followed by $\pi(K-1), \ldots, \pi(2), \pi(1)$, then the achievable region is (Vishwanath, et al., 2003)

$$R_{\pi(k)} \leq \log_2 \frac{\det\left(N_0\mathbf{I} + \mathbf{H}_{\pi(k)}\left(\sum_{j=1}^{k}\mathbf{Q}_{\pi(j)}\right)\mathbf{H}_{\pi(k)}^H\right)}{\det\left(N_0\mathbf{I} + \mathbf{H}_{\pi(k)}\left(\sum_{j=1}^{k-1}\mathbf{Q}_{\pi(j)}\right)\mathbf{H}_{\pi(k)}^H\right)}, \tag{7.93}$$

where $\mathbf{Q}_k$ is the covariance matrix of the transmit signal x_k destined to user k, and the transmit signal at the BS is given by $x = \sum_{k=1}^{K} x_k$. The dirty-paper achievable region is the convex hull of the union of the achievable vectors over all possible positive semi-definite matrices $\mathbf{Q}_1, \mathbf{Q}_2, \ldots, \mathbf{Q}_K$. It has been shown that this DPC achievable region is the capacity region (Weingarten *et al.*, 2006). Since the x_k are all independent, the covariance of the transmitted signal x is given by $\mathbf{Q} = \sum_{k=1}^{K} \mathbf{Q}_k$.

The sum capacity of the MIMO downlink channel can be upper-bounded if the receivers are arranged to cooperate such that the system operates as a single-user MIMO transmission. This cooperative bound can be tightened by using Sato's approach (Sato, 1978). The capacity region of the downlink channel depends only on the marginal distribution of the noise $P(\eta_1)$, $P(\eta_2), \ldots, P(\eta_K)$, not on the joint distribution $P(\eta_1, \eta_2, \ldots, \eta_K)$, because two downlink channels with the same marginal but different joint distribution can use the same encoder and decoder to maintain the same error probability (Yu and Cioffi, 2004). It is possible to tighten the cooperative bound by finding the least favorable noise that minimizes the bound over all possible joint distributions $P(\eta_1, \eta_2, \ldots, \eta_K)$. Consequently, the sum capacity upper-bound can be obtained by maximizing the input distribution and minimizing the noise distribution. In this case, there exists a saddle-point at which the max–min solution is equal to the min–max solution, and this saddle-point value corresponds to the sum capacity (Yu and Cioffi, 2004)

$$C_{\text{sum}} = \max_{\mathbf{Q}} \ \min_{\mathbf{Z}} \ \log \frac{\det(\mathbf{Z} + \mathbf{HQH}^{\text{H}})}{\det(\mathbf{Z})}. \tag{7.94}$$

Here, the covariance matrix of the signal, $\mathbf{Q}$, is a positive semi-definite matrix satisfying the power constraint $\text{Tr}(\mathbf{Q}) \leq P$; and the block-diagonal term of the noise covariance matrix, $\mathbf{Z}$, is given by $\mathbb{E}[\mathbf{Z}^{(k)}] = N_0 \mathbf{I}$ for $k = 1, 2, \ldots, K$.

7.3.4 Downlink–Uplink Duality

In order to get optimal performance in a MIMO system we can jointly encode messages at the transmitter and jointly decode them at the receiver. SVD (singular-value decomposition) operation and ML (maximum-likelihood) detection are good examples of joint encoding and decoding. In a single-user MIMO system, the transmitters (i.e., transmit antennas) are allowed to cooperate to encode messages to an input vector. Likewise, the receivers (i.e., receive antennas) are allowed to cooperate to detect the message jointly by using the whole receive vector, not each received symbol. However, in some cases, they are not allowed to cooperate, so that encoding and decoding should be done individually. For example, the receivers are not allowed to cooperate in the downlink while the transmitters are not allowed to cooperate in the uplink. Thus, in the downlink and the uplink, cooperation is allowed at only one side.

In a single-user MIMO system, the capacity is unchanged when the roles of the transmit and the receive antennas are interchanged (Telatar, 1999). Similarly, there is duality between the broadcast channel and the multiple-access channel such that the capacity region of the MAC is related to that of the BC. More specifically, the sum capacity of a Gaussian broadcast channel $r_k = \mathbf{H}_k x + \eta_k$, $k = 1, 2, \ldots, K$, and the sum capacity of a Gaussian dual multiple-access channel $r = \sum_{k=1}^{K} \mathbf{H}_k^{\text{H}} x_k + \eta$ are the same, as given by (7.94). We first consider the duality in the case of single-antenna BC and MAC, and then extend it to the MIMO BC and MAC in the following.

7.3.4.1 Downlink–Uplink Duality with Single Antenna

Consider the case where the base station and each mobile station have only one antenna. The channel gain between the BS and one MS is a scalar. In the broadcast channel, the BS transmits the independent information to each MS by broadcasting the signal to K different users to yield

$$r_k = h_k x + \eta_k. \tag{7.95}$$

In the multiple-access channel, K mobile stations send independent information to the same BS to yield

$$r = \sum_{k=1}^{K} h_k x_k + \eta. \tag{7.96}$$

The noise variances of the broadcast channel and the MAC are the same and given by N_0. The BC is dual of the MAC, and vice versa.

In the MAC, the optimal decoding is a successive decoding in which the receiver detects a signal after subtracting the previously detected signal from the received signal. Every decoding order in the successive decoding corresponds to the optimal corner point in the MAC capacity region. Given a decoding order $\pi(1)$, $\pi(2)$, ..., $\pi(K)$, meaning that user $\pi(1)$ is decoded first and user $\pi(K)$ is decoded last, the achieved rate is

$$R_{\pi(k)}^{MAC} = \log_2 \left(1 + \frac{|h_{\pi(k)}|^2 P_{\pi(k)}^{MAC}}{\sum_{j=k+1}^{K} |h_{\pi(j)}|^2 P_{\pi(j)}^{MAC} + N_0} \right), \tag{7.97}$$

where P_k^{MAC} is the power constraint of user k in the MAC. In the case of the BC, user messages are encoded in a successive way so that the dirty-paper coding presubtracts the previously encoded message. If we assume that the opposite encoding order is used for the BC (i.e., user $\pi(K)$ is encoded first and user $\pi(1)$ is encoded last), then the achievable rate is

$$R_{\pi(k)}^{BC} = \log_2 \left(1 + \frac{|h_{\pi(k)}|^2 P_{\pi(k)}^{BC}}{|h_{\pi(k)}|^2 \sum_{j=1}^{k-1} P_{\pi(j)}^{BC} + N_0} \right), \tag{7.98}$$

where P_k^{BC} is the power of user k in the BC and there is a total power constraint $\sum_{k=1}^{K} P_k^{BC} = P$.

Interestingly, the capacity region of the Gaussian BC is equal to the capacity region of the dual Gaussian MAC with the same sum power constraint, instead of individual power constraints (Jindal, et al., 2004). So, the uplink and the downlink channels differ in the sense only that individual power constraints exist at each mobile station. Every boundary point at the BC capacity region is a corner point of the dual MAC for some set of powers having the same sum power constraint. In addition, every corner point of the MAC is in the dual BC capacity region having the same sum power. So, there exists a power transformation that makes two achieved rates in the BC and the MAC equal; that is, $R^{\text{MAC}}_{\pi(k)} = R^{\text{BC}}_{\pi(k)}$. We can compute the BC powers from the MAC powers using the relations

$$
P^{\text{BC}}_{\pi(1)} = P^{\text{MAC}}_{\pi(1)} \frac{N_0}{N_0 + \sum_{k=2}^{K} |h_{\pi(k)}|^2 P^{\text{MAC}}_{\pi(k)}}
$$

$$
P^{\text{BC}}_{\pi(2)} = P^{\text{MAC}}_{\pi(2)} \frac{N_0 + |h_{\pi(2)}|^2 P^{\text{BC}}_{\pi(1)}}{N_0 + \sum_{k=3}^{K} |h_{\pi(k)}|^2 P^{\text{MAC}}_{\pi(k)}}
$$

$$
\vdots
$$

$$
P^{\text{BC}}_{\pi(K)} = P^{\text{MAC}}_{\pi(K)} \frac{N_0 + |h_{\pi(K)}|^2 \sum_{k=1}^{K-1} P^{\text{BC}}_{\pi(k)}}{N_0},
$$

(7.99)

and compute the MAC powers from the BC powers using the relations

$$
P^{\text{MAC}}_{\pi(K)} = P^{\text{BC}}_{\pi(K)} \frac{N_0}{N_0 + |h_{\pi(K)}|^2 \sum_{k=1}^{K-1} P^{\text{BC}}_{\pi(k)}}
$$

$$
P^{\text{MAC}}_{\pi(K-1)} = P^{\text{BC}}_{\pi(K-1)} \frac{N_0 + |h_{\pi(K)}|^2 P^{\text{MAC}}_{\pi(K)}}{N_0 + |h_{\pi(K-1)}|^2 \sum_{k=1}^{K-2} P^{\text{BC}}_{\pi(k)}}
$$

(7.100)

$$
\vdots
$$

$$
P^{\text{MAC}}_{\pi(1)} = P^{\text{BC}}_{\pi(1)} \frac{N_0 + \sum_{k=2}^{K} |h_{\pi(k)}|^2 P^{\text{MAC}}_{\pi(k)}}{N_0}.
$$

Using these transformations, we can prove that the capacity region of a Gaussian BC with power constraint P is equal to the union of the capacity region of the dual MAC with power constraints $P_1, P_2, \ldots, P_K$ with $\sum_{k=1}^{K} P_k = P$; that is (Jindal, et al., 2004)

$$
C_{\text{BC}}(P, (h_1, h_2, \ldots, h_K)) = \bigcup_{\sum_{k=1}^{K} P_k = P} C_{\text{MAC}}(P_1, P_2, \ldots, P_K, (h_1, h_2, \ldots, h_K)).
$$

(7.101)

7.3.4.2 Downlink–Uplink Duality with Multiple Antennas

Similarly, in the multiple-antenna downlink and uplink cases, the capacity region of the BC with power constraint P is equal to the union of the capacity regions of its dual MAC, where the union is taken over all individual power constraints that sum up to P (Vishwanath, et al., 2003):

$$C_{BC}(P, (\mathbf{H}_1, \mathbf{H}_2, \ldots, \mathbf{H}_K)) = \bigcup_{\sum_{k=1}^{K} P_k = P} C_{MAC}(P_1, P_2, \ldots, P_K, (\mathbf{H}_1, \mathbf{H}_1, \ldots, \mathbf{H}_K)). \quad (7.102)$$

There are also the MAC–BC and the BC–MAC covariance transformations that link the two capacity regions (Vishwanath, *et al.*, 2003). The MAC–BC duality is very useful because the capacity region of the BC leads to non-concave rate functions of the covariance matrices, whereas the rates in the dual MAC is a concave function of the covariance matrices. So, we can find the optimal covariance for the BC easily in the following way. We first find the optimal MAC covariance matrices by using the standard convex optimization algorithms. Then, we transform them to the corresponding optimal BC covariance matrices by using the MAC–BC covariance transformations. In the case of the sum-capacity, we can modify the iterative water-filling algorithm in Figure 7.16 to get the optimal BC covariance matrices (Jindal *et al.*, 2005; Yu, 2006).

7.3.5 Downlink Precoding Schemes

As discussed previously, in the downlink the signal for one user may interfere with the other user signals but the receivers cannot cooperate, so the overall performance is not good unless the interference is precancelled at the transmitter. *Beam-forming* is intended to maximize the power of the desired signals while minimizing (or nulling) the power of the interfering signals by using multiple transmit/receive antennas. In the context of the multiuser MIMO, the transmit beam-forming precancels the interference at the transmitter so that each user can receive nearly interference-free signals. The following discussion briefly reviews the linear and nonlinear precoding schemes for a MIMO downlink channel.

Consider the case where the base station has M transmit antennas and each user has a single antenna. The received signal at user k ($k = 1, 2, \ldots, K$) is given by

$$r_k = h_k x + \eta_k \quad (7.103)$$

where r_k is a received signal, h_k is a channel from the BS to user k, x is a transmitted symbol vector, and η_k is an additive zero-mean white Gaussian noise vector with variance N_0. In vector form, we get

$$\mathbf{r} = \mathbf{Hx} + \eta, \quad (7.104)$$

where $\mathbf{r} = [r_1, r_2, \ldots, r_K]^T$, $\mathbf{H} = [h_1^T, h_2^T, \ldots, h_K^T]^T$, and $\eta = [\eta_1, \eta_2, \ldots, \eta_K]^T$. Assume that the channel information $\mathbf{H}$ is known to the BS. By using downlink–uplink duality, the sum-capacity of this downlink channel is equal to the sum-capacity of its dual uplink channel, which is given by

$$C_{sum} = \max_{\mathbf{Q}} \log_2 \left(I + \frac{1}{N_0} \mathbf{H}^H \mathbf{Q} \mathbf{H} \right), \quad (7.105)$$

where $\mathbf{Q}$ is a $K \times K$ diagonal non-negative definite matrix. The sum-capacity grows in proportion to $\min [M, K] \log SNR$ for large SNR. So, if the number of users is sufficiently large, the capacity scales with the number of transmit antennas. This linear growth in sum-capacity means that more users can be accommodated simply by adding more antennas, without increasing the total transmitted power or bandwidth or lowering the rate to the existing users (Hochwald and Vishwanath, 2002).

A simple method to suppress interference at the transmitter is to invert the channel so that the received signal is free from interference. Such *channel inversion* can be done as if the *zero-forcing detection* were performed at the transmitter. The ZF at the transmitter can be performed in a manner similar to (7.35); that is

$$s = \mathbf{H}^H(\mathbf{HH}^H)^{-1}\mathbf{u} \tag{7.106}$$

where $\mathbf{u} = [u_1, u_2, \ldots, u_K]^T$ is the symbol vector to be transmitted physically. Since it is not normalized, we transmit the normalized signal

$$x = \frac{s}{\gamma}, \tag{7.107}$$

which is normalized by $\gamma = |s|^2$ or $\gamma = \mathbb{E}[|s|^2]$. This channel inversion not only makes the received signal interference-free, but also achieves a linear growth in the sum rate; that is, $K \log SNR$ when $M > K$. However, it is suboptimal because it achieves only about 80% of the sum-capacity at best (Hochwald and Vishwanath, 2002). Furthermore, it fails to achieve the linear growth when $M = K$ because the smallest eigenvalue of $\mathbf{HH}^H$ may possibly make γ tremendously large (Peel, *et al.*, 2005a).

In order to circumvent this drawback, we regularize the inversion by adding a multiple of the identity matrix before inverting the channel. Then the transmit signal vector changes to

$$s = \mathbf{H}^H(\mathbf{HH}^H + \alpha\mathbf{I})^{-1}\mathbf{u}. \tag{7.108}$$

Such *regularized inversion* brings in crosstalk among the received user signals, whose amount is determined by α. However, the addition of $\alpha\mathbf{I}$ may reduce γ, thereby mitigating the performance degradation caused by the smallest eigenvalue. So, it is necessary to carefully choose the value α. It is reported that $\alpha = KN_0$ is an optimal value that can maximize SINR for any eigenvalue distribution of $\mathbf{H}$ (Peel *et al.*, 2005a). So, regularized inversion looks like channel inversion by using a MMSE filter.

Even though regularized channel inversion improves the sum rate, there is still room for improvement. We perturb the information data in a data-dependent way (unknown to the receivers) in order to minimize the norm of the regularized transmit vector, $|s|^2$. Since the perturbation is not known to the receivers, we perturb each element of u by an integer so that the receiver can easily remove the perturbing vector by using a modulo operation. A modulo operation is used to perturb the input vector such that

$$\mathbf{u}' = \mathbf{u} + \tau l, \tag{7.109}$$

where τ is a positive real constant, and l is a K-dimensional complex vector $a + jb$, for integer numbers a and b. Then the transmit vector is given by

$$s = \mathbf{H}^H(\mathbf{HH}^H)^{-1}(\mathbf{u} + \tau l), \tag{7.110}$$

where the perturbing vector is chosen to minimize $\gamma = |s|^2$; that is (Peel *et al.*, 2005b)

$$l = \arg \min_l |\mathbf{H}^H(\mathbf{HH}^H)^{-1}(\mathbf{u} + \tau l)|^2. \tag{7.111}$$

This is a K-dimensional integer-lattice least-squares problem, and the Fincke–Pohst algorithm is one of the well-known to solve this (Fincke and Pohst, 1985). The algorithm was first used for the lattice code decoder in space–time modulation and it has been called a *sphere decoder* (Damen *et al.*, 2000). In the downlink channel, the integer-lattice least-squares is performed at the transmitter, and thus is called a *sphere encoder*. The Fincke–Pohst algorithm avoids exhaustive search over all possible integers in the lattice by limiting the search space to a sphere of some given radius centered on the starting point. In our problem, the center point is the vector $\mathbf{u}$ (Peel *et al.*, 2005b).

If we use the regularized channel inversion in (7.108) in addition, then we get the *regularized perturbation* scheme in which the transmit vector is

$$\mathbf{s} = \mathbf{H}^{\mathrm{H}}(\mathbf{H}\mathbf{H}^{\mathrm{H}} + \alpha\mathbf{I})^{-1}(\mathbf{u} + \tau l) \tag{7.112}$$

and the perturbing vector is determined to be (Peel *et al.*, 2005b)

$$l = \arg\min_{l} |\mathbf{H}^{\mathrm{H}}(\mathbf{H}\mathbf{H}^{\mathrm{H}} + \alpha\mathbf{I})^{-1}(\mathbf{u} + \tau l)|^2. \tag{7.113}$$

In this case the optimal value of α is N_0.

Finally, compare the error performance of the various transmit precoding schemes for the downlink. Assume that the base station has 10 antennas and there are 10 users, each having a single antenna. The modulation used in the simulation is QPSK. Figure 7.22 plots the bit error performance of the channel inversion in (7.106), the regularized inversion in (7.108), the

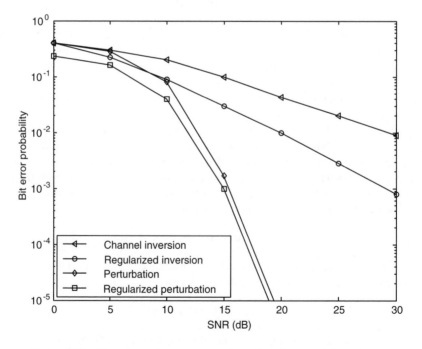

Figure 7.22   Bit-error performance of various downlink precoding schemes

vector perturbation in (7.110), and the regularized perturbation in (7.112). Note that the vector perturbation greatly reduces the error probability. The regularized inversion outperforms the channel inversion at high SNR. The regularized perturbation scheme performs better than the vector perturbation scheme but the gain is not significant when compared with the gain of the regularized inversion over the channel inversion.

References

Alamouti, S. (1998) A simple transmit diversity techniques for wireless communications. *IEEE Journal on Selected Areas in Communications*, **16**, 1451–1458.

Caire, G. and Shamai, S. (2003) On the achievable throughput of a multi-antenna Gaussian broadcast channel. *IEEE Transactions on Information Theory*, **49**, 1691–1706.

Costa, M.H.M. (1983) Writing on dirty paper. *IEEE Transactions on Information Theory*, **29**, 439–441.

Cover, T.M. and Thomas, J.A. (1991) *'Elements of Information Theory'*, John Wiley & Sons.

Damen, M.O., Chkeif, A. and Belfiore, J.-C. (2000) Lattice code decoder for space–time codes. *IEEE Communications Letters*, **4**, 161–163.

Erez, U., Shamai, S. and Zamir, R. (2005) Capacity and lattice-strategies for cancelling known interference. *IEEE Transactions on Information Theory*, **51**, 3820–3833.

Fincke, U. and Pohst, M. (1985) Improved methods for calculating vectors of short lengths in a lattice, including a complexity analysis. *Mathematics of Computation*, **44**, 463–471.

Foschini, G.J. (1996) Layered space-time architecture for wireless communication in a fading environment when using multiple antennas. *Bell Labs Technical Journal*, **1** (2), 41–59.

Hochwald, B.M. and Vishwanath, S. (2002) Space–time multiple access: Linear growth in the sum rate. In: *Proceedings Allerton Conference on Computing, Communications, Control*, 387–396.

Jindal, N., Vishwanath, S. and Goldsmith, A. (2004) On the duality of Gaussian multiple-access and broadcast channels. *IEEE Transactions on Information Theory*, **50**, 768–783.

Jindal, N., Rhee, W., Vishwanath, S., Jafar, S.A. and Goldsmith, A. (2005) Sum power iterative water-filling for multi-antenna Gaussian broadcast channels. *IEEE Transactions on Information Theory*, **51**, 1570–1580.

Lozano, A., Tulino, A. and Verdú, S. (2006) Multiantenna capacity: myths and realities. In: *'Space–Time Wireless Systems: From Array Processing to MIMO Communications'*, Cambridge University Press.

Onggosanusi, E.N., Dabak, A.G. and Schmidl, T.M. (2002) High rate space-time block coded scheme: Performance and improvement in correlated fading channels. In: *Proceedings WCNC*, Orlando, FL, 194–199.

Peel, C.B., Hochwald, B.M. and Swindlehurst, A.L. (2005a) A vector-perturbation technique for near-capacity multiantenna multiuser communication-part I: channel inversion and regularization. *IEEE Transactions on Communications*, **53**, 195–202.

Peel, C.B., Hochwald, B.M. and Swindlehurst, A.L. (2005b) A vector-perturbation technique for near-capacity multiantenna multiuser communication-Part II: Perturbation. *IEEE Transactions on Communications*, **53**, 537–544.

Proakis, J.G. (2001) *'Digital Communications'*, 4th edn, McGraw-Hill.

Sato, H. (1978) An outer bound on the capacity region of broadcast channels. *IEEE Transactions on Information Theory*, **24**, 374–377.

Tarokh, V., Seshadri, N. and Calderbank, A.R. (1998) Space-time codes for high data rate wireless communication: performance criterion and code construction. *IEEE Transactions on Information Theory*, **44**, 744–765.

Tarokh, V., Seshadri, N. and Calderbank, A.R. (1999) Space-time block codes from orthogonal designs. *IEEE Transactions on Information Theory*, **45**, 1456–1467.

Telatar, E. (1999) Capacity of multi-antenna Gaussian channel. *European Transactions on Telecommunications*, **10**, 585–595.

Tse, D.N.C., Viswanath, P. and Zheng, L. (2004) Diversity-multiplexing tradeoff in multiple-access channels. *IEEE Transactions on Information Theory*, **50**, 1859–1874.

Varanasi, M. and Guess, T. (1997) Optimum decision feedback multiuser equalization with successive decoding achieves the total capacity of the Gaussian multiple access channel. In: *Asilomar Conference on Signals, Systems and Computers*, Pacific Grove, CA, 1405–1409.

Vishwanath, S., Jindal, N. and Goldsmith, A. (2003) Duality, achievable rates, and sum-rate capacity of MIMO broadcast channels, *IEEE Transactions on Information Theory*, **49**, 2658–2668.

Weingarten, H., Steinberg, Y. and Shamai, S. (2006) The capacity region of the Gaussian multiple-input multiple-output broadcast channel. *IEEE Transactions on Information Theory*, **52**, 3936–3964.

Yu, W. (2006) Sum-capacity computation for the Gaussian vector broadcast channel via dual decomposition. *IEEE Transactions on Information Theory*, **52**, 754–759.

Yu, W. and Cioffi, J.M. (2004) Sum capacity of Gaussian vector broadcast channels. *IEEE Transactions on Information Theory*, **50**, 1875–1892.

Yu, W., Rhee, W., Boyd, S. and Cioffi, J.M. (2004) Iterative water-filling for Gaussian vector multiple access channels. *IEEE Transactions on Information Theory*, **50**, 145–152.

Zheng, L. and Tse, D. (2003) Diversity and multiplexing: A fundamental tradeoff in multiple-antenna channels. *IEEE Transactions on Information Theory*, **49**, 1073–1096.

Further Reading

Biglieri, E. and Taricco, G. (2004) Transmission and reception with multiple antennas: Theoretical foundations. *Foundation and Trends in Communications and Information Theory*, **1** (2), 183–332.

Gesbert, D., Shafi, M., Shiu, D.-S., Smith, P.J. and Naguib, A. (2003) From theory and practice: an overview of MIMO space-time coded wireless systems. *IEEE Journal on Selected Areas in Communications*, **21**, 281–302.

Goldsmith, A., Jafar, S.A., Jindal, N. and Vishwanath, S. (2003) Capacity limits of MIMO channels. *IEEE Journal on Selected Areas in Communications*, **21**, 684–702.

Hottinen, A., Tirkkonen, O. and Wichman, R. (2003) *'Multi-antenna Transceiver Techniques for 3G and Beyond'*, John Wiley & Sons.

Paulraj, A., Nabar, R. and Gore, D. (2003) *'Introduction to Space–Time Wireless Communications'*, Cambridge University Press.

Tse, D. and Viswanath, P. (2004) *'Fundamentals of Wireless Communication'*, Cambridge University Press.

Viswanath, P. and Tse, D.N.C. (2003) Sum capacity of the vector Gaussian broadcast channel and uplink–downlink duality. *IEEE Transactions on Information Theory*, **49**, 1912–1921.

8

Inter-cell Resource Management

If a single base station (BS) were to serve all the mobile stations (MS) in an entire service area, it would not be possible to support all service requirements simultaneously. It would require very high transmission power for those mobile stations located far away from the BS, which would cause severe interference among the users. Moreover, the single BS would become a bottleneck as the limited bandwidth available would have to be shared by all the users. This problem can be resolved if the entire service area is divided into multiple smaller *cell* coverage areas, with a BS serving all the users belonging to each cell and reusing the wireless bandwidth repeatedly in each cell.

So, the concept of cellular systems was developed to provide quality services to a large number of users spread over the entire service area using finite wireless resources. The basic idea is to divide the service area into multiple smaller service areas, and deploy one BS at each cell. Within each cell, the distance from the BS to each MS, as well as the number of served mobiles, is reduced significantly, making it possible to provide the desired level of service quality using a much smaller amount of transmission power and bandwidth.

This cellular concept introduces some new problems, such as *inter-cell interference management* and *handoff management*.

Consider first inter-cell interference management. If two neighboring cells use the same frequency bandwidth simultaneously, a mobile station in one cell, located at the border zone of the two cells, would suffer from strong interference from the other cell. This interference corrupts the user signals and degrades quality of service at the cell boundary. One possible way of mitigating inter-cell interference is to divide the entire bandwidth into multiple disjoint sets and *reuse* a particular set only at base stations spaced far enough apart. Inter-cell interference can be effectively suppressed by adopting such a spectrum- (or frequency-) reuse scheme, but at a price; the bandwidth for use in each cell diminishes. As a consequence, a tradeoff exists between the level of inter-cell interference and the bandwidth used by each BS. Section 8.1 discusses inter-cell interference management in detail, focusing on how to exploit this tradeoff effectively.

The other important issue is handoff management. Since a cellular system separates the coverage of each BS geographically, a mobile station crossing a boundary has to redirect its connection seamlessly to a new BS. *Handoff* refers to the process of transferring the channel of

Wireless Communications Resource Management Byeong Gi Lee, Daeyoung Park, and Hanbyul Seo
© 2009 John Wiley & Sons (Asia) Pte Ltd

an ongoing connection associated with one BS to another connection associated with another BS. It is widely accepted that sudden termination of an ongoing connection is much more undesirable than blocking a new connection. In the context of resource management, this means that each BS should be equipped with a *prioritization scheme* in allocating wireless resources such that a handoff connection is prioritized over a new one. Section 8.2 discusses handoff management methods developed to balance the QoS of the handoff and the new connection arrivals.

8.1 Inter-cell Interference Management

A frequency-reuse scheme helps to mitigate inter-cell interference by reusing a common frequency band only at base stations spaced far apart. This approach relies on the fact that the power of a transmitted signal decreases rapidly as the signal propagates through space.

A set of cells which share the same channel set is called a *co-channel set*; the minimum distance among the members of the co-channel set is called the *reuse distance*; and the number of channel sets that divides the total bandwidth is called the *reuse factor*. There is a one-to-one correspondence between the reuse factor and the reuse distance; that is, once the reuse factor is determined, the reuse distance is calculated uniquely. So two terms, reuse factor and reuse distance, can be used interchangeably. In a sense, inter-cell interference management converges to the issue of determining the reuse distance in consideration of the QoS requirement of the system, as the reuse distance dictates the inter-cell interference level and the bandwidth used by each BS.

Channel allocation refers to the design task of determining which channels are to be used by which cells, and in what reuse distance to attain maximal frequency-reuse efficiency. Among a large number of schemes proposed to date, we deal here with those that strive to maximize the performance metrics while satisfying given QoS requirements. The schemes are categorized by two criteria – the heterogeneity of the reuse distance, and the flexibility of channel assignment. Four schemes are covered:

1. *fixed channel allocation* for which the channels assigned to each cell are fixed during the run time;
2. *dynamic channel allocation* that assigns channels dynamically to each cell while employing an explicit homogeneous reuse distance;
3. *channel allocation based on SINR measurement* that dynamically decides the reusability of the channels based on SINR measurement;
4. *channel allocation with inter-cell power control* that permits each cell to control its power allocated to each channel continuously.

8.1.1 Fixed Channel Allocation

In *fixed channel allocation* (FCA), the available channels are divided into multiple disjoint channel sets, with each set having its own reuse distance and each cell taking the channel set that satisfies its required reuse distance. If the reuse distance is the same for all channel sets, then it is called an *FCA with homogeneous reuse distance*. Otherwise, it is called an *FCA with heterogeneous reuse distance*.

8.1.1.1 FCA with Homogeneous Reuse Distance

For FCA with the same reuse distance among all channels sets, the resource manager needs to determine only two factors: the size of the reuse distance, and the specific channel allocation among the cells. This task is called *cell planning*. Figure 8.1 illustrates the result of cell planning

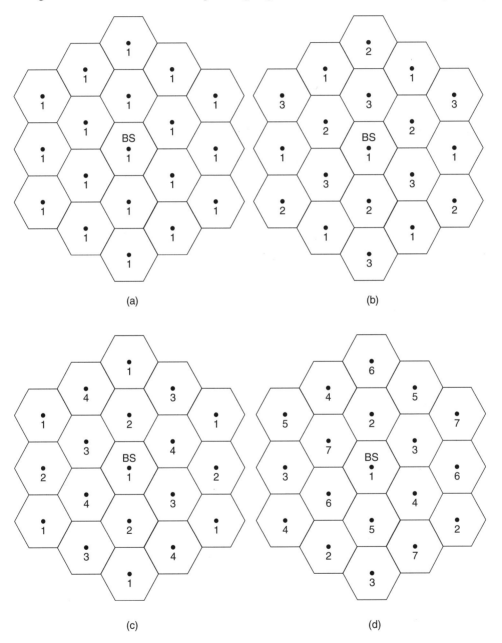

Figure 8.1 Homogeneous cell planning with (a) reuse factor 1, (b) reuse factor 3, (c) reuse factor 4, and (d) reuse factor 7

with reuse factors of $1, 3, 4$, and 7. The number in each cell denotes the index of the channel set allocated to that cell.

The suitability of the determined reuse distance is dictated by the QoS of the cell boundary users as they are likely to suffer from inter-cell interference more than other users in the cell. Thus, the resource manager determines the reuse distance that can maximize the system objective while satisfying the target QoS of the boundary users.

For example, consider a downlink system that has the objective of maximizing the bandwidth used by each BS. Assume that there are 100 channels in total and each user requires the system to guarantee the minimum SINR of 6 dB. For simplicity, assume that the effects of shadowing and short-term fading are negligible and each cell uses the same transmit power. Then, the interference level from the other cells may be determined only by the distances between the target user and the interfering cells. Since the interference level decays very rapidly with the distance (see Section 2.1), it is dominated by the closely located cells. Thus, we can assume that inter-cell interference is predominantly caused by the *first-tier* cells within the co-channel set, if the cells are categorized into multiple tiers according to the distance. Figure 8.2 illustrates the first-tier cells in cases with reuse factors 1, 3, and 4.

Figure 8.2 indicates, for each reuse factor value, the distance from the center of each first-tier cell to the boundary point P by a dashed line. This helps to derive the SINR expression of the user at the boundary point P. Assuming a path-loss exponent of 3, the SINR expression for a reuse factor 1, $SINR_1$, is

$$\text{SINR}_1 \approx \frac{R^{-3}}{2R^{-3} + 2(2R)^{-3} + 2(\sqrt{7}R)^{-3}} = \frac{1}{2 \cdot 1^{-3} + 2 \cdot 2^{-3} + 2 \cdot \sqrt{7}^{-3}} \approx 0.424 = -3.73 \,(\text{dB}).$$

$$(8.1)$$

The SINR value in this setting does not satisfy the QoS requirement of 6 dB, so the system manager would give up the case of reuse factor 1 although it would offer the largest bandwidth to each BS. Repeating the calculation for reuse factors 3 and 4 yields

$$\text{SINR}_3 \approx \frac{1}{2^{-3} + 4^{-3} + 2 \cdot \sqrt{7}^{-3} + 2 \cdot \sqrt{13}^{-3}} \approx 3.43 = 5.36 \,(\text{dB}), \qquad (8.2)$$

$$\text{SINR}_4 \approx \frac{1}{2 \cdot \sqrt{7}^{-3} + 2 \cdot \sqrt{13}^{-3} + 2 \cdot \sqrt{19}^{-3}} \approx 5.72 = 7.57 \,(\text{dB}). \qquad (8.3)$$

From these results, the system manager would select the reuse factor 4 as it meets the QoS requirement of the boundary user. In this case, the given 100 channels will be divided into four different channel sets, making each BS use 25 channels exclusively.

The SINR experienced by a cell boundary user changes depending on the path-loss exponent. Although we simply put the exponent value to 3 in the above example, it varies in general within a certain range (e.g., from 2 in free space to 5 in an urban area) according to the communication environment (refer to Section 2.2.1). Figure 8.3 depicts how the SINR changes as the path-loss

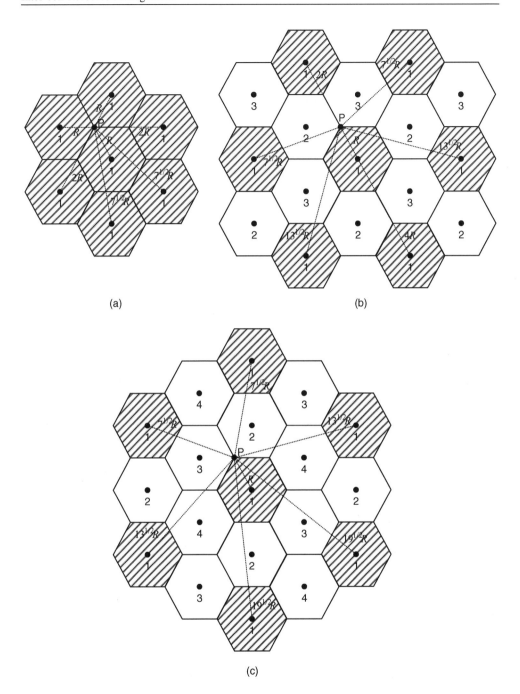

(a)

(b)

(c)

Figure 8.2   The first-tier cells in the cases of (a) reuse factor 1, (b) reuse factor 3, and (c) reuse factor 4

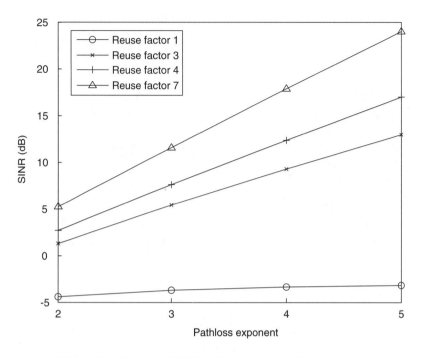

Figure 8.3   Change of SINR with respect to path-loss exponent

exponent varies for reuse factors 1, 3, 4, and 7. Observe that the SINR increases almost linearly on a log scale for all cases. This happens because, as the path-loss exponent increases, the interfering signals which travel over the distances relatively longer than the desired signal attenuate more rapidly than the desired signal. Note too that the slope of the curve is steeper for a larger reuse factor, which happens because the system with a smaller reuse factor has a smaller reuse distance. These observations lead us to conclude that the system manager may allocate channels more densely, or take a lower reuse factor, when the path-loss exponent is larger. If we revisit the above example, the system manager may now take the reuse factor 3, instead of 4, if the path loss exponent is 4, as it guarantees an SINR over 6 dB to the cell boundary users, according to Figure 8.3.

After the reuse factor is determined, the system manager decides which channels to allocate to which cell. This process may be done in an arbitrary way, as each cell would not have any preference on the channels. Once the channel allocation is done, each cell possesses the same number of channels for its exclusive use. Such a fixed type of channel allocation would yield an optimal means for handling the traffic as long as the traffic distribution of the system is uniform. However, the traffic distribution is not uniform, in general, and varies depending on time and space. If the future traffic pattern can be predicted in advance, the channel allocation can be modified such that the channels assigned to a lightly loaded cell can be borrowed by a heavily loaded adjacent cell located farther than the reuse distance. This borrowing arrangement may be done during the cell planning process in such a way that the borrowed channels then belong to the heavily loaded cell permanently. Such a borrowing strategy, called the *static borrowing* (Anderson, 1973), will be treated in Section 8.1.3 as a special case of dynamic channel allocation.

8.1.1.2 FCA with Heterogeneous Reuse Distance

For FCA with heterogeneous reuse distance, the system manager partitions the total bandwidth to multiple different cell plans having different reuse distances. Each cell is divided into multiple concentric subcells (see Figure 8.4), with each subcell being associated with each cell plan. Thus the innermost subcell is associated with the cell plan having the smallest reuse distance and the outermost subcell with the cell plan having the largest reuse distance. Beyond that, the channel allocation for each cell plan is similar to the case of FCA with homogeneous reuse distance. In contrast, though, heterogeneous FCA involves multiple cell plans with different reuse distances, so it is called *reuse partitioning*. It is sometimes referred to as a *fractional frequency-reuse scheme* in the context of an OFDMA system (IEEE 802.20, 2005).

The underlying principle behind multiple cell plans is that the power level required to achieve a given SINR requirement may be much lower in the inner subcell than in the outer subcell. This is because the inner subcells are closer to the BS which is located at the center of the cell. As a consequence, a user in the inner subcell can get the required QoS even when the inner subcell uses a cell plan with a much smaller reuse factor. This indicates that a channel can be reused more frequently provided it is assigned to the inner subcells, so reuse partitioning can exploit the tradeoff relation between reuse factor and bandwidth in a more efficient way.

For reuse partitioning, the system manager has to determine how to partition each cell into subcells and what number of channels to assign to each subcell. We shall revisit the example considered in the homogeneous arrangement to illustrate these arrangements. For simplicity,

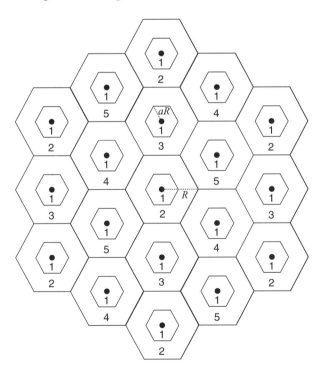

Figure 8.4 Cell structure of FCA with heterogeneous reuse distance (two-subcell case with reuse factors 1 and 4)

consider the case of partitioning each cell into two subcells, with each subcell taking reuse factor 1 or 4. Then we divide the whole frequency spectrum into five channel sets 1–5, among which we assign channel set 1 to the inner subcells with reuse factor 1 and assign channel sets 2–5 to the outer subcells with reuse factor 4. This arrangement is illustrated in the cell structure diagram in Figure 8.4.

The next task is to determine the radius of the inner subcell, aR, $0 \leq a \leq 1$. The SINR experienced by a boundary user of the inner cell may be approximated by

$$
\text{SINR}_1 \approx \frac{(aR)^{-3}}{2R^{-3} + 2(2R)^{-3} + 2(\sqrt{7}R)^{-3}} = a^{-3} \frac{1}{2 \cdot 1^{-3} + 2 \cdot 2^{-3} + 2 \cdot \sqrt{7}^{-3}} \tag{8.4}
$$
$$
= 0.424a^{-3} = -30 \log_{10} a - 3.73 \,(\text{dB}).
$$

Recalling that the SINR requirement is 6 dB, we determine a such that

$$
-30 \log_{10} a - 3.73 \geq 6, \tag{8.5}
$$

which yields $a \leq 10^{-9.73/30} = 0.474$. Noting that a larger area of the inner cell enables allocating more channels to each BS, we take $a = 0.474$.

The final task of the system manager is to determine the number of channels to assign to each subcell in consideration of the traffic intensity. Assume, for simplicity, that users are distributed uniformly and each user has the same offered load; then the traffic intensity of each subcell will be proportional to the area of the subcell. Let c_1 and c_2 be the number of channels assigned to the inner and outer subcells, respectively. Then we have the relations

$$
c_1 : c_2 = 0.474^2 : 1 - 0.474^2, c_1 + 4c_2 = 100. \tag{8.6}
$$

The second relation reflects that the reuse factor associated with the outer subcell is 4. Solving the equations, we obtain $c_1 = 8$ and $c_2 = 23$ as the most suitable choice. This means that each base station under the FCA scheme with heterogeneous reuse distance can use 31 channels exclusively, which is higher than the number of channels that a BS under the FCA scheme with homogeneous reuse distance can use (i.e., 25).

Figure 8.5 compares the resulting channel allocations of FCA schemes with homogeneous and heterogeneous reuse distances, listing the allocated channel numbers in serial numbers. When a user requests a connection to a base station, the BS simply assigns a channel among those assigned to the subcell where the user resides.

8.1.2 Dynamic Channel Allocation (DCA)

FCA schemes can do cell planning to provide a guaranteed QoS to cell boundary users but cannot adjust the cell planning according to short-term temporal and spatial variations of traffic. Due to this limitation, it can happen that a connection request to a BS under the FCA schemes is blocked even though there are many idle channels in the adjacent cells. This limitation is an obstacle in attaining high channel efficiency and, thus, has motivated development of *dynamic channel-allocation* (DCA) methods that can adapt the channel allocation to the traffic variation.

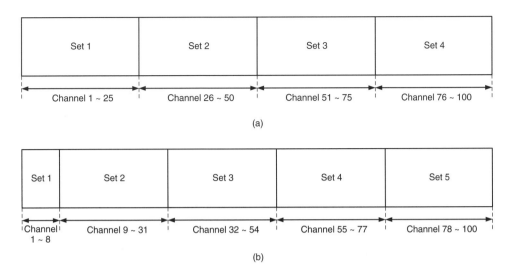

Figure 8.5 Comparison of channel allocations by FCA with (a) homogeneous, and (b) heterogeneous reuse distances

From the various types of DCA scheme, we consider in this section those using the *explicit reuse-distance* concept. The term "explicit reuse distance" means that the reuse distance is incorporated into the DCA scheme in such an explicit manner that a channel used in a cell cannot be reused in the neighboring cells at a distance less than a certain reuse distance. We differentiate such DCA schemes from other DCA schemes that operate without any explicit reuse distance, which will be discussed in the next sections.

The main goal of DCA schemes with implicit reuse distance is to allow a cell that temporally suffers from heavy traffic to use more channels than were originally assigned under the FCA principle. Clearly, such dynamic allocation of channels will help to avoid the blocking problems that might have occurred otherwise at heavily loaded cells under the FCA scheme. However, this dynamic operation may possibly lead to inefficient channel reuse unless the dynamic channel allocation is done very carefully, since any rearrangement of the channels assigned by FCA is likely to break the maximal channel reuse incorporated by FCA under the given reuse distance constraint. For example, consider a cellular system with reuse factor 3. Suppose that a channel that was originally assigned to channel set 1 under FCA is reallocated under DCA to another cell using channel set 2. Then, this rearrangement can no longer maintain the maximal channel reuse unless the reallocated channel is also used at all the cells that used channel set 2 under FCA in the entire system. However, such a DCA scheme is too complicated to implement since all the cells in the system should adjust the channel assignment simultaneously and the signaling overhead required for that simultaneous reassignment among all cells may be exceedingly large. Even if the DCA scheme is implemented, the channel efficiency may drop, since some of the cells with channel set 2 might be readily satisfied without the additional channel assignment.

Hence, any practical DCA scheme is unable to maintain maximal channel reuse under a given reuse distance constraint. If it is desired to guarantee the separation of the co-channel set farther than the reuse distance in a DCA, it is sometimes unavoidable to determine the distance among co-channels to be farther than the reuse distance, giving up the possibility of the maximal channel reuse attained in FCA. Thus, a hastily done DCA may severely deteriorate the overall channel reuse efficiency. In developing a DCA scheme, it is necessary to arrange such that each channel is reused as many times as possible and allocated dynamically to each cell under a given reuse distance constraint. This section will deal with two different types of DCA schemes that meet those guidelines. One uses *nominal channel assignment and borrowing*, in which each channel is first assigned by the FCA and temporally lent to an adjacent cell when needed. The other uses *channel sharing with reuse compaction*, in which each channel is shared by the entire cells in the system and reused in a compact way while guaranteeing the reuse distance.

8.1.2.1 DCA with Nominal Channel Assignment and Borrowing

The system manager first assigns nominal channels to each cell at the cell planning stage, as for the FCA schemes. The nominal channels are for use in connection setup within the assigned cells. However, if a cell is overloaded, it can borrow a nominal channel of a neighboring cell in such a way that the borrowed channel does not interfere with the existing connections. The cell that borrows a channel is called the *acceptor cell* and the neighboring cell that lends the channel is called the *donor cell*. After the acceptor cell terminates the connection using the borrowed channel, it returns the channel to the donor cell. This type of DCA scheme improves the channel reuse efficiency without degrading the channel efficiency as it makes the nominal channels available as the first candidate within each cell. Note that a nominal channel is assigned through the FCA cell planning which is optimized for channel reuse.

The key issue of the nominal channel allocation and borrowing scheme is how to determine the donor cell. The *borrowing-from-the-richest* (BR) scheme is a criterion for the donor cell proposed in (Anderson, 1973). In the BR scheme, if all nominal channels are busy, an attempt is made to borrow a channel from one of the adjacent cells. If the adjacent cell having a channel available for borrowing is more than one, the cell having the largest number of available channels is selected as the donor cell. As it is difficult, in general, to exactly determine whether or not the borrowed channel would interfere with existing connections, the channel that is free in the donor cell as well as in other interfering co-donor cells is selected. An "interfering co-donor cell" refers to a cell that is located within the reuse distance from the acceptor cell and its nominal channels are the same as the co-channel set of the donor cell.

Figure 8.6 illustrates the case where cell A1 wants to borrow channel x from cell D1 in a cellular system with reuse factor 7. In this example, cells D2 and D3 are the interfering co-donor cells since their distance to A1 is shorter than the reuse distance and their nominal channels are equal to those of D1. As the use of channel x may interfere with cells D2, D3, and D1, channel x becomes available if and only if it is not occupied in those cells.

Once a channel is lent to an acceptor cell, the channel becomes locked in the donor cell and the interfering co-donor cells. A channel locked in a particular cell is considered to be excluded from control of its nominal cell. That is, both the donor cell and the interfering co-donor cells cannot use the channel for connection setup or lend it to any other cells until it is returned from the acceptor cell. This is a stringent constraint, but is necessary as otherwise the existing connections would be interfered with. In the case of Figure 8.6, for example, if cell B borrows the locked channel x from cell D3, one of the interfering co-donor cells, it will certainly corrupt the corresponding connection in A1.

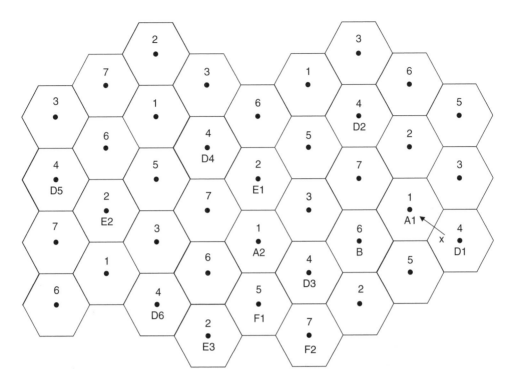

Figure 8.6   The BR scheme in a cellular system with reuse factor 7

Since the BR scheme is designed to borrow a cell from the richest adjacent cell, it tends to equalize the probability that no channel is assigned to a new connection in each cell, and thus reduces the overall connection blocking probability. However, the BR scheme has two serious problems. First, the availability requirement of the BR scheme is overly stringent. In Figure 8.6, for example, if channel x is borrowed to A1 from D1, cell A2 cannot borrow channel x from cell D3 as it is locked in D3 but, in reality, A2 may use channel x without interfering with the connection in A1 because the distance between A1 and A2 is larger than the reuse distance. As another example, cell E1 cannot borrow channel x from cell D4 because this borrowing requires channel x to be free in D4, D3, and D2 simultaneously but, in reality, cell E1 may borrow channel x from D4 because the distance between E1 and A1 is larger than the reuse distance. Second, the BR does not take account of channel locking when choosing a candidate channel for borrowing. Once the richest adjacent cell is determined, it randomly selects one among the available channels belonging to the cell, thereby incurring unnecessarily many locked channels in the system. In Figure 8.6, for example, if cell E2 borrows channel y from cell D5 after cell A1 borrows channel x from D1, then cell E3 cannot borrow either channel x or channel y from cell D6 since x is locked in D3 and y is locked in D6. In reality, however, this problem would not happen if E2 borrows channel x from D5 as E3 is allowed to borrow channel y from D6 in this case.

In order to overcome the above problems, *borrowing with directional channel locking* (BDCL) was proposed (Zhang, 1989). When a channel is borrowed in this scheme, its locking in the co-channel cells is restricted only to those affected by this borrowing. This approach helps to mitigate the first problem of the BR scheme, the unnecessary conservativeness of the channel locking mechanism. As a result, the number of channels available for borrowing becomes much larger than with the BR scheme. BDCL specifies the "lock direction" for each locked channel in order to determine if it is available for a new borrowing attempt. This directivity of channel locking can be illustrated by the following example. When cell A1 borrows channel x from cell D1, channel x in cell D3 needs to be locked in the upper-right directions only. In other words, cells A2, F1, and F2 may borrow channel x since this borrowing would not interfere with the corresponding connection in A1. Hence, in this case, cell A2 can borrow channel x from D3 as long as channel x is not locked in D4 and D6. This directional locking loosens the availability requirement without violating the reuse distance, thereby enhancing the adaptability to the traffic variations.

To overcome the second problem of BR – the inefficiency caused by random selection of the borrowed channel – BDCL introduces the *channel-ordering* and *channel-reallocation* concepts.

- *Channel ordering* sorts all the nominal channels into a predetermined order such that the first channel has the highest priority for being assigned to the local call and the last channel has the highest priority for being borrowed by the adjacent cells. In other words, the borrowed channel is chosen not only from the adjacent cell having the maximum number of available channels but also from the last ordered one in the list of this adjacent cell satisfying the availability for borrowing condition. By operating this way, each cell in the co-channel set tends to lend the same channel to its adjacent cell, and the unnecessary channel locking problem caused by the random selection can be diminished.
- *Channel reallocation* means that the system reallocates a channel released by a terminating connection to another ongoing connection as necessary. This concept helps resolving the unnecessary channel locking problem on one side and contributes to minimizing the occurrence of channel borrowing on the other.

When a connection terminates, a channel is reallocated according to the following rules:

1. When a connection on a nominal channel terminates and there is another connection on a higher order nominal channel in the same cell, then the connection in the higher order nominal channel is reallocated to the newly released lower order one.
2. When a connection on a borrowed channel terminates and there is another connection on a lower order borrowed channel, the connection carried on the lower order borrowed channel is switched to this channel.
3. When a connection on a nominal channel terminates and there is another connection on a borrowed channel, the call on the borrowed channel is switched to the nominal channel and the borrowed channel is released to the original owner.

Rules (1) and (2) are necessary to prevent the unnecessary channel locking problem as they support smooth operation of channel ordering. By rule (3), the number of the borrowed channels reduces when compared with BR, and each cell becomes more probable to use its own nominal channel in supporting a new connection request. This property is useful in attaining maximal channel reuse in the DCA context.

8.1.2.2 DCA with Channel Sharing with Reuse Compaction

In this type of DCA scheme, there is no fixed relationship between channels and cells; and even the nominal channel concept does not apply. All the channels are kept in a pool that is shared by all cells and are assigned dynamically to each cell as new connections arrive in the system. After a connection is completed, its channel is returned to the central pool. Since all the channels are shared by all cells in the system, any cell is eligible to use any channel, provided that it is used in none of the neighboring cells at a distance less than the predetermined reuse distance. Then, the problem is how to devise a decision criterion for selecting a channel among the channels in the pool to support a new connection request. The devised selection strategy affects the efficiency of channel reuse, which was pointed out as the problem of the DCA schemes above. It is obvious that more efficient channel reuse can be attained as the distance among the co-channel sets gets smaller. Therefore, DCA schemes in this category focus on finding a way to reuse the channels in a more compact way while maintaining the reuse distance. *Reuse compaction* addresses this point.

The simplest scheme in this sharing strategy is the *first available* (FA) scheme proposed in Cox and Reudink (1972). When a new connection arrives at a cell, the system searches the central pool and the first available channel is assigned to the connection. Once a channel is assigned to a cell, it is considered unavailable in the neighboring cells at a distance less than the reuse distance. The FA scheme minimizes system computational time, but does not care about reuse compaction.

Some other algorithms such as the *mean-square* (MSQ) and *nearest-neighbor* (NN) schemes have been developed to implement reuse compaction (Cox and Reudink, 1972). When a new request arrives at a cell, the MSQ scheme first gathers from the channel pool the candidate channels that are not occupied in the cells closer than the reuse distance. Then, the activity of each candidate channel is checked at the cells located in the interval between one and two times the reuse distance away from the assigned cell. The MSQ scheme selects the channel which minimizes the criterion

$$\underset{m=1,2,\ldots,M}{\text{minimize}} \frac{1}{N_m} \sum_{i=1}^{N_m} D_{im}^2, \text{ for } D_R \leq D_{im} \leq 2D_R, \tag{8.7}$$

where m is the channel index in the channel pool, D_R denotes reuse distance, N_m is the number of cells using the candidate channel m within this interval, and D_m indicates the distance between the assigned cell and the cell using the candidate channel m.

The NN scheme is the same as the MSQ except for the minimization metric. It selects for allocation the channel that is in use in the cell nearest to the assigned cell but still at least a reuse distance away. This implies that the NN scheme minimizes the distance to the nearest cell that uses the same channel while guaranteeing the reuse distance. By minimizing the distance among the co-channel sets, both the MSQ and NN schemes try to perform reuse compaction; that is, they try to reuse each channel as many times as possible.

8.1.2.3 Comparison between FCA and DCA

It turns out that DCA performs best under low traffic intensity, but FCA becomes superior in a heavily loaded environment. The following example explains the reason for this conclusion.

For simplicity we shall use BDCL as representative of the DCA schemes. Consider again the situation depicted in Figure 8.6. When a user in cell A1 requests a connection but there is no channel available in cell A1, the base station of cell Al chooses one of two different decisions: one is to borrow a channel from the neighboring cell (in this example, to borrow channel x from cell D1) to accept the request, and the other is just to block the request. Clearly, a system using DCA (in this example, BDCL) will take the former policy while a system using FCA will take the latter one. We assume that the system uses DCA and examine what will happen after the system decides to borrow channel x from cell D1.

At the point of borrowing channel x, DCA has a definite performance gain over FCA as a DCA system can accept the request that an FCA system would block. However, it also has a weak point that channel x is locked in cells D2 and D3 until the connection is terminated. If no connection request occurs in cells D2 and D3 during this interval, or if enough channels are available to support new requests in cells D2 and D3, the locking of channel x does not affect the performance at all. However, if connection requests occur in cells D2 and D3 during this interval, with no channel being available in cells D2, D3, and their neighbors, the locking of channel x can severely degrade the performance. This is because the system using DCA has to block those connection requests, whereas a system using FCA would be able to allocate channel x to those requests. Since two cells lock their channels to support only one connection request in DCA, in this example, borrowing channel x may possibly lead to blocking two requests in the cells where the channel is locked. It is clear that the former situation where DCA takes superiority over FCA corresponds to the lightly loaded traffic case and the latter one where FCA is superior to DCA corresponds to the heavily loaded traffic case.

The above phenomenon is an inherent characteristic which is commonly observed in various DCA schemes. Figure 8.7 compares the BDCL scheme with that of FCA in terms of the blocking probability which is analytically derived under the assumption of a one-dimensional cellular system with reuse factor 2.[1] A and B in Figure 8.7(a) denote the nominal channel sets consisting of m channels each. Figure 8.7(b) illustrates the analyzed blocking probabilities of the BDCL and the FCA schemes with $m = 5$. Here, the traffic intensity is defined by the average number of connection attempts normalized by the number of channels and the average call length. Observe that the blocking probability of BDCL is lower than that of FCA for traffic intensity less than 0.9, but FCA begins to outperform BDCL for a higher traffic load. This result is consistent with the above comment that DCA performs better under low traffic intensity but FCA becomes superior in heavily loaded cases.

8.1.3 Channel Allocation based on SINR Measurement

So far this section has discussed inter-cell resource management schemes that are established based on the explicit constraint of reuse distance. Each channel was allocated such that two cells using the same channel can be geometrically separated more than the required reuse distance. Such channel allocation was possibly implemented in static or dynamic manner.

Schemes based on the explicit reuse distance concept, however, share a common limitation that they may lead to overly conservative channel reuse, with the users near to the base station getting over-satisfied QoS. The constraint on the reuse distance is imposed to guarantee the QoS requirement even in the worst case, providing users with much better channels than are

---

[1] Refer to Yeung and Yun (1995) for the assumptions and analytical results in detail.

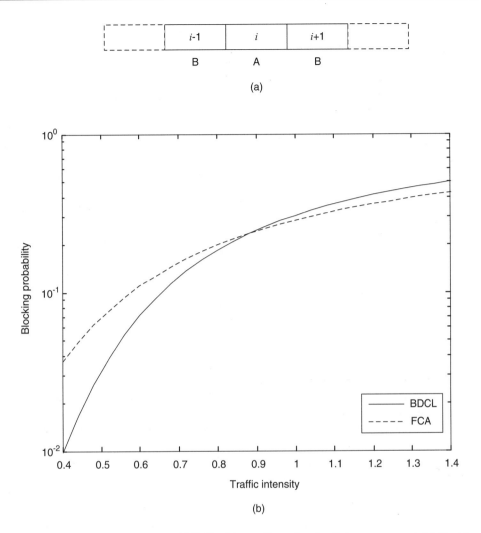

Figure 8.7 Comparison of FCA and BDCL: (a) one-dimensional cellular system, and (b) blocking probability

required. The reuse partitioning method was devised as a method to overcome the inefficiency problem, but unfortunately limitation arose again in the cell planning of subcells. Moreover, since the number of channels assigned to subcells is relatively smaller, it may suffer from the flexibility problem in dynamic channel allocation.

As an effort to resolve the problem of conservative channel allocation completely, some SINR-based approaches have been presented that allocate channels dynamically based on the SINR measurement. Each user measures the level of co-channel interference and then determines the reusability of the channel. This means that a channel is considered available to any user only if the SINR that the user measured at the channel is greater than the requirement. The salient difference of SINR-based schemes is that the reusability of each channel is determined by the relation among the users that use the channel, not between the cells. Such a user-centric channel reuse helps to avoid the bandwidth inefficiency issue caused by QoS over-satisfaction. SINR measurement-based schemes are necessarily dynamic as the user position and channel condition may vary in time, and the channel allocation is performed regardless of the distances among the co-channel sets as long as the SINR requirement can be satisfied.

A potential problem with measurement-based allocation is that the QoS of the existing connections is vulnerable to the interference added by new channel allocations. QoS degradation by a new channel allocation is called *service interrupt*. Since a channel is allocated to a connection request if the predicted SINR of the corresponding user is above the requirement, it is not possible to exactly predict how the channel allocation would affect the SINR and the reusability of the ongoing connections of other users. It can happen that a new channel allocation degrades an ongoing connection's SINR below its requirement, thereby interrupting the service. In this case, if the interrupted connection cannot find an acceptable new channel immediately, its service will be terminated prematurely, which is not acceptable. Even if the interrupted connection finds an acceptable channel immediately, setting up a link by using the new channel may cause an interruption of another established link. It is noteworthy that this problem is the cost of removing the explicit constraint on the reuse distance, which used to make channel allocations free of service interrupt.

Therefore, in devising a dynamic channel allocation scheme based on SINR measurement, the system manager should take care of the following two problems simultaneously: to reuse the channel more efficiently, and to avoid service interrupt.

One method proposed to implement SINR-based DCA is *autonomous reuse partitioning* (ARP) (Kanai, 1992). The objective of ARP is to generate an efficient frequency reuse pattern which is similar to that of the reuse partitioning in Section 8.1.1. In ARP, an ordering is made on the entire channels that are shared by all the cells. When a connection request arrives at a base station, it searches channels in this order and allocates the first unused channel satisfying the SINR requirement to the connection. Since every BS searches the channels in the same order, a channel with a higher order is used more frequently and, as a result, suffers from a higher interference level. Consequently, a channel with a higher order is likely to be allocated to a user who needs a strong signal, probably a user near the BS. On the other hand, a channel with a lower order is less affected by the inter-cell interference, thus is more probable to be allocated to a user close to the cell boundary. Due to such characteristics, ARP tends to make channel allocation that is quite similar to reuse partitioning. Nevertheless, ARP can provide a better channel reuse that reuse partitioning as it does not assume a discrete subcell structure and static channel assignment.

The reuse efficiency can be further improved by another channel allocation called *flexible reuse* (FRU) (Onoe and Yasuda, 1989). In FRU, when a user requests a connection to a cell, the channel with the smallest SINR margin among the available channels is allocated. This operation also yields a channel allocation similar to ARP. A user near a base station will select a channel that is reused by many neighboring cells. This happens because it has a comparatively higher interference than a rarely reused one but the user has enough received power to overcome the high interference. Thus, a user near a BS tends to be allocated a channel reused frequently. On the other hand, a user at the cell boundary tends to be allocated a channel reused by only a few cells as the frequently reused channel suffers from too high interference to satisfy the SINR requirement by a weak desired signal power. FRU tries to reuse each channel in the most aggressive way by selecting the SINR with the smallest margin. This operation renders more chance for a channel to be reused but also has a more risk of service interrupt by the smallest SINR margin itself. This means that an ongoing connection becomes more probable to be corrupted by a future channel allocation.

In order to resolve this service interrupt problem, a DCA scheme called *maximum SINR* (MSIR) (Serizawa and Goodman, 1993) was proposed that allocates the channel with the largest SINR, in contrast to FRU. Since this scheme first assigns an unused or the most lightly loaded channel to a new connection, each channel has much more SINR margin than in ARP and FRU. However, the benefit of efficient reuse partitioning does not arise much in MSIR as even a user near a base station prefers an unused channel and does not try to reuse a channel used in the neighboring cells. Thus, this MSIR scheme is more vulnerable to blocking a connection request than ARP and FRU.

In essence, there exists a tradeoff between the number of channel reuses and the possibility of service interrupt. If we try to reuse a channel more aggressively, we have no choice but to reduce the SINR margin in selecting a channel for allocation, and this small margin yields more chances of service interrupt. The FRU scheme is the extreme case in this direction. On the other hand, if we try to improve the vulnerability to service interrupt, we have to set enough SINR margin, which may disallow a compact reuse of each channel.

The channel reuse efficiency (i.e., the blocking probability and the service interrupt possibility) of FRU and MSIR can be demonstrated by an example. Consider the situation depicted in Figure 8.8. User A represents a user in the vicinity of the base station, so it experiences less interference compared with user B located near the cell boundary. There are two channels available in this cell, x and y. Assume that channel x is not used around the cell and has relatively low interference compared with channel y. Accordingly, assume that the SINRs of channels x and y, experienced by user A, are 10 dB and 8 dB, respectively, and those seen by user B are 8 dB and 4 dB, respectively. Assume also that the minimum SINR requirement is 6 dB. Under these assumptions, there is no difference between FRU and MSIR provided user B requests a channel before user A does. Now consider what will happen if user A requests a connection to the cell before user B does. As explained above, if the system adopts MSIR, user A who has requested first will take channel x as the SINR margin of channel x is larger than that of channel y. In this case, the only channel user B can get is y, but it cannot satisfy the 6 dB requirement of user B. As a result, the request of user B will be blocked. However, if the system adopts FRU, user A will take channel y rather than x, as FRU allocates the channel with the smallest SINR margin among the available channels to the connection request. Then, the system can allocate the remaining channel x to user B. This example illustrates that FRU yields a smaller blocking probability than MSIR does. This result originates from the fact that FRU tries to reuse channels in a more compact way by assigning the channel with the least SINR margin. In the above example, FRU assigns channel y, the more interfered with but still satisfactory one, to user A, the user near the base station.

Figure 8.8 Locations of users A and B within a cell

To compare the service interrupt probability, assume that several new connection requests arrive in the neighboring cells and the neighboring cells allocate channels x or y to them. These new connections cause additional interference in the channels readily allocated to users A and B. For simplicity, assume that the SINRs of those channels seen by both users decrease equally by 3 dB. If the system uses FRU, both of the ongoing connections of users A and B will be interrupted by those new calls because their SINR is now 5 dB, which is insufficient to satisfy the required minimum SINR. However, if the system uses MSIR, it will not undergo service interruption as channel x allocated to user A will still have 1 dB SINR margin. Thus, we see that MSIR is more robust to the service interruption than FRU. Therefore, as to the tradeoff relation between channel utilization and service interrupt, we may conclude as follows. An approach that tries to maximize channel reuse, like FRU, yields improved channel utilization but is more vulnerable to service interrupt. In contrast, a conservative allocation, such as MSIR, provides more robust QoS at the cost of lowered channel utilization.

8.1.4 Channel Allocation with Inter-cell Power Control

The channel allocation schemes discussed so far share several properties. First, the SINR requirement of users is the most important criterion in allocating channels to cells. Second, the transmission rate achieved by users is not an important issue as long as their SINR requirement is met. Third, it is of great concern whether a channel is used or not, but it is of no concern how to determine the transmission power of each channel. Those properties are well aligned for serving voice calls in that the QoS requirement is very strict and no flexibility is allowed to manipulate the QoS of each user in different ways. In fact, it is most important in serving voice calls to maintain the voice quality above a certain level. In addition, devising mechanisms to maximize the transmission rate is not important in serving voice calls, as voice is not the kind of source that generates enormous traffic.

In the case of data traffic, it is desirable to take different approaches in developing inter-cell resource management schemes. In general, data traffic has an elastic characteristic, so maximization of the total throughput of users is of high importance. In contrast, the maintenance of the minimum SINR is of less importance as long as each user is provided with a sufficient transmission rate. The SINR requirement may not be strictly observed, as the channel quality can be reinforced by the support of HARQ and AMC techniques. In addition, two channels should be distinguished by quantitative SINR values, not by the SINR requirement satisfaction, as more data can be delivered through the higher SINR channel. In this context, it makes more sense, in dealing with data traffic, to arrange the channel allocation algorithm in conjunction with the transmission power control. Therefore, this section discusses different types of channel allocation schemes that consider the channel allocation issue in conjunction with power control, under different system objectives and constraints.

8.1.4.1 Water-filling Power Allocation as Inter-cell Interference Management

Among several different power allocation techniques, the *water-filling* method is known to be optimal in the sense that it maximizes overall throughput. The following two characteristics of the water-filling method in particular are important here. First, it tends to allocate almost equal power to each channel when the SINRs of all the channels are high enough. Second, when the overall SINR is low, it selectively allocates power such that more power is allocated to better channels and, sometimes, allocates no power to the channels with very poor SINR. These characteristics share similarity with the principle of channel allocation with explicit reuse distance in the following respects. If two cells are located closely, severely interfering each other, water-filling-based power allocation would allocate separate channels to them; that is, each cell will be allocated with the channels on which the other cell does not put any transmission power. This corresponds to the case where two cells within the reuse distance are provided with different channel sets. On the other hand, if two cells are separated far apart, all the channels would have low interference and thus be used by both cells. This corresponds to the case where two cells located off the reuse distance are allowed to use the same channels.

The following illustrates the operation of water-filling power allocation between two interfering cells. Assume that each cell has 12 units of transmission power to allocate over the six channels shared by both cells. The channel gains of the six channels are $0, -1, 0, -2, 1$, and 2 dB for cell 1, and $1, 2, -2, 0, -1$, and 0 dB for cell 2. We assume a noise density of one unit. Consider two different cases: near-interferer and far-interferer. The two cases are differentiated in terms of the interference gain; that is, the channel gain for the inter-cell interference from the other cell.

First, Figure 8.9 illustrates water-filling power allocation for the near-interferer case where the two cells are located close to each other and the interference gain is 0 dB. In the beginning, the two cells distribute their transmission power equally. Each cell performs water-filling power allocation based on the SINR measured after the previous iteration of power allocation. If cell 1 allocates some additional transmission power to a channel, cell 2 observes an increase in inter-cell interference at that channel. This observation makes cell 2 reduce the power allocation at that channel in the next iteration (e.g., the channels 3, 5, and 6). Then, cell 1 observes inter-cell interference decreasing, so allocates even more transmission power to that channel in the next iteration. After repeated iterations, the power allocation converges to an equilibrium point where the two cells use two independent sets of channels, as shown in Figure 8.9(d). A cell allocates no transmission power to the channels on which the other cell puts a high transmission power. This operation corresponds to the channel allocation with a higher reuse factor.

Second, Figure 8.10 illustrates water-filling power allocation for the far-interferer case where the two cells are located far from each other and the interference gain is -10 dB. The initial power allocation is the same as for the above case but the power allocation behavior changes because the additional power allocation at a cell does not affect much the SINR of the channel. This is because the interference gain is set to a low value and thus the inter-cell interference is not a dominating factor when compared with the noise density. As a consequence, the power allocation does not change much from the initial allocation even though each cell adjusts its transmission power according to the water-filling power allocation principle as above. Observe from the equilibrium state in Figure 8.10(d) that each cell utilizes the entire channels by allocating transmission power of similar level. This operation corresponds to the universal channel reuse.

8.1.4.2 Game-theoretic Channel and Power Allocation: Single-user Case

Based on the above implication, an inter-cell interference management scheme was developed by employing a distributed non-cooperative game approach (Han *et al.*, 2004). The objective of this scheme is to minimize the overall transmitted power under the constraints of

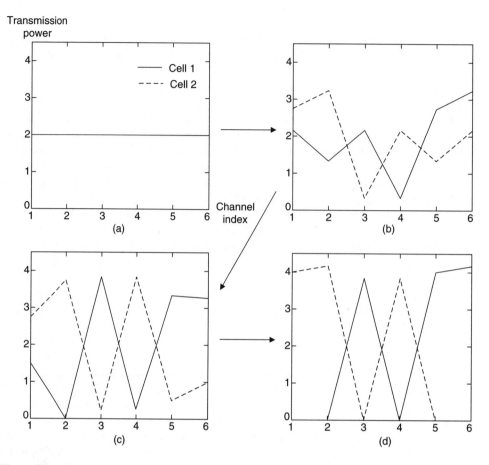

Figure 8.9 Illustration of water-filling power allocation between two closely located cells: (a) initial allocation, (b) allocation after first iteration, (c) allocation after second iteration, and (d) allocation at an equilibrium point

the maximal power and minimal throughput requirement of each cell. For simplicity, assume that each cell has only one user. For N cells sharing L channels, the optimization problem is formulated by

$$\text{minimize} \quad \sum_{n=1}^{N}\sum_{l=1}^{L}P_n^l$$

$$\text{subject to} \quad \sum_{l=1}^{L}R_n^l - R_n^{\text{req}} \geq 0, \forall n,$$

$$\sum_{l=1}^{L}P_n^l - P_{\max} \leq 0, \forall n,$$

$$r_n^l, P_n^l \geq 0, \forall n, l,$$

(8.8)

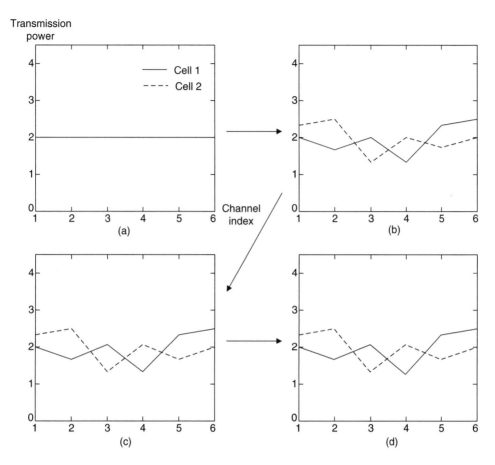

Figure 8.10 Illustration of water-filling power allocation between two faraway cells: (a) initial allocation, (b) allocation after first iteration, (c) allocation after second iteration, and (d) allocation at an equilibrium point

where r_n^l and P_n^l are the rate and power of channel l in cell n, which are related by

$$r_n^l = W \log_2 \left(1 + \frac{P_n^l G_{nn}^l}{\Gamma \left(\sum_{i \neq n} P_i^l G_{in}^l + \sigma^2 \right)} \right) \tag{8.9}$$

for bandwidth W, SNR gap r, noise variance σ^2, and channel gain G_{in}^l, from the transmitter at cell i to the receiver at cell n in channel l. Here, R_n^{req} and P_{max} are the minimal rate and the maximal power constraints of cell n. Define the power vector to represent the whole transmission power by $\mathbf{P} = [P_1^1, P_1^2, \ldots, P_1^L, P_2^1, \ldots, P_N^L]$.

The problem in (8.8) is hard to solve by centralized constrained nonlinear integer optimization, because the complexity and communication overhead grow very fast as the number of cells increases. For an alternative solution, a distributed algorithm with limited controls was devised by using the non-cooperative game approach. In the non-cooperative game, each player (i.e., each base station in this case) chooses the best strategy to maximize its own utility (i.e., minimize the total transmission power in the cell in this case) without cooperating with other players. This non-cooperative game is necessarily an iterative one because the strategy of each cell depends on that of the other cells. This implies that the channel and power allocation of a cell should be modified if another cell changes its resource allocation. In this sense, the power allocation may be regarded as an inter-cell power control that tries to enable a harmonious coexistence of the constituent cells by properly regulating the inter-cell interference. This iteration of the power control process will be balanced if the following situation occurs. If the strategy of the other player is given, each player cannot change its own strategy (i.e., given the channel and power allocation of the other cell, no cell can reduce its transmitted power alone by changing its rate allocation to different channels). Game theory calls this status the *Nash equilibrium*. If a system falls into Nash equilibrium, the system becomes stable but it does not guarantee optimality in general.

This inter-cell resource management algorithm operates as follows. At first, the system manager initializes the channel and power allocation so that the constraints are satisfied. Suppose that the initial power allocation vector is given by $\mathbf{Q} = [Q_1^1, Q_1^2, \ldots, Q_1^L, Q_2^1, \ldots, Q_N^L]$ before starting iterations. Then, each cell minimizes its own transmitted power for a given normalized interference level

$$I_n^l = \Gamma \frac{\sum_{i \neq n} Q_i^i G_{in}^l + \sigma^2}{G_{nn}^l}, \tag{8.10}$$

in a distributed way. The operation which base station n plays is given by

$$\begin{aligned} \text{minimize} \quad & \sum_{l=1}^{L} P_i^l \\ \text{subject to} \quad & \sum_{l=1}^{L} R_i^l = \sum_{l=1}^{L} W \log_2 \left(1 + \frac{P_n^l}{I_n^l} \right) \leq R_n^{\text{req}}. \end{aligned} \tag{8.11}$$

The above problem is nothing but the power allocation problem over parallel channels (see Section 6.2). In this case, it is easy to show that the optimal strategy that each cell takes is given by

$$P_n^l = \left[\frac{1}{\lambda} - I_n^l \right]^+, \tag{8.12}$$

where λ is a threshold determined by

$$\sum_{l=1}^{L} W \log_2 \left(1 + \frac{[\frac{1}{\lambda} - I_n^l]^+}{I_n^l} \right) = R_n^{\text{req}}. \tag{8.13}$$

Observe that this power allocation is the well-known water-filling method. Each cell updates its transmission power according to the inter-cell interference. Note that the interference level given in (8.10) may vary as the iteration proceeds. It is because the other cells also update their transmission power at every iteration step. This iterative operation can be expressed explicitly by writing the transmission power determination in (8.12) and (8.13) by $\mathbf{P}(\mathbf{Q})$, indicating that the power determination after an iteration step is a function of the power vector before that step. This iteration is repeated until a Nash equilibrium point is attained.

In general, the system has multiple Nash equilibrium points, including some undesirable ones. To avoid those undesirable equilibria, we can judge the desirability of the attained equilibrium and, if it turns out undesirable, restart the whole power update iteration at a different environment. For example, we may exclude a cell that has a very low SINR at a channel when allocating that channel.[2] The problem of undesirable equilibrium points can be mitigated in a multiuser situation where each cell has multiple users in it and channel allocation is done user-wise, not cell-wise.

8.1.4.3 Game-theoretic Channel and Power Allocation: Multiuser Case

Another channel and power allocation scheme based on the non-cooperative game that considers multiple users in each cell is proposed in Kwon and Lee (2006). Here, the cell n is assumed to have K_n users in it and the channel gain of the user k in the cell n is represented by $G_{in}^l(k)$ from the transmitter at cell i in channel l. In this scheme, the utility function represents the sum of the data rates and the power consumption in a cell. More specifically, the objective of each cell is given by

$$\text{maximize} \sum_{l=1}^{L} R_n^l - cP_n^l \qquad (8.14)$$

for the price per unit power, c. This objective incorporates the concept of balancing the cell throughput and the total power consumption by a properly designed parameter c. It can be interpreted as the net throughput achieved by resource allocation when the cost is paid in proportion to the total power consumption at the rate c. Each base station in the cell is treated as a player, and the strategy chosen by each player determines the channel assignment and power allocation.

The algorithm in Kwon and Lee (2006) is similar to that in Han $et\ al.$ (2004), but each cell also performs the channel allocation among the users in it. The user-wise channel allocation corresponds to the maximal rate scheduling that assigns each channel to the user who has the largest SINR on it. This is a natural result from the objective of the power minimization (or the throughput maximization, equivalently). Thus, for each channel, the normalized interference is expressed by

$$I_n^l = \min_{k=1,\dots,K_n} \Gamma \frac{\sum_{i \neq n} Q_i^l G_{in}^l(k) + \sigma^2}{G_{nn}^l(k)}, \qquad (8.15)$$

for a given power allocation vector $\mathbf{Q} = [Q_1^1, Q_1^2, \dots, Q_1^L, Q_2^1, \dots, Q_N^L]$. Since the SINR is determined at each channel after the user assignment, the BS can operate the water-filling

---

[2] See Han $et\ al.$ (2004) for more detailed description about avoiding undesirable equilibrium points.

algorithm as the solution of (8.14).[3] The transmission power of each channel is given by the water-filling method

$$P_n^l = \left[\frac{1}{\lambda} - I_n^l\right]^+, \qquad (8.16)$$

for $\lambda = c \ln 2/W$. As in the single-user case above, we can represent this power update by $\mathbf{P}(\mathbf{Q})$. Note that each cell updates not only the transmission power but also the channel allocation among the users at every iteration step.

By the virtue of the multiuser diversity effect, a multiuser environment is helpful in avoiding the problem of undesirable equilibrium points.[4] This means that, if there are many users in each cell, it is highly improbable that each channel is assigned to a user with a poor SINR. Thus, in a multiuser environment, the channel allocation in each cell performs the operation of removing the bad channel users who may yield undesirable equilibrium points in the single-user environment.

Interestingly, it is known that water-filling-based channel allocation has a unique Nash equilibrium point under some conditions (Kwon and Lee, 2006). We assume the following two assumptions:

- The power allocation holds $\mathbf{P}(\mathbf{Q}) > 0$ for any power vector $\mathbf{Q}$.
- The noise density is much smaller than the inter-cell interference; that is, $N_0/Q_n^l \to 0$ for any power vector $\mathbf{Q}$.

The first assumption is not unrealistic as there may exist at least one channel that has a very high ratio of channel gain to interference gain in a multiuser environment. Therefore, a positive level of transmission power would be set to each channel even though the other cells send strong interference. The second assumption is reasonable in the system whose performance is limited by inter-cell interference. Then, the power allocation rule $\mathbf{P}(\mathbf{Q})$ becomes type-II standard as discussed below, which is known to have a unique equilibrium point.[5]

First, the positivity holds directly by the first assumption. Second, the type-II monotonicity holds as follows. Suppose power is allocated such that $\mathbf{Q}' \geq \mathbf{Q}$. Then, the normalized interference in (8.15) monotonically increases for all the channels and all the users. Then, since the water-filling algorithm allocates less power for a higher interference, the power allocation gets reduced component-wise. In other words, we have $I_n^l(\mathbf{Q}') \geq I_n^l(\mathbf{Q})$, and by (8.16) we get

$$P_n^l(\mathbf{Q}') = \left[c' - I_n^l(\mathbf{Q}')\right]^+ \leq \left[c' - I_n^l(\mathbf{Q})\right]^+ = P_n^l(\mathbf{Q}), \qquad (8.17)$$

which implies the type-II monotonicity.

---

[3] The problem in (8.14) corresponds to the unconstrained version of the throughput maximization problem under a constraint on the power consumption. Note that price c plays the role of Lagrange multiplier.
[4] Refer to Chapter 5 for detailed discussion on the multiuser diversity effect.
[5] Refer to Section 6.1.3 for detailed discussion on the type-II standard power control.

Third, the type-II scalability holds as follows. Suppose that the transmission power $\mathbf{Q}$ is rescaled by $\alpha\mathbf{Q}$ for $\alpha > 1$. Then, by applying the second assumption to (8.15), the normalized interference is also rescaled by

$$I_n^l(\alpha\mathbf{Q}) = \min_{k=1,\ldots,K_n} \Gamma \frac{\sum_{i\neq n}\alpha Q_i^l G_{in}^l(k)}{G_{nn}^l(k)} = \alpha I_n^l(\mathbf{Q}) \qquad (8.18)$$

Then by (8.16) we have

$$\begin{aligned}
P_n^l(\alpha\mathbf{Q}) - \frac{1}{\alpha}P_n^l(\mathbf{Q}) &= c' - \alpha I_n^l(\mathbf{Q}) - \frac{1}{\alpha}(c' - I_n^l(\mathbf{Q})) = \frac{\alpha-1}{\alpha}c' - \frac{\alpha^2-1}{\alpha}I_n^l(\mathbf{Q}) \\
&= \frac{\alpha-1}{\alpha}(c' - (\alpha+1)I_n^l(\mathbf{Q})),
\end{aligned} \qquad (8.19)$$

where the first equality holds due to the first assumption. We consider another power allocation $P_n^l((\alpha+1)\mathbf{Q})$, which is equal to the term in the parentheses in the last equality of (8.19). Then, this power allocation is also positive by the first assumption, so we have

$$P_n^l(\alpha\mathbf{Q}) - \frac{1}{\alpha}P_n^l(\mathbf{Q}) = \frac{\alpha-1}{\alpha}P_n^l((\alpha+1)\mathbf{Q}) > 0, \qquad (8.20)$$

which implies type-II scalability.

Based on this property, the inter-cell power control in which each cell iteratively performs a water-filling power allocation renders an effective inter-cell resource management algorithm especially in a multiuser scenario.

8.2 Handoff Management[6]

The functionality of handoff is needed for wireless cellular systems such that services are provided seamlessly even when the users make geographical movement over the cells. *Handoff* (or *handover*) refers to the process of transferring the ongoing connection in a channel associated with a base station to another channel associated with another BS, as illustrated in Figure 8.11. Suppose that a mobile station with a connection in progress is moving away from the coverage of the associated BS. Then, the quality of the connection will degrade, so the user needs to get associated with another BS in order to keep the channel quality above the sustainable level. Then the handoff mechanism redirects the mobile station's connection to a new one that provides a more favorable communication channel. As such, handoff is a means of maintaining the QoS of an ongoing connection in reaction to crossing a cell boundary.

In the context of resource management, handoff raises the issue of resource reservation and admission control. As noted above, handoff is a means of maintaining the QoS of the ongoing connection at a pre-specified level. This task requires the reserving of a certain amount of resource as otherwise handoff may fail due to lack of available channels. Resource reservation

---

[6] The contents of the front part of this section are reproduced by permission from Lee and Choi, *Broadband Wireless Access and Local Networks,* Norwood, MA: Artech House, Inc., 2008. © 2008 by Artech House, Inc.

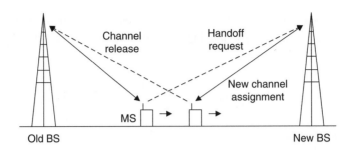

Figure 8.11 Handoff operation (Reproduced with permission from B.G. Lee and S. Choi, *Broadband Wireless Access and Local Networks: Mobile WiMAX and WiFi*, Artech House, Norwood, MA. © 2008 Artech House, Inc.)

for handoff is tied in with the admission control of the newly originating connections, as the total resource has to be shared by the handoff connections and the newly generated connections. In this sharing, it is reasonable to set a higher priority on the handoff connections since a forced termination of the ongoing connections makes much severer impact on the connection QoS, as well as customer satisfaction, than the denial of access (i.e., blocking) of newly originating connections.

From the resource manager's point of view, the above prioritization policy may be interpreted as an assignment of different admission priorities to two different types of traffic arrival – handoff arrivals and newly originating arrivals – with a higher priority set to the former. This admission priority is directly related to the issue of reserving the resource to the two different types of traffic arrivals. Therefore the essence of handoff management is how to divide the resource to the two different traffic arrivals such that it can satisfy the QoS requirement of the handoff connections and, at the same time, provide a reasonable level of performance to the new connections. This section discusses various approaches of resource reservation and admission control in handoff management.

8.2.1 Handoff Procedure and Performance

In preparation for the detailed discussions of handoff management schemes, we first introduce the handoff process and a mathematical model for calculating performance metrics.

8.2.1.1 Handoff Procedure

Handoff occurs when the signal strength of the existing connection from an MS to the BS is not good enough to maintain the QoS, whereas the signal strength of a new connection being considered to a new BS is better. Thus, the handoff procedure is initiated based on the mobile's measurement of the signal strengths of multiple base stations. There are several types of criteria for initiating handoff, depending on the signal strengths of the old and new connections.

One simple criterion of handoff is that the MS redirects its connection to the BS with the highest signal strength. According to Figure 8.12, the MS may initiate the handoff procedure at position A at which the signal strength of the new BS becomes stronger than that of the serving BS. This method will surely guarantee that the MS is always associated with the BS having the strongest channel, but at the same time it can make a handoff attempt even when the current connection has acceptable quality – and thereby induce unnecessary handoffs.

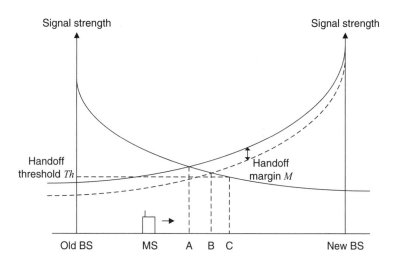

Figure 8.12   Handoff initialization criteria (Reproduced with permission from B.G. Lee and S. Choi, *Broadband Wireless Access and Local Networks: Mobile WiMAX and WiFi*, Artech House, Norwood, MA. © 2008 Artech House, Inc.)

Another criterion for handoff initiation is to redirect the connection to a new BS only when the signal strength of the new BS is higher than the current signal strength by a predetermined margin. According to Figure 8.12, this criterion makes the MS initiate the handoff at point B at which the new signal strength is higher than the existing one by the margin M. This method can prevent the aforementioned ping-pong effect, or repeated handoffs between two base stations, which may possibly occur at the cell boundary. In this case a larger value of margin may be taken to shift the handoff point closer to the new BS (e.g., to point C), thereby reducing the handoff attempts further, but an excessive margin could delay handoff initialization and consequently deteriorate the connection quality.

A third handoff criterion may be to initiate handoff only when the current signal strength falls below a given threshold and the new BS sends a signal with a higher strength. According to Figure 8.12, this criterion makes handoff initiated at point C at which the current signal strength drops below the threshold Th. In this criterion, the decision of the threshold level governs the handoff initiation. If the threshold is too high, the signal strength of the current BS would always be below the threshold at the cell boundary; and if it is too low, handoff may not take place even when the current connection falls into a poor state. Therefore the threshold should be set adequately by considering the above three possible criteria, so as to prevent unnecessary handoffs while maintaining the connection quality above an acceptable level.

In the last two cases, the coverage of the old and new cells overlaps and thus the moving MS may be associated with the old BS for a certain duration even after the signal strength to the new BS becomes stronger. In Figure 8.12, the signal strength becomes stronger to the new BS beyond point A but the handoff procedure is initiated later at points B and C respectively for the two cases. Thus, the area between point A and the point where handoff takes place (B or C) belongs to the overlapping area of the two base stations, in which a newly generated connection is associated with the new BS while communications continue with the old BS for the handoff connection.

As such, signal strength measurement and the handoff decision are two key handoff procedures that may take place in the old and new base stations and/or the moving MS.

Signal strength may be measured by the base stations using the uplink transmission or by the MS using the downlink transmission. The handoff decision may be made at the base stations according to their own measurements or the feedback from the MS, or may be decided at each MS in a decentralized manner.

8.2.1.2 Performance Metrics

Consider the traffic flow diagram in Figure 8.13. This shows that a handoff traffic that requests for a handoff connection may be either dropped or admitted. A new traffic flow that requests for a new connection may be either blocked or admitted. The ongoing connections may be either completed or handed off to a new connection.

There are several performance metrics involved in handoff management, such as dropping probability, forced termination probability, and blocking probability. *Dropping probability*, P_D, is the probability that a handoff attempt becomes unsuccessful due to the lack of resources in the new cell. It indicates how well the QoS of a readily admitted connection is maintained by the communication system. *Forced termination probability*, P_F, is the probability that a connection that was not blocked at the connection initiation is eventually forced to terminate by handoff failure before it completes its connection. *Blocking probability*, P_B, is the probability that the communication system rejects a newly originating connection. It reflects how the employed resource reservation scheme affects the QoS of newly originating connections.

In order to determine the above performance metrics, handoff management schemes are often formulated as a Markov chain under the assumption of memoryless connection arrival

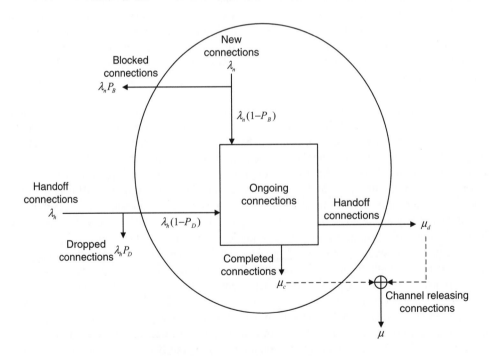

Figure 8.13 Traffic flow diagram of a cell

and departure (Hong and Rappaport, 1986). The new and handoff call arrivals are assumed to be a Poisson process with average arrival rates λ_n and λ_h, respectively. Assume that the *connection time*, T_c, or the time duration of connection, is an exponential random variable with mean $1/\mu_c$. This means that the CDF of the channel connection time is given by

$$F_c(t) = \Pr[T_c \le t] = 1 - e^{-\mu_c t} \tag{8.21}$$

We also assume that the *dwell time*, T_d, or the time duration spent in a given cell until handoff, is also an exponential random variable with mean $1/\mu_d$. In addition, we assume that all the above processes are independent of one another. This implies that all the inter-arrival and departure times are exponentially distributed and thus memoryless. In the case of the connection time, for example, the remaining connection time is independent of the time elapsed after its origination. In other words, the probability that a connection terminates after a certain time is always the same no matter how long the connection has already lasted.

With the above assumptions, we can determine the distribution of the *channel holding time*, T_{ch}, or the time duration that a connection holds a channel of a given cell. Since an ongoing connection releases its channel either when its communication is completed or when it is handed off to another cell, the channel holding time of a connection is the smaller of the connection time and the dwell time; that is

$$T_{ch} = \min(T_c,\ T_d). \tag{8.22}$$

So the CDF of the channel holding time is given by

$$\begin{aligned} F_{ch}(t) &= 1 - \Pr[T_{ch} > t] = 1 - \Pr[T_c > t] \times \Pr[T_d > t] \\ &= 1 - e^{-\mu_c t} e^{-\mu_d t} = 1 - e^{-(\mu_c + \mu_d)t}. \end{aligned} \tag{8.23}$$

This implies that the channel holding time is also an exponential random variable with mean $1/\mu = 1/(\mu_c + \mu_d)$ and has the memoryless property.

Since handoff occurs when the connection time T_c is longer than the dwell time T_d, we can determine the probability that a connection will attempt handoff before completion, P_H, by

$$P_H = \Pr[T_c > T_d] = \int_0^\infty \Pr[T_c > t | T_d = t] f_d(t) \mathrm{d}t = \int_0^\infty e^{-\mu_c t} \mu_d e^{-\mu_c t} \mathrm{d}t = \mu_d/(\mu_c + \mu_d), \tag{8.24}$$

for the probability density function of the dwell time, $f_d(t)$. The last equality holds due to the independence of T_c and T_d.

Now, we relate the dropping probability P_D with the forced termination probability P_F. A connection is dropped at the first handoff attempt with probability $P_H P_D$ and is dropped at the second attempt with probability $P_H^2(1 - P_D)P_D$. The forced termination probability is the summation of all these dropping probabilities:

$$P_F = P_H P_D + P_H^2(1 - P_D)P_D + P_H^3(1 - P_D)^2 P_D + \cdots = \frac{P_H P_D}{1 - P_H(1 - P_D)}. \tag{8.25}$$

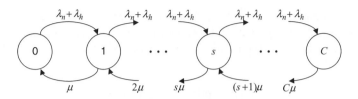

Figure 8.14   State transition diagram for the non-prioritized policy

Now that the forced termination probability can be obtained from the dropping probability, we may focus on determining the dropping and blocking probabilities from the Markov chain formulation.

The problem can be formulated by a Markov chain model for the case where no prioritization is applied between the handoff and new arrivals. Assume that the base station has C channels which are shared by both the new and handoff arrivals. We denote the state of the system by the number of channels occupied by the ongoing connections. Then, we can model the system by a Markov chain, whose state transition occurs as shown in Figure 8.14. At state s, an additional user arrives at the cell with the rate $\lambda_n + \lambda_h$, and this user occupies an additional channel, thereby causing a state transition to state $(s + 1)$. Also, in this state, each user departs the cell with the rate μ, so the system transits to state $(s - 1)$ with the rate $s\mu$. Assume that the time interval for state transition is very short and the probability that more than two users arrive and depart simultaneously is negligible.

At the steady state, the incoming flow to a given state is balanced with the outgoing flow from that state. If we denote by P_s the steady-state probability of state s, we get from Figure 8.14 the balance equation

$$(\lambda_n + \lambda_h)P_{s-1} = s\mu P_s, \quad 1 \leq s \leq C. \tag{8.26}$$

Applying the relation recursively, we get

$$P_s = \frac{(\lambda_n + \lambda_h)^s}{s!\mu^s}P_0 = \frac{(\rho_n + \rho_h)^s}{s!}P_0, \quad 0 \leq s \leq C, \tag{8.27}$$

where $\rho_n = \lambda_n/\mu$ and $\rho_h = \lambda_h/\mu$. Since the sum of all the state probabilities is 1, that is

$$\sum_{s=0}^{C} P_s = 1, \tag{8.28}$$

we get the probability at state 0 by

$$P_0 = \frac{1}{\displaystyle\sum_{j=0}^{C} \frac{(\rho_n + \rho_h)^j}{j!}}. \tag{8.29}$$

All the steady-state probabilities can be obtained by inserting (8.29) into (8.27). In this non-prioritized scheme, the dropping and blocking probabilities are not distinguishable; that is

$$P_{\mathrm{D}} = P_{\mathrm{B}} = P_{\mathrm{C}} \tag{8.30}$$

since the resource is not separated for the handoff and new traffics. A handoff or new connection is dropped or blocked if all the C channels are occupied when it enters the given cell.

8.2.2 Resource Reservation via Guard Channel Policy

A guard channel policy always reserves a certain number of channels for handoff arrivals. This subsection first formulates this policy to calculate its performance and derives its optimal property. Then it considers an extended guard channel policy for a gradual tradeoff between the dropping and blocking probabilities, and discusses the Pareto optimality of this policy.

8.2.2.1 Guard Channel Policy

The prioritization policy which puts a higher priority on handling the handoff traffic than any new traffic may be implemented in such a way that a fixed number of channels, called the *guard channels* (GC), are reserved for the handoff arrivals. This *GC policy* or *cutoff prioritization policy* may be operated by setting a channel threshold T as depicted in Figure 8.15. Any handoff connection is admitted as long as a channel is available, but a new connection is admitted only when the number of ongoing connections is less than the threshold T. This implies that, out of the total resource of C channels, the GC policy always reserves $(C - T)$ channels for admitting the handoff connections.

With the GC policy, the state transition diagram changes to the shape shown in Figure 8.16. It differs from the non-prioritized policy case in that the handoff rate is involved in increasing the system state number after state T.

From the state transition diagram, we get the balance equations

$$\begin{aligned}
(\lambda_{\mathrm{n}} + \lambda_{\mathrm{h}})P_{s-1} &= s\mu P_s, \, 1 \leq s \leq T, \\
\lambda_{\mathrm{h}} P_{s-1} &= s\mu P_s, \, T+1 \leq s \leq C.
\end{aligned} \tag{8.31}$$

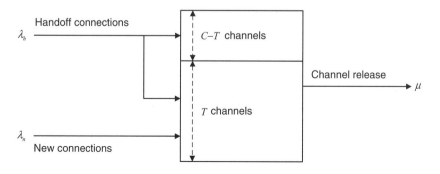

Figure 8.15 Modeling of the guard channel policy

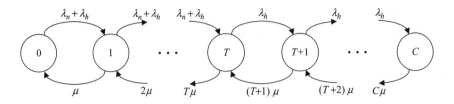

Figure 8.16   State transition diagram for the guard channel policy

and the steady-state probabilities

$$
P_s =
\begin{cases}
\dfrac{(\rho_h + \rho_n)^s}{s!} P_0, & \text{if } 0 \le s \le T, \\[4mm]
\dfrac{(\rho_h + \rho_n)^T}{s!} \rho_h^{\,s-T} P_0, & \text{if } T+1 \le s \le C,
\end{cases}
\tag{8.32}
$$

where

$$
P_0 = \frac{1}{\displaystyle\sum_{j=0}^{T} \frac{(\rho_h + \rho_n)^j}{j!} + \sum_{j=T+1}^{C} \frac{(\rho_h + \rho_n)^T}{j!} \rho_h^{\,j-T}}.
\tag{8.33}
$$

Since a handoff connection may be dropped only at state C, the dropping probability is given by $P_D = P_C$. A new connection may be blocked when the system is at states $T, (T+1), \ldots, C$, so the blocking probability is given by

$$
P_B = \sum_{n=T}^{C} P_n.
\tag{8.34}
$$

The tradeoff relation between the dropping and blocking probabilities can be exploited by adjusting the reservation threshold T. The dropping probability decreases and the blocking probability increases as the threshold decreases, because more channels are reserved for the handoff connections.

Clearly, a tradeoff relation exists between P_D and P_B. The above discussion indicates that the two probabilities cannot be reduced simultaneously due to this tradeoff. In order to balance the two performance metrics, we need to formulate a linear objective function as a weighted sum of the two probabilities; that is, $F = \alpha_B P_B + \alpha_D P_D$, for some positive constants α_B and α_D. Assume that $\alpha_B/\lambda_n < \alpha_D/\lambda_h$. This assumption is reasonable since a higher priority is given on the handoff connections in formulating the objective (i.e., $\alpha_D > \alpha_B$) and a new connection is generated at higher frequency than a handoff connection (i.e., $\lambda_n > \lambda_h$). Then, the system objective becomes minimization of the objective function F for the given number of channels C. In view of this optimization problem, the GC policy is optimal as it minimizes the given linear objective function (Ramjee *et al.*, 1997); therefore the GC policy renders a simple but efficient operation while optimally exploiting the tradeoff between blocking and dropping probabilities.

The optimality of the above problem can be shown by the sequential optimization method discussed in Section 4.2. Suppose that the system is observed at every Δt time and the observation interval Δt is small enough to assume that no two different arrivals or departures occur simultaneously at each observation. We observe the system M times and count the number of dropped and blocked connections during the observation. We define by $1_B(m)$ the indicator which becomes 1 if a new connection is blocked at the mth observation, and becomes zero otherwise. We define by $1_D(m)$ in the same manner to be the indicator for handoff dropping.

The blocking and dropping probabilities are expressed by using the above notations. When the number of observations, M, increases to infinity, the number of new connection attempts converges to $\lambda_n \Delta t M$. Thus, from the ergodicity of the Markov formulation, the blocking probability is equal to the average ratio of the number of the blocked connections to the whole newly originated connections; that is

$$P_B = \lim_{M \to \infty} \mathbb{E}\left[\frac{\sum_{m=1}^{M} 1_B(m)}{\lambda_n \Delta t M}\right]. \tag{8.35}$$

For the dropping probability, we similarly get

$$P_D = \lim_{M \to \infty} \mathbb{E}\left[\frac{\sum_{m=1}^{M} 1_D(m)}{\lambda_h \Delta t M}\right]. \tag{8.36}$$

Then, the minimization of a linear objective function can be posed as the following objective minimization problem:

$$\text{minimize } \lim_{M \to \infty} \frac{1}{M \Delta t} \mathbb{E}\left[\sum_{m=1}^{M} \frac{\alpha_B 1_B(m)}{\lambda_n} + \sum_{m=1}^{M} \frac{\alpha_D 1_D(m)}{\lambda_h}\right]. \tag{8.37}$$

So we consider the cost minimization problem with finite observations in the form

$$\text{minimize } \mathbb{E}\left[\sum_{m=1}^{M} A'_B 1_B(m) + \sum_{m=1}^{M} A'_D 1_D(m)\right]. \tag{8.38}$$

for $A'_B \equiv \alpha_B / \lambda_n$ and $A'_D \equiv \alpha_D / \lambda_h$. If we find the optimal policy which minimizes (8.38) regardless of the number of observations, M, it becomes the optimal policy that minimizes the linear objective function $F = \alpha_B P_B + \alpha_D P_D$.

Define by $V_m(s)$ the optimal cost after m observations starting at state s. At the first observation, a connection leaves the cell with probability $s \mu \Delta t$ and the expected cost becomes $V_{m-1}(s-1)$. A new connection arrives with probability $\lambda_n \Delta t$, and if we admit this connection, the cost does not change but the state increases to $(s+1)$. In this case, the expected cost becomes $V_{m-1}(s+1)$. If we reject this new connection, the state remains unchanged while the cost increases by A'_B. The expected cost for this case is $A_B + V_{m-1}(s)$. The optimal policy chooses the action which has the smaller expected cost between admission and rejection. We apply an analogous relation to a handoff arrival. Finally, no event occurs with probability $(1 - (\lambda_n + \lambda_h + s\mu)\Delta t)$ and the system remains the same while making the expected cost $V_{m-1}(s)$. By combining all the above elements, we get the following recursive relation of the optimal cost

$$
\begin{aligned}
V_m(s) = {}& \lambda_n \Delta t \min\{V_{m-1}(s+1), A_B + V_{m-1}(s)\} + \lambda_h \Delta t \min\{V_{m-1}(s+1), A_D + V_{m-1}(s)\} \\
& + s\mu \Delta t V_{m-1}(s-1) + (1 - (\lambda_n + \lambda_h + s\mu)\Delta t)V_{m-1}(s),
\end{aligned}
\tag{8.39}
$$

for $1 \leq s \leq C - 1$. We can determine all $V_m(s)$ by applying $V_0(s) = 0$ for $0 \leq s \leq C$ and $V_m(s) = \infty$ for $s > c$. For notational simplicity, we rewrite (8.39) by

$$
\begin{aligned}
V_m(s) = {}& \lambda_n \min\{U(s+1), A'_B + U(s)\} \\
& + \lambda_h \min\{U(s+1), A'_D + U(s)\} + sW(s-1) + (B-s)W(s),
\end{aligned}
\tag{8.40}
$$

where $U(s) \equiv \Delta t V_{m-1}(s)$, $A_B \equiv A'_B \Delta t$, $A_D \equiv A'_D \Delta t$, $W(s) \equiv \mu \Delta t V_{m-1}(s)$, and $B \equiv (1 - (\lambda_n + \lambda_h)\Delta t)/\mu \Delta t$.

We first show by induction that $V_m(S)$ is a non-decreasing function of s for a given m. The first step is obvious since $V_m(S) = 0$ for $0 \leq s \leq C$. We assume $V_{m-1}(s)$ is non-decreasing and consider the difference $V_m(s) - V_m(s-1)$. The first term of the difference is

$$
\lambda_n[\min\{U(s+1), A_B + U(s)\} - \min\{U(s), A_B + U(s-1)\}] \geq 0,
\tag{8.41}
$$

since $U(s+1) \geq U(s)$ and $A_B + U(s) \geq A_B + U(s-1)$ by the assumption. Similarly, the second term of the difference is also larger than or equal to zero. The third and forth terms are given by

$$
\begin{aligned}
& sW(s-1) + (B-s)W(s) - (s-1)W(s-2) - (B-(s-1))W(s-1) \\
& = (s-1)\{W(s-1) - W(s-2)\} + (B-s)\{W(s) - W(s-1)\} \geq 0,
\end{aligned}
\tag{8.42}
$$

by the assumed non-decreasing property of $W(s)$. Hence, we get $V_m(s) - V_m(s-1) \geq 0$ and $V_m(s)$ is non-decreasing.

Second, we show by induction that $V_m(s)$ is a convex function of s for a given m. The initial step is trivial again since $V_0(s) = 0$ for $0 \leq s \leq C$. For the induction, we assume that $V_{m-1}(s)$ is convex. Then, we can show the convexity of $V_m(s)$ by showing that the difference $V_m(s) - V_m(s-1)$ is non-decreasing. Consider the first term of the difference which is given by

$$
\lambda_n D_1(s) \equiv \lambda_n[\min\{U(s+1), A_B + U(s)\} - \min\{U(s), A_B + U(s-1)\}].
\tag{8.43}
$$

By the non-decreasing property and convex assumption of $U(s)$, there exists an s^* such that

$$
\begin{aligned}
U(s) &< A_B + U(s-1), \quad s < s^* \\
U(s) &\geq A_B + U(s-1), \quad s \geq s^*.
\end{aligned}
\tag{8.44}
$$

Then we get

$$D_1(s) = \begin{cases} U(s+1)-U(s), & s < s^*-1, \\ A_B, & s = s^*-1, \\ U(s)-U(s-1), & s \geq s^*. \end{cases} \tag{8.45}$$

By the convexity of $U(s)$, we get

$$D_1(s) \geq D_1(s-1), 1 \leq s \leq s^*-2 \text{ or } s \geq s^*+1. \tag{8.46}$$

Moreover, by (8.44) we get

$$D_1(s^*-1) = A_B > U(s^*-1)-U(s^*-2) = D_1(s^*-2), \tag{8.47}$$

$$D_1(s^*) = U(s^*)-U(s^*-1) \geq A_B + U(s^*-1)-U(s^*+1) = A_B = D_1(s^*-1). \tag{8.48}$$

By combining the inequalities in (8.46)–(8.48), we can prove that the first term of the difference $V_m(s) - V_m(s-1)$ is non-decreasing. Similarly, we can prove the second term also is non-decreasing. The third and fourth terms are given by

$$D_{3,4}(s) \equiv (B-s)W(s) + (2s+1-B)W(s-1)-(s-1)W(s-2). \tag{8.49}$$

Then we get

$$\begin{aligned}
D_{3,4}(s) &- D_{3,4}(s-1) \\
&= (B-s)W(s)+(2s+1-B)W(s-1)-(s-1)W(s-2) \\
&- \{(B-s+1)W(s-1)+(2s-1-B)W(s-2)-(s-2)W(s-3)\} \\
&= (B-s)W(s)+(3s-2B)W(s-1)-(3s-2-B)W(s-2)+(s-2)W(s-3) \\
&= (B-s)(W(s)-2W(s-1)+W(s-2))+sW(s-1)-2(s-1)W(s-2)+(s-2)W(s-3) \\
&\geq sW(s-1)-2(s-1)W(s-2)+(s-2)W(s-3) \\
&\geq (s-1)(W(s)-2W(s-1)+W(s-2)) \geq 0.
\end{aligned} \tag{8.50}$$

In (8.50), the first and third inequalities hold by applying Jensen's inequality to $W(s)$, and the second inequality holds due to the non-decreasing property of $W(s)$. The result of (8.50) implies that the third and fourth terms of the difference $V_m(s) - V_m(s-1)$ are also non-decreasing. Therefore, it is proven that the optimal cost function $V_m(s)$ is convex.

Now we return to the recursive relation of $V_m(s)$ in (8.39) with its non-decreasing and convex properties. When a new connection arrives at a cell with s ongoing connections, the optimal policy admits the connection if $V_{m-1}(s + 1) < A_B + V_{m-1}(s)$, and rejects it otherwise. Since the cost function $V_m(s)$ is non-decreasing and convex, there exists a state s_n^* which satisfies

$$\begin{aligned}
V_{m-1}(s) &< A_B + V_{m-1}(s-1), s < s_n^*, \\
V_{m-1}(s) &\geq A_B + V_{m-1}(s-1), s \geq s_n^*.
\end{aligned} \tag{8.51}$$

Equation (8.51) indicates that the optimal criterion for new connection admission is a threshold-type policy. That is, the optimal policy admits the connection if the current state is less than s_n^* and rejects it if there are more than s_n^* ongoing connections. Likewise, the optimal criterion for handoff connection admission is also a threshold-type for the threshold state s_h^*. However, since $A_B < A_D$, the threshold state of the handoff connection case is larger than or equal to that of the new connection case; that is, $s_n^* \leq s_h^*$. It is obvious that $s_h^* = C$ since any different $s_h^* (< C)$ will result in unnecessarily idle channels. Therefore, we can conclude that the optimal admission policy which minimizes the linear objective $F = \alpha_B P_B + \alpha_D P_D$ is a guard channel policy with a properly defined threshold T.

8.2.2.2 Fractional Guard Channel Policy

The guard channel policy is an effective method of channel reservation for handoff connections as it minimizes a linear combination of the dropping and blocking probabilities. However, since the number of reserved channels can only be an integer number in the GC policy, the tradeoff between the two probabilities can be exploited only in a discrete manner. Even when the current dropping probability is slightly larger than the required level, the GC policy has no choice but to increase the number of the reserved channel by one. However, this increment is likely to over-satisfy the dropping probability requirement, while blocking newly originating connections excessively.

As a solution to the above problem, the *fractional guard channel* (FGC) policy was developed (Cruz-Perez *et al.*, 1999; Markoulidakis *et al.*, 2000), which allows reservation of a real number (i.e., non-integer number) of channels based on probability-based admission. The FGC policy normally operates in the same way as the GC policy except at state T. A handoff connection is admitted if there is an available channel, and a new connection is admitted if the number of ongoing connections is less than the threshold T. If a new connection arrives when T connections exist in the cell, then the new connection is admitted with a probability β in [0, 1]. By this random behavior, the FGC policy reserves $(C - T - \beta)$ channels *on average* for the handoff connections, and the number of reserved channels is controlled by the probability β. Figure 8.17 depicts the system model of the FGC policy, which is a *generalization* of the GC policy.

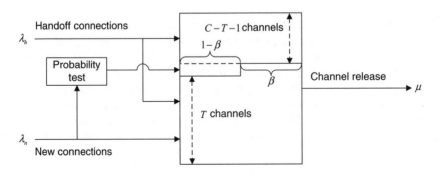

Figure 8.17   System model for the fractional guard channel policy

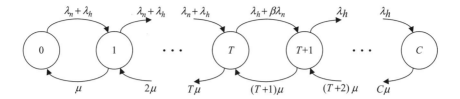

Figure 8.18   State transition for the FGC policy

The state transition diagram of the FGC policy is given in Figure 8.18, which differs from Figure 8.16 only in the arriving rate at state T. So we can get the steady-state probability

$$
P_s = \begin{cases} \dfrac{(\rho_h + \rho_n)^s}{s!} P_0, & \text{if } 0 \le s \le T, \\[3mm] \dfrac{(\rho_h + \rho_n)^T}{s!}(\rho_h + \rho_n\beta)\rho_h^{s-(T+1)} P_0, & \text{if } T+1 \le s \le C, \end{cases}
\tag{8.52}
$$

where

$$
P_0 = \frac{1}{\displaystyle\sum_{j=0}^{T} \frac{(\rho_h + \rho_n)^j}{j!} + \sum_{j=T+1}^{C} \frac{(\rho_h + \rho_n)^T}{j!}(\rho_h + \rho_n\beta)\rho_h^{j-(T+1)}}.
\tag{8.53}
$$

The dropping probability is given by $P_D = P_C$ and the blocking probability is expressed by

$$
P_B = (1-\beta)P_T + \sum_{j=T+1}^{C} P_j.
\tag{8.54}
$$

We can interpret the FGC policy with the threshold T and the probability β (which we denote by $f(T, \beta)$ as a combination of the GC policies with the thresholds T and $(T + 1)$ (which we denote by $g(T)$ and $g(T + 1)$, respectively). This means that the FGC policy $f(T, \beta)$ employs the GC policy $g(T)$ with a probability $(1 - \beta)$ and employs the GC policy $g(T + 1)$ with a probability $(\pi$. One can easily show that the performance of $f(T, \beta)$ falls in between $g(T)$ and $g(T + 1)$; that is, the dropping probability satisfies the inequalities

$$
P_D(g(T)) \le P_D(f(T,\beta)) \le P_D(g(T+1)),
\tag{8.55}
$$

and the blocking probability satisfies the inequalities

$$
P_B(g(T)) \ge P_B(f(T,\beta)) \ge P_B(g(T+1)).
\tag{8.56}
$$

In the following, we consider how the performance metrics of $f(T, \beta)$ are expressed in terms of the performance metrics of $g(T)$ and $g(T + 1)$.

First consider the dropping probability. By (8.55), we can write the dropping probability of the FGC policy $f(T, \beta)$ by

$$(1-\beta_D)P_D(g(T)) + \beta_D P_D(g(T+1)) = P_D(f(T,\beta)). \tag{8.57}$$

for a constant $\beta_D(\in[0, 1])$. For notational simplicity, we put

$$x = \frac{(\rho_h + \rho_n)^T}{C!}\rho_h{}^{C-(T+1)}, \quad y = \sum_{j=0}^{T}\frac{(\rho_h+\rho_n)^j}{j!}, \quad z = \sum_{j=T+1}^{C}\frac{(\rho_h+\rho_n)^T}{j!}\rho_h{}^{j-(T+1)}. \tag{8.58}$$

Then, the left-hand side of (8.57) is expressed by

$$(1-\beta_D)P_D(g(T)) + \beta_D P_D(g(T+1)) = (1-\beta_D)\rho_h x\frac{1}{y+\rho_h z} + \beta_D(\rho_h+\rho_n)x\frac{1}{y+(\rho_h+\rho_n)z}$$

$$= \frac{\beta_D\rho_n xy}{(y+\rho_h z)(y+(\rho_h+\rho_n)z)} + \frac{\rho_h x}{y+\rho_h z}, \tag{8.59}$$

and the right-hand side is given by

$$P_D(f(T,\beta)) = \frac{(\rho_h+\rho_n\beta)x}{y+(\rho_h+\rho_n\beta)z}. \tag{8.60}$$

By equating the two equations, we get

$$\beta_D = \frac{(y+\rho_h z)(y+(\rho_h+\rho_n)z)}{\rho_n y} \times \frac{\rho_n\beta y}{(y+(\rho_h+\rho_n\beta)z)(y+\rho_h z)} = \frac{y+(\rho_h+\rho_n)z}{y+(\rho_h+\rho_n\beta)z}\beta. \tag{8.61}$$

Now consider the blocking probability, which takes the expression

$$(1-\beta_B)P_B(g(T)) + \beta_B P_B(g(T+1)) = P_B(f(T,\beta)). \tag{8.62}$$

Since each blocking probability is expressed by a sum of the steady-state probabilities, we separate out the components of the blocking probability by that related to state T and those related to the states higher than T. As a new connection is blocked at state T with probabilities 1, 0, and $(1-\beta)$ respectively by the policies $g(T)$, $g(T+1)$, and $f(T,\beta)$, (8.62) at state T can be written as

$$(1-\beta_B) \times 1 \times P(g(T))_T + \beta_B \times 0 \times P(g(T+1))_T = (1-\beta) \times P(f(T,\beta))_T, \tag{8.63}$$

where $P(\alpha)_s$ denotes the steady-state probability of the state s when the policy α is used. If we put

$$w = \frac{(\rho_h+\rho_n)^T}{T!}, \tag{8.64}$$

the left-hand and right-hand sides of (8.63) can be expressed by

$$(1-\beta_B)P(g(T))_T + \beta_B \times 0 = (1-\beta_B)w\frac{1}{y+\rho_h z}, \tag{8.65}$$

$$(1-\beta)P(f(T,\beta))_T = (1-\beta)w\frac{1}{y+(\rho_h+\rho_n\beta)z}. \tag{8.66}$$

By equating them together, we get

$$\beta_B = \frac{y+(\rho_h+\rho_n\beta)z-(1-\beta)(y+\rho_h z)}{y+(\rho_h+\rho_n\beta)z} = \frac{y+(\rho_h+\rho_n)z}{y+(\rho_h+\rho_n\beta)z}\beta, \tag{8.67}$$

which is identical to β_D in (8.61). Since a connection is blocked by all the three policies at the states higher than T, one can easily get

$$(1-\beta_D)P(g(T))_s + \beta_D P(g(T+1))_s = P(f(T,\beta))_s \tag{8.68}$$

for $T+1 \leq s \leq C$. This means that the parameter β_B that relates the blocking probability by (8.63) is equal to the parameter β_D that relates the dropping probability by (8.57). Consequently, the performance metrics of the FGC policy is represented by a convex combination of those of the two neighboring GC policies.

The Pareto optimality of the FGC policy can be shown by using the above-mentioned property. Consider a two-dimensional performance vector $[P_D(\cdot), P_B(\cdot)]$ for the admission control policy depicted in Figure 8.17. Then, we plot the performance vectors for the GC policy by increasing the threshold T one by one. Due to the tradeoff relation, the dropping probability increases and the blocking probability decreases as the threshold increases. Now consider two adjacent GC policies and a line that connects their performance vectors. Denote the normal vector of this line by $[\alpha_D, \alpha_B]$. Then, any performance vector located in the left-lower side of the line is infeasible. If such a vector were feasible, that vector would render a linear objective $F = \alpha_B P_B + \alpha_D P_D$ lower than any GC policy, which contradicts the property that the optimal policy minimizing any linear objective is a GC policy. Therefore, we can achieve at best the segment between the two adjacent GC policies. Any performance vector on the segment can be achieved by the FGC policy since its performance is written by

$$[P_D(f(T,\beta)), P_B(f(T,\beta))] = (1-\beta')[P_D(g(T)), P_B(g(T))] + \beta'[P_D(g(T+1)), P_B(g(T+1))] \tag{8.69}$$

for a constant $\beta' = \beta_D = \beta_B (\in[0, 1])$. This implies that the FGC policy is Pareto-optimal: there exists no policy that renders both dropping and blocking probabilities lower than the FGC policy simultaneously (see Figure 8.19).

It is noteworthy that the dropping and blocking probabilities of an FGC policy are a convex combination of two GC policies but the combining parameter β' is different from β, the admission probability at state T. For example, an FGC with $\beta = 0.5$ does not necessarily render

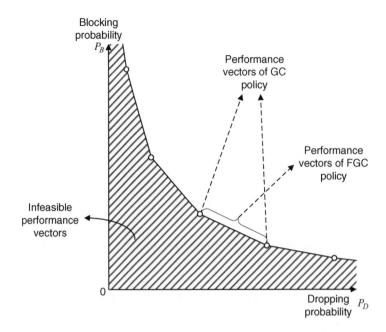

Figure 8.19 Dropping and blocking probabilities of the GC and FGC policies

the performance vector which is located at the center of the line segment connecting the two adjacent GC policies.

8.2.3 Handoff Request Queuing and Soft Handoff

In analyzing the system performance in Section 8.2.2, it was assumed that a handoff connection is dropped if the new cell has no available channel at the very moment when the handoff request arrives. This implies that an ongoing connection sends the handoff request to the new base station and releases the channel of the old cell at the same time. Figure 8.20(a) depicts this operation, in which the handoff initiation and the channel release occur simultaneously. As depicted, the coverage of the two neighboring cells is separated into two disjoint regions, allowing no buffering zone in between. In this case, the handoff is requested at the moment the mobile station crosses the cell boundary, and connectivity cannot be maintained unless the request is admitted immediately.

In some cases, however, the channel of the old BS does not need to be released at the moment of handoff initialization. An MS that initiated a handoff request may maintain the old channel with an acceptable quality as long as the inter-cell interference can be managed well. The old channel may be released after the handoff request is admitted and a channel is allocated by the new BS. This implies that the instance of the channel release is separated from the handoff initialization instance. Figure 8.20(b) shows the operation and cell structure of this case. There exists a buffering zone between the two neighboring cells, which is usually called the *handoff region*. An MS entering the handoff region initiates handoff but holds the channel of the old cell until a channel is allocated by the new cell. When all the channels of the new BS are occupied, the handoff request is *queued* while waiting for an available channel. The handoff is completed

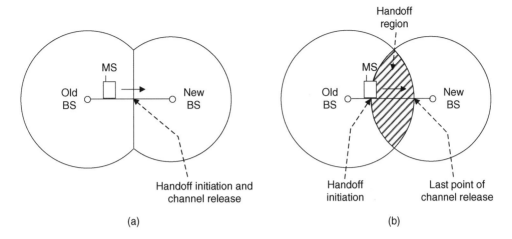

Figure 8.20 Handoff initialization and channel release for two neighboring cells: (a) without handoff region, and (b) with handoff region

successfully if a channel is allocated before the MS moves out of the handoff region. Since the current connection is maintained until a reliable connection is established or the MS gets out of the handoff region, the probability that the handoff connection is terminated due to the unavailability of channels is considerably lowered. Thus, the existence of the handoff region makes it possible to save some handoff connections that would be dropped off instantly otherwise.

Once the handoff request is successfully admitted by a new BS, the MS may release the channel of the old BS for other MSs to use, but it is also possible to arrange things such that the MS keeps the old channel to communicate with the old and the new BSs simultaneously while staying in the handoff region. In this handoff procedure, the MS can open multiple channels with two or more BSs near the cell boundary, in contrast to the above single BS association case. This type of handoff is described by the term "make-before-break" and is called *soft handoff*, in contrast to *hard handoff* which releases the old channel when a new channel is allocated. As will be discussed later, soft handoff provides an effective means of improving the quality of connections near the cell boundary.

8.2.3.1 Handoff Request Queuing

In formulating handoff management equipped with handoff request queuing, we assume the guard channel policy with the threshold T. As an MS moves away from the currently associated BS, the received power from the BS decreases. When the signal strength goes below the employed handoff criteria, the handoff procedure is initiated. The MS enters the handoff region and the handoff request is sent to a new BS. If the new BS has an available channel, it is allocated to the MS and the handoff attempt is completed successfully. If all the channels are occupied by the ongoing connections, the handoff request is queued in *first-in first-out* (FIFO) manner. If any channel is released while the MS is in the handoff area, the handoff attempt at the head of the queue is accomplished successfully. If the MS is not assigned a channel until it moves out of

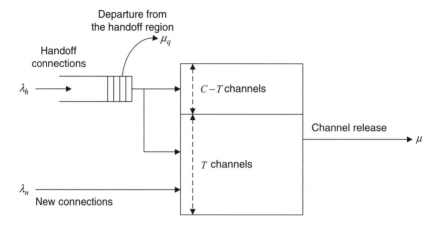

Figure 8.21 System model for handoff management with handoff request queuing

the handoff area, its connection is dropped. Figure 8.21 shows the system model of such handoff management with handoff request queuing.

The above handoff management method can also be formulated by a Markov chain (Hong and Rappaport, 1986). For an easy formulation, assume that a connection that has requested handoff is not completed while it is waiting for a channel allocation in the queue. Denote by T_q the dwell time of an MS in the handoff region. Assume that the dwell time is independent of the connection time T_c and the cell dwell time T_d. In addition, assume that it is an exponential random variable with mean $1/\mu_q$. Then we get the state transition diagram in Figure 8.22. The system operates the same way as the GC policy from state 0 to state C. The states $(C + k)$ means that k handoff requests are queued and waiting for a channel release. Since the system moves from state $(C + k)$ to state $(C + k + 1)$ if an additional handoff request arrives, the transition rate is λ_h. The system moves from state $(C + k)$ to $(C + k - 1)$ if one of the C ongoing connections releases its channel or one of the k queued requests leaves the handoff region. Thus, the transition rate is $C\mu + k\mu_q$.

With this diagram, we get the balance equation

$$
\begin{aligned}
(\lambda_n + \lambda_h)P_{s-1} &= s\mu P_s, 1 \le s \le T, \\
\lambda_h P_{s-1} &= s\mu P_s, T+1 \le s \le C, \\
\lambda_h P_{s-1} &= (C\mu + (s-C)\mu_q)P_s, s \ge C+1,
\end{aligned}
\tag{8.70}
$$

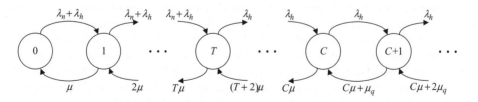

Figure 8.22 State transition diagram for handoff request queuing

and thus the steady-state probabilities

$$
P_s =
\begin{cases}
\dfrac{(\rho_h + \rho_n)^s}{s!} P_0, & \text{if } 0 \le s \le T, \\[3ex]
\dfrac{(\rho_h + \rho_n)^T}{s!} \rho_h^{s-T} P_0, & \text{if } T+1 \le s \le C, \\[3ex]
\dfrac{(\rho_h + \rho_n)^T \rho_h^{C-T}}{C!} \dfrac{\lambda_h^{s-C}}{\prod_{j=1}^{s-c}(C\mu + j\mu_q)} P_0, & \text{if } s \ge C+1,
\end{cases}
\tag{8.71}
$$

where

$$
P_0 = \left[\sum_{j=0}^{T} \frac{(\rho_h + \rho_n)^j}{j!} + \sum_{j=T+1}^{C} \frac{(\rho_h + \rho_n)^T}{j!} \rho_h^{j-T} + \sum_{j=C+1}^{\infty} \frac{(\rho_h + \rho_n)^T \rho_h^{C-T}}{C!} \frac{\lambda_h^{s-C}}{\prod_{i=1}^{s-c}(C\mu + i\mu_q)} \right]^{-1}.
\tag{8.72}
$$

The blocking probability P_B takes the same form as in the GC policy; that is

$$
P_B = \sum_{n=T}^{\infty} P_n.
\tag{8.73}
$$

However, the dropping probability P_D takes a quite different form. Suppose that there exist k handoff requests in the queue when a target handoff request arrives at the new BS. This case happens with the probability P_{C+k} in (8.71). Then, the target handoff connection is dropped if its dwell time in the handoff region is shorter than the time taken to move to the head of the request queue and get a channel. Denote by $P_{D|k}$ the probability of handoff dropping when a handoff arrival finds k requests in the queue. Then, the dropping probability is given by

$$
P_D = \sum_{k=0}^{\infty} P_{C+k} P_{D|k}.
\tag{8.74}
$$

We briefly digress to determine the conditional dropping probability $P_{D|k}$. Denote by $P(i|i+1)$ the probability that the target handoff request in the $(i+1)$th position of the queue advances to the ith position before the MS leaves the handoff region. This target request fails to advance if its remaining dwell time in the handoff region, $T_{q,i+1}$, is smaller than all the remaining dwell times in the queue, $T_{q,1}, T_{q,2}, \ldots, T_{q,i}$, and the minimum channel holding time of the C ongoing connections, X. Note that X is an exponential random variable with mean $1/C\mu$. Thus we have

$$
1 - P(i|i+1) = \Pr[T_{q,i+1} \le \min(X, T_{q,1}, T_{q,2}, \ldots, T_{q,i})] = \Pr[T_{q,i+1} \le Y_i],
\tag{8.75}
$$

for $Y_i = \min(X, T_{q,1}, T_{q,2}, \ldots, T_{q,i})$. By the independency of X and $T_{q,1}, T_{q,2}, \ldots, T_{q,i}$, the CDF of Y_i is given by

$$F_{Y_i}(t) = 1 - (1 - F_X(t))(1 - F_{T_{q,1}}(t))(1 - F_{T_{q,2}}(t)) \cdots (1 - F_{T_{q,i}}(t)) = 1 - \exp(-(C\mu + i\mu_q)). \tag{8.76}$$

The second equality holds because X and $T_{q,1}, T_{q,2}, \ldots, T_{q,i}$ are independent exponential random variables. Then we have

$$
\begin{aligned}
P(i|i+1) &= \Pr[T_{q,i+1} > Y_i] = \int_0^\infty \Pr[T_{q,i+1} > t | Y_i = t] f_{Y_i}(t) dt \\
&= \int_0^\infty \Pr[T_{q,i+1} > t](C\mu + i\mu_q) \exp(-(C\mu + i\mu_q)t) dt \\
&= \int_0^\infty \exp(-\mu_q t)(C\mu + i\mu_q) \exp(-(C\mu + i\mu_q)t) dt = \frac{C\mu + i\mu_q}{C\mu + (i+1)\mu_q}.
\end{aligned}
\tag{8.77}
$$

By the memoryless property of the dwell and channel holding times, the probability that the handoff request at the $(k+1)$th position to be the head of the queue is given by

$$\prod_{i=1}^k P(i|i+1) = \prod_{i=1}^k \frac{C\mu + i\mu_q}{C\mu + (i+1)\mu_q} = \frac{C\mu + \mu_q}{C\mu + (k+1)\mu_q}. \tag{8.78}$$

After the handoff request arrives at the head of the queue, it will get a channel if its remaining dwell time $T_{q,1}$ exceeds X. With a method similar to (8.77), this probability is given by

$$\Pr[T_{q,1} > X] = \frac{C\mu}{C\mu + \mu_q}. \tag{8.79}$$

As a result, the conditional dropping probability is given by

$$
\begin{aligned}
P_{D|k} &= 1 - \Pr[\text{reach the head of the queue and get a channel at the head}] \\
&= 1 - \left[\prod_{i=1}^k P(i|i+1) \right] \Pr[T_{q,1} > X] = 1 - \frac{C\mu + \mu_q}{C\mu + (k+1)\mu_q} \times \frac{C\mu}{C\mu + \mu_q} = \frac{(k+1)\mu_q}{C\mu + (k+1)\mu_q}.
\end{aligned}
\tag{8.80}
$$

If we insert the above conditional dropping probability into (8.74), we get the dropping probability with handoff request queuing.

The handoff dropping probability can be reduced further if an intelligent queue management scheme is employed instead of FIFO. When an MS moves from the old BS to the new one, the signal strength from the old BS gets gradually reduced and eventually becomes too weak to support the connection. For a successful handoff, a channel should be allocated before the MS leaves the handoff region. So it is desirable to assign a channel to the particular MS that will leave the handoff region first among all the MSs with queued handoff requests. In terms of signal strength, if there exists an MS whose current received signal strength approaches the minimum tolerable level, it is desirable to allocate an available channel to that MS first. Based on this observation, it is proposed in Tekinay and Jabbari (1992) to sort the queued handoff requests in ascending order of the signal strength from the old BS. This scheme puts a higher priority to the MS with a weaker power and dynamically changes the positions of the queued requests according to the current priority of the MSs. With this scheme employed, the most urgent handoff requests that are about to leave the handoff region will be shifted toward the head of the queue and get channels prior to others, and a smaller dropping probability will be achieved.

8.2.3.2 Soft Handoff

As discussed above in relation to the handoff region in Figure 8.19, soft handoff takes place in such a way that an MS maintains its old channel with the old BS even after a new channel is assigned by a new BS by associating with two or more BSs simultaneously until the MS leaves the handoff region. Figure 8.23 illustrates the soft handoff procedure.

The advantage of soft handoff is that an MS in the process can exploit a diversity effect by communicating with two or more base stations. In the downlink case, multiple copies of a single signal arrive at the MS and are combined into a more reliable signal. In the uplink case, the signal transmitted at the MS is received by each associated BS and all the received signals are gathered via the network that connects the neighboring BSs. The diversity effect achieved by soft handoff is sometimes called the *macro diversity*, in contrast to the other diversity schemes which are performed in geometrically small scales. It is noteworthy that this macro diversity is exploited only by the MSs that are located near the cell boundary where the signal strength is weak. The MSs near the BS will not need such macro diversity, as the high channel gain will easily satisfy the required QoS. So, soft handoff helps to balance the QoS of the constituent users regardless of their locations, effectively providing continuous services to the mobile stations near the cell boundary.

Theoretically, soft handoff is applicable in any type of multiple-access technology, but there are some implemental challenges. FDMA, TDMA, and OFDMA usually use a frequency reuse factor greater than one. This means that two neighboring cells use different frequency bands and, as a consequence, an MS cannot perform soft handoff unless it is capable of operating over multiple frequency bands simultaneously. In fact, those technologies separate the frequency bands among the neighboring cells to avoid interference, as they do not have any other

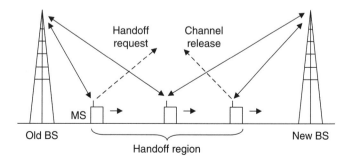

Figure 8.23 The soft handoff procedure

mechanism to suppress inter-cell interference. In the case of TDMA and OFDMA, soft handoff may be possibly applied if the two neighboring cells allocate the same channel (i.e., the same time slot or the same subchannel). In principle, inter-cell interference can be removed in this case as the neighboring cells transmit the same signal over the channel that the MS watches. In reality, it requires very strict synchronization of the signal arrivals. This means that the neighboring base stations should be precisely synchronized, and the propagation delay from/to the BSs should be completely compensated, as otherwise a signal from/to one BS would not be orthogonal to the signal of the other BS and thus would be seen as interference. However, this strict synchronization is a big burden in practical implementation. In contrast, CDMA has some properties that enable an easy implementation of soft handoff. First, separation of frequency bands is not required, as inter-cell interference is suppressed automatically during the despreading process. Second, strict synchronization is not required because the time difference among the neighboring BSs can be regarded as the delay spread caused by multipaths and the interference caused by this difference can be mitigated by despreading each of the received signals.

Handoff dropping and new call blocking in soft handoff can be formulated by the model used in handoff request queuing. The only difference is that soft handoff does not release the old channel for a certain time after the channel allocation, but this affects only the statistics of the cell dwell time in the old BS. Therefore, the soft handoff can be formulated and analyzed based on the Markov model in this subsection with slight modifications (Su et al., 1996; Chang and Sung, 2001).

8.2.4 Advanced Handoff Management Schemes

This subsection considers some advanced topics in handoff management. It first considers policies that take the adjacent cell information into account in admitting a new connection arrival. Then it considers how the simple channel reservation schemes can be extended to cases where multiple traffic classes exist with each having different priority.

8.2.4.1 Admission using Adjacent Cell Information

The admission policies discussed so far used only the target cell information (i.e., the number of occupied channels) in deciding the admission of a newly originating connection. However, it would be much more beneficial to use the information of the adjacent cells as well. For example, suppose two neighboring cells have different populations such that cell 1 has many ongoing connections but cell 2 does not. If this information can be utilized in an admission policy, it would help to adjust the number of reserved channels dynamically according to the number of ongoing connections in the adjacent cell as follows.

A new connection arriving at cell 2 may be admitted by the conventional policies but this admission can cause an avoidable handoff dropping while resulting in an unnecessarily high dropping probability. This is because the connection admitted to cell 2 is highly likely to get dropped when it is handed off to cell 1 which suffers from a heavy load. In addition, the connection admission in cell 2 reduces the amount of channel reservation for the potentially heavy handoff connections from cell 1, thereby increasing the handoff probability. On the other hand, new connections at cell 1 may be admitted in spite of the heavy load in cell 1,

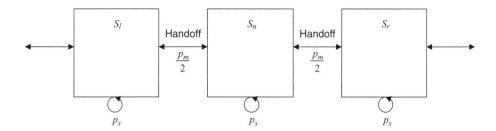

Figure 8.24   One-dimensional cell structure

since it is not necessary to reserve many channels for the handoff from cell 2 which has a small population.

An admission policy that adjusts the admission threshold by using the information of the adjacent cells was proposed in (Naghshineh and Schwartz, 1996). We discuss this policy using the one-dimensional cellular system model depicted in Figure 8.24. Denote by S_n the cell in which a new connection arrives and by S_r and S_l two adjacent cells of S_n. We assume that each cell has C channels and the number of the occupied channels are n, r, and l respectively for cells S_n, S_r, and S_l. In order to estimate the effect of handoff connections, we adopt a stochastic model for each MS's mobility. We simply assume that, after time t lapses, each MS in a cell remains in the same cell with probability p_s and hands off to one of the adjacent cells with probability $p_m/2$. In addition, we assume that the probability that a connection hands off twice or more during time t is negligible.

Under the above system model, we can consider the following admission criterion. At time t in the future, the probability that the population of each cell exceeds the capacity C should be less than the predetermined level p_{req}. Obviously, the connections that hand off from the adjacent cells should be taken into consideration in calculating the population after time t. If a new connection is admitted according to this criterion, the probability that a cell suffers from an overload situation can be maintained below the required level. For this operation, we need to estimate the population after time t as follows.

First consider the future population of S_n, $\tilde{n}$. This population consists of three components: m_n connections remaining in S_n; and m_r and m_l handoff connections respectively from S_r and S_l. The three components are binomial random variables with distributions $\text{B}(m_n, n, p_s)$, $\text{B}(m_r, r, p_m/2)$, and $\text{B}(m_l, l, p_m/2)$, respectively, where $\text{B}(m, k, p)$ denotes the probability that m connections are selected out of k connections with selection probability p. This probability is given by

$$\text{B}(m,k,p) = \frac{k!}{m!(k-m)!}p^m(1-p)^{k-m}, \tag{8.81}$$

and has mean kp and variance $kp(1-p)$. Since $\tilde{n} = m_n + m_r + m_l$, the distribution of the future population can be represented by using a convolution sum of the three binomial random variables, but this yields very complicated distribution of $\tilde{n}$. Thus, based on the central limit theorem, we approximate the binomial random variable in (8.81) to a Gaussian random variable with the same mean and variance. Then, $\tilde{n}$ becomes a sum of three independent Gaussian random variables; that is, a Gaussian random variable with the distribution

$$\tilde{n} \sim N(np_s + (l+r)p_m/2, np_s(1-p_s) + (l+r)p_m(1-p_m/2)/2). \qquad (8.82)$$

From this probability distribution, we can obtain the future overload probability P_{OL}, or the probability that the future population exceeds the capacity, by

$$P_{OL} = \sum_{i=C+1}^{n+r+L} \Pr[\tilde{n} = i] \approx Q\left(\frac{C-(np_s + (l+r)p_m/2)}{\sqrt{np_s(1-p_s) + (l+r)p_m(1-p_m/2)/2}}\right), \qquad (8.83)$$

where $Q(\cdot)$ is the integral over the tail of a normalized Gaussian distribution given by

$$Q(x) = \int_x^\infty \frac{1}{\sqrt{2\pi}} \exp(-t^2) dt. \qquad (8.84)$$

Then we get

$$C-(np_s + (l+r)p_m/2) = Q^{-1}(p_{req})\sqrt{np_s(1-p_s) + (l+r)p_m(1-p_m/2)/2}. \qquad (8.85)$$

On rearranging this for n, we can find an admission threshold n_1 which satisfies the admission criterion for S_n. We can repeat the same process to obtain the admission thresholds n_2 and n_3 by focusing on the future populations of the cells S_r and S_i, respectively, and arranging the equations for n. Then the final admission threshold is given by

$$T = \min(n_1, n_2, n_3). \qquad (8.86)$$

Here T implies the finally determined admission threshold; that is, the maximum number of connections that can be admitted to the cell S_n such that the future overload probability can be maintained below the predetermined level.

In the above dynamic admission policy, a homogeneous mobility is assumed for all the mobile stations in calculating the future population. This means that all the MSs are assumed to have the same handoff probability. In addition, each MS is assumed to visit each cell in its adjacencies with equal probability. However, this assumption is not true in practice. An MS with a higher speed is much more likely to hand off than an MS with a lower speed. Thus, the probability that an MS will remain in the cell or will hand off to a neighboring cell is different for different users. Moreover, the cell that a handoff MS will visit is strongly related to the current direction of the MS's mobility. If an MS moves to a new direction, it is much probable that the MS will visit a cell located in that direction.

If this heterogeneity of mobility can be taken into consideration, a more efficient resource reservation in handoff management can be achieved. Suppose an MS moves eastward with high speed. Then, the cells in this direction should make a channel reservation for this MS while the other cells may ignore the effect of this MS on their channel reservation. The mobility of each MS can be represented by the probability that the MS will reside in each cell in the future, as illustrated in Figure 8.25. In this figure, a darker cell means that it is more probable for the MS

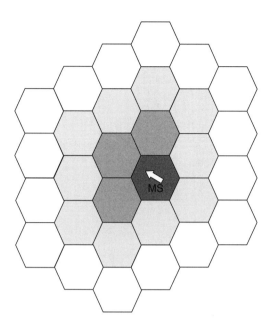

Figure 8.25 Mobility and residential probability at each cell (Reproduced with permission from D.A. Levine, I.F. Akyildiz, and M. Naghshineh, "A resource estimation and call admission algorithm for wireless multimedia networks using the shadow cluster concept," *IEEE/ACM Transactions on Networking*, 5, February, 1–12, 1997. © 1997 IEEE)

will reside in after a given time t. It is obvious that the cells along the direction of an MS have a higher residential probability.[7] If this information is available, the future population of each cell can be estimated more precisely. Several dynamic admission policies have been developed based on stochastic estimation of the mobility of each MS (Levine *et al.*, 1997; Choi and Shin, 2002). The mobility can be estimated from information such as the current location and velocity, as well as the historical record of movement. Then, the amount of channel reservation can be estimated for each cell and the admission threshold can be adjusted based on this estimation.[8]

If the location and velocity of each MS can be estimated more accurately, it is also possible to send a channel reservation message to the approaching cell (Chiu and Bassiouni, 2000). A typical operation is shown in Figure 8.26. An MS in the cell S_1 and approaching cell S_2 sends a reservation request to the cell at point A based on its direction and velocity. If it hands off to S_2 as predicted, it occupies the reserved channel (i.e., point B). If the MS changes its direction to cell S_3, a cancellation message is sent to S_2 and another reservation request is sent to S_3 at point C. This operation is also a dynamic reservation policy since each BS adjusts the admission threshold according to the reservation and cancellation messages. It distinguishes itself from the above dynamic reservation policies in that it sends an explicit reservation request and cancellation message while the others determine the number of channel reservations based on stochastic information.

---

[7] In Levine *et al.* (1997), the cells having considerable residential probabilities are said to form a *shadow cluster*.

[8] A heuristic algorithm is proposed in each reference for adjustment of the admission threshold. See the references for more detailed operation of each admission policy.

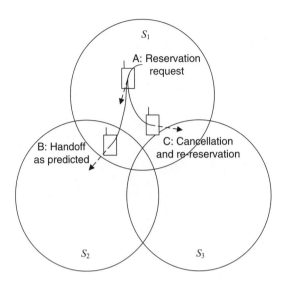

Figure 8.26   Channel reservation and cancellation requests (Chiu and Bassiouni, 2000)

8.2.4.2 Multiple Traffic Classes

So far we have considered handoff management under the assumption that there exists only a single traffic class. However, handoff management needs to be extended to encompass multiple traffic classes as future wireless systems are expected to support various different types of traffics simultaneously. Since the main objective of handoff management is to maintain the QoS of each user at an acceptable level, handoff management in multiple traffic classes is required to be able to deal with the different QoS requirements of each class.

The guard channel policy developed for the single traffic class can be extended to support multiple traffic classes by using different admission thresholds for different traffic classes. A lower threshold is employed for the traffic with a lower priority such that a connection with a low priority can be admitted only when there are enough available channels at the base station. Since each traffic class has two kinds of arrivals (i.e., new origination and handoff), different admission thresholds may be employed for each of them. In the following, we first introduce the simplest case with two different classes, and then discuss some of its variations.

First consider two different classes of traffic, each of which requires one channel per connection (Yin *et al.*, 2000). Assume that class 1 traffic has a higher priority in resource reservation than class 2 traffic.[9] Denote by λ_{n1}, λ_{h1}, λ_{n2}, and λ_{h2} the arrival rates of the new connection and handoff connections of classes 1 and 2, respectively. Similarly, denote by μ_1 and μ_2 the departure rates of the two classes. Assume that the arrival and departure processes are memoryless, as in the previous cases.

---

[9] In many studies, the two traffic classes are considered as voice and data traffic, respectively.

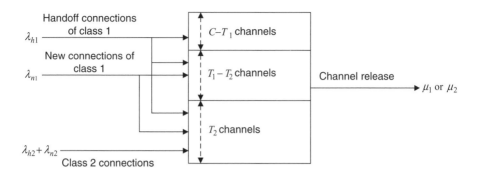

Figure 8.27   System model for the dual-threshold reservation policy

The employed handoff management scheme, called *dual-threshold reservation* (DTR), divides the C channels in the cell into three regions by two thresholds T_1 and T_2 ($T_1 > T_1$). When the number of the occupied channels is less than T_2, both classes can be admitted into the system; and otherwise a class 2 connection is rejected regardless of its arrival type. When the number of the occupied channels exceeds T_1, only a handoff connection of class 1 is admitted. The handoff connection of class 1 will be dropped only when no channel is available. Figure 8.27 illustrates the channel reservation of this DTR scheme.

The DTR scheme can be formulated as a two-dimensional Markov chain. The system state is denoted by (S_1, S_2) when there are S_1 class 1 connections and S_2 class 2 connections in the cell. Its state transition diagram is quite complicated to depict, but we can obtain the steady-state balance equations by calculating the incoming and outgoing rates at each state, as follows. First, for $s_2 \neq T_2$, we have

$$
P_{s_1,s_2} = \begin{cases}
0, & \text{if } s_1 + s_2 > C, \\[2mm]
\dfrac{\lambda_{h1} P_{s_1-1,s_2}}{s_1\mu_1 + s_2\mu_2}, & \text{if } s_1 + s_2 = C, \\[4mm]
\dfrac{(s_1+1)\mu_1 P_{s_1+1,s_2} + \lambda_{h1} P_{s_1-1,s_2} + (s_2+1)\mu_2 P_{s_1,s_2+1}}{\lambda_{h1} + s_1\mu_1 + s_2\mu_2}, & \text{if } T_1 < s_1 + s_2 < C, \\[4mm]
\dfrac{(s_1+1)\mu_1 P_{s_1+1,s_2} + (\lambda_{h1}+\lambda_{n1}) P_{s_1-1,s_2} + (s_2+1)\mu_2 P_{s_1,s_2+1}}{\lambda_{h1} + s_1\mu_1 + s_2\mu_2}, & \text{if } s_1 + s_2 = T_1, \\[4mm]
\dfrac{(s_1+1)\mu_1 P_{s_1+1,s_2} + (\lambda_{h1}+\lambda_{n1}) P_{s_1-1,s_2} + (s_2+1)\mu_2 P_{s_1,s_2+1}}{\lambda_{h1} + \lambda_{n1} + s_1\mu_1 + s_2\mu_2}, & \text{if } T_2 < s_1 + s_2 < T_1, \\[4mm]
\dfrac{(s_1+1)\mu_1 P_{s_1+1,s_2} + (\lambda_{h1}+\lambda_{n1}) P_{s_1-1,s_2} + (s_2+1)\mu_2 P_{s_1,s_2+1} + (\lambda_{h2}+\lambda_{n2}) P_{s_1,s_2-1}}{\lambda_{h1} + \lambda_{n1} + s_1\mu_1 + s_2\mu_2}, & \text{if } s_1 + s_2 = T_2, \\[4mm]
\dfrac{(s_1+1)\mu_1 P_{s_1+1,s_2} + (\lambda_{h1}+\lambda_{n1}) P_{s_1-1,s_2} + (s_2+1)\mu_2 P_{s_1,s_2+1} + (\lambda_{h2}+\lambda_{n2}) P_{s_1,s_2-1}}{\lambda_{h1} + \lambda_{n1} + s_1\mu_1 + s_2\mu_2 + \lambda_{h2} + \lambda_{n2}}, & \text{if } 0 < s_1 + s_2 < T_2, \\[4mm]
\dfrac{(s_1+1)\mu_1 P_{s_1+1,s_2} + (s_2+1)\mu_2 P_{s_1,s_2+1}}{\lambda_{h1} + \lambda_{n1} + \lambda_{h2} + \lambda_{n2}}, & \text{if } s_1 = s_2 = 0, \\[4mm]
0, & \text{if } s_1 < 0 \text{ or } s_2 < 0.
\end{cases}
$$

$$(8.87)$$

If $s_2 = T_2$, we have

$$
P_{s_1,s_2} = \begin{cases}
0, & \text{if } s_1 > C - T_2, \\[1mm]
\dfrac{\lambda_{h1} P_{s_1-1,s_2}}{s_1\mu_1 + s_2\mu_2}, & \text{if } s_1 = C - T_2, \\[2mm]
\dfrac{(s_1+1)\mu_1 P_{s_1+1,s_2} + \lambda_{h1} P_{s_1-1,s_2}}{\lambda_{h1} + s_1\mu_1 + s_2\mu_2}, & \text{if } T_1 - T_2 < s_1 < C - T_2, \\[2mm]
\dfrac{(s_1+1)\mu_1 P_{s_1+1,s_2} + (\lambda_{h1} + \lambda_{n1}) P_{s_1-1,s_2}}{\lambda_{h1} + s_1\mu_1 + s_2\mu_2}, & \text{if } s_1 = T_1 - T_2, \\[2mm]
\dfrac{(s_1+1)\mu_1 P_{s_1+1,s_2} + (\lambda_{h1} + \lambda_{n1}) P_{s_1-1,s_2}}{\lambda_{h1} + \lambda_{n1} + s_1\mu_1 + s_2\mu_2}, & \text{if } 0 < s_1 < T_1 - T_2, \\[2mm]
\dfrac{(s_1+1)\mu_1 P_{s_1+1,s_2} + (\lambda_{h2} + \lambda_{n2}) P_{s_1-1,s_2}}{\lambda_{h1} + \lambda_{n1} + s_2\mu_2}, & \text{if } s_1 = 0, \\[2mm]
0, & \text{if } s_1 < 0.
\end{cases}
\tag{8.88}
$$

These equations are obtained by equalizing the incoming rate to a state with the outgoing rates from it. For example, consider a state (s_1, s_2) with $s_2 \neq T_2$ and $s_1 + s_2 = T_1$. The system leaves this state if a handoff of class 1 arrives (at rate $\lambda_{h1} P_{s_1,s_2}$) or one of the connection releases the channel (at rate $(s_1\mu_1 + s_2\mu_2) P_{s_1,s_2}$). An arrival of class 2 or a new connection of class 1 is rejected as T_1 channels are occupied currently. They are rejected and the system does not leave this state. On the other hand, the system enters this state if a class 2 connection releases its channel at state $(s_1 + 1, s_2)$ (at rate $(s_1 + 1)\mu_1 P_{s_1+1,s_2}$), a class 1 connection arrives at state $(s_1 - 1, s_2)$ (at rate $(\lambda_{h1} + \lambda_{n1}) P_{s_1-1,s_2}$), or a class 2 connection leaves state $(s_1, s_2 + 1)$ (at rate $(s_2 + 1)\mu_2 P_{s_1,s_2+1}$). An arrival of class 2 is rejected at state $(s_1, s_2 - 1)$ as $s_1 + s_2 - 1 = T_1 - 1 \geq T_2$, so the system does not enter this state in this case. Then, by equating the incoming and outgoing rates, we obtain the balance equation in the fourth case of (8.87).

Differently from the single-class case, the steady-state probability for two traffic classes cannot be solved in closed form. It can be obtained by solving the set of linear equations numerically. After obtaining the steady-state probability, we can calculate the system performance. The dropping probability of the class 1 traffic is given by

$$
P_{D,1} = \sum_{s_1 + s_2 = C} P_{s_1,s_2},
\tag{8.89}
$$

and its blocking probability is given by

$$
P_{B,1} = \sum_{s_1 + s_2 \geq T_1}^{s_1 + s_2 = C} P_{s_1,s_2}.
\tag{8.90}
$$

For the class 2 traffic, the dropping and blocking probabilities are the same; that is

$$
P_{D,1} = P_{B,1} = \sum_{s_1 + s_2 \geq T_2}^{s_1 + s_2 = C} P_{s_1,s_2}
\tag{8.91}
$$

Additionally, several research results on handoff management for multiple traffic classes have been reported which analyze the system under slightly different system models. Li *et al.* (2004) dealt with the case where a single connection of class 2 requires more than one channel. It also analyzed the system assuming that the bandwidth of class 2 is elastic, which means that the class 2 connections can release some channels they occupy in order to prevent dropping a class 1 handoff connection. Zeng and Agrawal (2000) considered handoff request queuing for two classes and analyzed the system performance. Zeng and Agrawal (2001) analyzed the system under the assumption that a class 1 handoff connection is allowed to preempt a channel of a class 2 connection if there is no channel available. Ureta *et al.* (2003) extended the channel reservation scheme to the case where more than two traffic classes exist.

References

Anderson, L. (1973) A simulation study of some dynamic channel assignment algorithms in high capacity mobile telecommunications system. *IEEE Transactions on Vehicular Technology*, **22**, 210–217.

Chang, J.W. and Sung, D.K. (2001) Adaptive channel reservation scheme for soft handoff in DS-CDMA cellular suystems. *IEEE Transactions on Vehicular Technology*, **50**, 341–353.

Chiu, M.-H. and Bassiouni, M.A. (2000) Predictive schemes for handoff prioritization in cellular networks based on mobile positioning. *IEEE Journal on Selected Areas in Communications*, **18**, 510–522.

Choi, S. and Shin, K.G. (2002) Adaptive bandwidth reservation and admission control in QoS-sensitive cellular networks. *IEEE Transactions on Parallel and Distributed Systems*, **13**, 882–897.

Cox, D.C. and Reudink, D.O. (1972) A comparison of some channel assignment strategies in large-scale mobile communications systems. *IEEE Transactions on COM-20*, **2**, 190–195.

Cruz-Perez, F.A. *et al.* (1999) Fractional channel reservation in mobile communication systems. *IEEE Electronic Letters*, **35**, 2000–2002.

Han, Z., Ji, Z. and ray Liu, K.J. (2004) Power minimization for multi-cell OFDM networks using distributed non-cooperative game approach. In: Proceedings IEEE Globecom, Dallas, TX, 3742–3747.

Hong, D. and Rappaport, S.S. (1986) Traffic model and performance analysis for cellular mobile radio telephone systems with prioritized and nonprioritized handoff procedures. *IEEE Transactions on Vehicular Technology*, **35**, 77–92. See also the comments on this paper in *IEEE Transactions on Vehicular Technology*, **49**, 2037–2039.

IEEE/Qualcomm (2005) IEEE 802.20. QFDD and QTDD: Proposed Draft Air Interface Specification, Qualcomm.

Kanai, T. (1992) Autonomous reuse partitioning in cellular systems. In: Proceedings IEEE Vehicular Technology Conference, Denver, CO, 782–785.

Kwon, H. and Lee, B.G. (2006) Distributed resource allocation through noncooperative game approach in multi-cell OFDMA systems. In: Proceedings IEEE ICC, Istanbul, Turkey, 4345–4350.

Lee, B.G. and Choi, S. (2008) '*Broadband Wireless Access and Local Networks: Mobile WiMAX and WiFi*', Artech House.

Levine, D.A., Akyildiz, I.F. and Naghshineh, M. (1997) A resource estimation and call admission nalgorithm for wireless multimedia networks using the shadow cluster concept. *IEEE/ACM Transactions on Networking*, **5**, 1–12.

Li, B., Li, L., Bo, Li. *et al.* (2004) Call admission control for voice/data integrated cellular networks: Performance analysis and comparative study. *IEEE Journal on Selected Areas in Communications*, **22**, 706–718.

Markoulidakis, J.G. *et al.* (2000) Optimal system capacity in handoff prioritised schemes in cellular mobile telecommunication systems. *Computer Communications Journal*, **23**, 462–475.

Naghshineh, M. and Schwartz, Mischa (1996) Distributed call admission control in mobile/wireless networks. *IEEE Journal on Selected Areas in Communications*, **14**, 711–717.

Onoe, S. and Yasuda, S. (1989) Flexible re-use for dynamic channel assignment in mobile radio systems. In: Proceedings IEEE ICC, Boston, MA, 472–476.

Ramjee, R. *et al.* (1997) On optimal call admission control in cellular networks. *Wireless Networks*, **3**, 29–41.

Serizawa, M. and Goodman, D. (1993) Instablity and deadlock of distributed dynamic channel allocation. In: Proceedings IEEE Vehicular Technology Conference, Secaucus, NJ, 528–531.

Su, S.-L. *et al.* (1996) Performance analysis of soft handoff in CDMA cellular systems. *IEEE Journal on Selected Areas in Communications*, **14**, 1762–1769.

Tekinay, S. and Jabbari, B. (1992) A measurement-based prioritization scheme for handoffs in mobile cellular networks. *IEEE Journal on Selected Areas in Communications*, **10**, 1343–1350.

Ureta, H.H., Perez, F.A.C. and Guerrero, L.O. (2003) Capacity optimization in multiservice mobile wireless networks with multiple fractional channel reservation. *IEEE Transactions on Vehicular Technology*, **52**, 1519–1538.

Yeung, K.L. and Yun, T.-S.P. (1995) Cell group decoupling analysis of a dynamic channel assignment strategy in linear microcellular radio systems. *IEEE Transactions on Communications*, **43** (2), 1289–1292.

Yin, L., Zhang, Z. and Lin, Y.-B. (2000) Performance analysis of dual-threshold reservation scheme for voice/data integrated mobile wireless networks. In: Proceedings IEEE WCNC, Chicago, IL, 258–262.

Zeng, Q.-A. and Agrawal, D.P. (2000) Performance analysis of a handoff scheme in integrated voice/data wireless networks. In: *Proceedings IEEE VTC*, Boston, MA, 1986–1992.

Zeng, Q.-A. and Agrawal, D.P. (2001) Modeling and analysis of preemptive priority based handoffs in integrated wireless networks. In: Proceedings IEEE VTC, Atlantic City, NJ, 281–285.

Zhang, M. (1989) Comparisons of channel assignment strategies in cellular mobile telephone systems. *IEEE Transactions on Vehicular Technology*, **38**, 211–215.

Index